Research Foundations for Psychotherapy Practice

The Mental Health Foundation

As the UK's only charity concerned with both mental illness and learning disabilities, the Mental Health Foundation plays a vital role in pioneering new approaches to prevention, treatment and care. The Foundation's work includes allocating grants for research and community projects, contributing to public debate, educating policy-makers and health care professionals and striving to reduce the stigma attached to mental illness and learning disabilities.

The Foundation's work in supporting research and community projects is overseen by five specialist committees: General Projects, Research, Learning Disabilities, Substance Abuse and Mentally Disordered Offenders. The members of all the committees are leading medical, academic, legal and health care professionals who donate their time and expertise voluntarily to select the most effective and innovative proposals from an ever increasing number.

The Mental Health Foundation held a conference on Research Foundations for Psychotherapy Practice at Balliol College, Oxford in September 1993. The conference brought together an exciting programme of international speakers with the aim of stimulating good research projects for future funding by the Research Committee of the Foundation.

Research Foundations for Psychotherapy Practice

Edited by
Mark Aveline
Nottingham Psychotherapy Unit, Nottingham, UK

David A. Shapiro
University of Sheffield, Sheffield, UK

JOHN WILEY & SONS
in association with

THE
MENTAL
HEALTH
FOUNDATION

JOHN WILEY & SONS
Chichester · New York · Brisbane · Toronto · Singapore

Copyright © 1995, The Mental Health Foundation
and Individual Contributors

Published 1995 by John Wiley & Sons Ltd,
Baffins Lane, Chichester,
West Sussex P019 1UD, England

Telephone: National   Chichester (01243) 779777
            International + 44 1243 779777

Other Wiley Editorial Offices

John Wiley & Sons, Inc., 605 Third Avenue,
New York, NY 10158-0012, USA

Jacaranda Wiley Ltd, 33 Park Road, Milton,
Queensland 4064, Australia

John Wiley & Sons (Canada) Ltd, 22 Worcester Road,
Rexdale, Ontario M9W 1L1, Canada

John Wiley & Sons (SEA) Pte Ltd, 37 Jalan Pemimpin #05–04,
Block B, Union Industrial Building, Singapore 2057

British Library Cataloguing in Publication Data

A catalogue record for this book is available from the British Library

ISBN 0 471 95219 2

Typeset in 10/12pt Palatino by
Mathematical Composition Setters Ltd, Salisbury, Wiltshire
Printed and bound in Great Britain by
Biddles Ltd, Guildford

For Anna and Diana

Contents

Contributors

Mark Aveline
Nottingham Psychotherapy Unit, 114 Thornywood Mount, Nottingham NG3 2PZ, UK

Michael Barkham
MRC/ESRC Social and Applied Psychology Unit, Department of Psychology, University of Sheffield, Sheffield S10 2TN, UK

M. Berelowitz
Royal Free Hospital, London, UK.

Ivy-Marie Blackburn
Newcastle Cognitive Therapy Centre, Newcastle Mental Health NHS Trust, Collingwood Clinic, St Nicholas Hospital, Gosforth, Newcastle upon Tyne NE3 3XT, UK

Chess Denman
The Cassel Hospital, 1 Ham Common, Richmond, Surrey TW10 7JF, UK

Robert K. Elliott Jnr
Department of Psychology, University of Toledo, Toledo, OH 43606-3390, USA

Chris Freeman
Royal Edinburgh Hospital, Morningside Terrace, Edinburgh EH10 5DT, Scotland, UK

Zaida Hall
University of Southampton Department of Psychiatry, Royal South Hants Hospital, Graham Road, Southampton SO9 4PE, UK

Gillian E. Hardy
Social and Applied Psychology Unit, Department of Psychology, University of Sheffield, Sheffield S10 2TN, UK

Ken Howard
Department of Psychotherapy, Northwestern University, 2029 Sheridan Road, Evanston, IL 60208-2710, USA

Michael King
Department of Psychiatry, University of London, Royal Free Hospital School of Medicine, Pond Street, London NW3 2QG, UK

I. Kolvin
Tavistock Clinic, 120 Belsize Lane, London NW3 5BA, UK.

Robert J. Luegar
Department of Psychology, Marquette University, Milwaukee, WI 53233, USA

Mark Mullee
Medical Statistics and Computing, University of Southampton, Southampton General Hospital, Southampton, UK

David E. Orlinsky
5555 South Everett Avenue, Chicago, IL 60637 USA

Glenys Parry
Department of Health, Richmond House, 79 Whitehall, London SW1A 2NS, UK

Michael J. Peckham
Department of Health, Richmond House, 79 Whitehall, London SW1A 2NS, UK

Anne Rees
MRC/ESRC Social and Applied Psychology Unit, Department of Psychology, University of Sheffield, Sheffeld S10 2TN, UK

Shirley Reynolds
University of East Anglia, UK

Michael Rohrbaugh
Department of Psychology, University of Arizona, Tucson, AZ 85718, USA

Paul M. Salkovskis
University of Oxford Department of Psychiatry, Warneford Hospital, Oxford OX3 7JX, UK

David A. Shapiro
MRC/ESRC Social and Applied Psychology Unit, Department of Psychology, University of Sheffield, Sheffeld S10 2TN, UK

Varda Shoham
Department of Psychology, University of Arizona, Tucson, AZ 85718, USA

Mike Startup
University College of North Wales, Bangor, Wales, UK

Chris Thompson
University of Southampton Department of Psychiatry, Royal South Hants Hospital, Graham Road, Southampton SO9 4PE, UK

Judith A. Trowell
Tavistock Clinic, 120 Belsize Lane, London NW3 5BA, UK

Preface

Every second year, the Mental Health Foundation organizes a research conference. This residential, invitation-only event is an occasion for senior researchers and practitioners and those earlier in their careers to gather and debate the state of knowledge and research in their field. For two days in September 1993, an international group of 90 met at Balliol College in Oxford to consider the research foundations for psychotherapy practice.

The Mental Health Foundation, a charity founded in 1948, devotes half of its resource to funding research in mental health. The aim of this conference was to promote high quality research in psychotherapy, a priority issue for the Foundation. Contributors were asked to focus on issues in methodology and evaluation, to clarify how particular research designs might answer specific psychotherapy service questions, to draw out guidelines for good practice through examining exemplars of good research practice, and to consider how inevitable problems in implementation can be anticipated and overcome. Contributors were enjoined to avoid being caught up in claim and counter-claim of efficacy. Our intention was to arrive at 'state of the art' research design guidelines which would be of broad interest to the research community, purchasers and providers of psychotherapy services, and funding bodies in both public and private sectors.

The chapters in this book broadly follow the structure of the conference. Part I opens with two position statements, one from an outcome researcher and the other a process researcher. Though the treatment effects are difficult to demonstrate, personal aptitude as a pre-disposing factor has face validity and is a promising area for research inquiry; how might this interaction be teased out? Next comes the neglected topic of the local organizational issues which need to be attended to if the research is to be adopted and supported by the clinicians and others in the context where the work is to be done. The section closes with an overview of the new framework for setting research priorities and allocating funding in the National Health Service.

These essentials set in place, the four chapters in Part II consider individual therapy, this being the most common form of psychotherapy. Can the management of the waiting-list in a NHS Psychotherapy Department be improved by a randomized controlled trial? How may duration effects be investigated? What are the obstacles to researching long-term therapy? How may the specific effects of cognitive-behavioural therapy be demonstrated and their active ingredients identified?

The focus broadens in Part III. How to differentiate and quantify the effects of pharmacotherapy and psychotherapy? Two papers address the assessment of the benefits of intervention with the sexually abused. Psychological and economic indices as well as the elucidation of the critical therapeutic ingredients are considered. Finally, the formal activity of counselling in General Practice is undergoing rapid growth. How might its clinical benefit be researched?

The volume concludes with a synthesis of a presentation made at the end of the conference by a panel of four and the subsequent discussion. Informed by the conference papers and discussion, the chapter addresses the opportunities and challenges facing psychotherapy research. Ways forward are identified. The chapter begins with the concerns of a provider of clinical service and a possible purchaser, it being essential for researchers to address the concerns of these constituents in order to maximize relevance and the chance of securing public funding. Detailed advice is given on research *desiderata*, strategy, design and method.

We hope that the book will help researchers construct a variety of well-founded research projects and carry them through successfully to the mutual benefit of all concerned and the advancement of the science of psychotherapy.

Mark Aveline
David A. Shapiro

Part I

ELEMENTS OF DESIGN, ORGANIZATION AND FUNDING

1 The Design of Clinically Relevant Outcome Research: Some Considerations and an Example

KENNETH I. HOWARD*, DAVID E. ORLINSKY[a] AND ROBERT J. LUEGER[b]

Northwestern University, [a]University of Chicago and [b]Marquette University, USA

ABSTRACT   One key to impacting clinical practice is to provide information relevant to the current case in treatment. How can we design such psychotherapy research? The time-honoured approach is the randomized clinical trial, but for this to be informative to a practitioner (or policy maker), the study sample must represent some recognizable (to a clinician) patient population and main effects must be unambiguous. Since we are very unlikely to ever mount a clinical trial with a generalizable patient sample, can rarely avoid attrition, outcome distributions will overlap, and there will be reliable within-cell variation, this approach seems fruitless in terms of clinical relevance.

The approach that we have taken is to develop a theoretical framework (the dosage and phase models), operationalize and test the concepts, and develop a system for feedback to the clinician. We advocate a naturalistic strategy that is an outgrowth of the case study method. This method entails the use of objective data, continuous assessment, a model of problem stability, diverse and heterogeneous samples of patients, and clear evidence of an effect that can be quantified and used to modify treatment. We utilize a three-phase conception of psychotherapy—remoralization, remediation and rehabilitation—and present examples from our study of psychotherapy utilization.

INTRODUCTION

Does psychotherapy 'work?' For 30 years this question dominated the attention of psychotherapy researchers. Finally, Smith *et al*. (1980) and Shapiro and Shapiro (1982) laid this question to rest; at this point in time there are over 500 studies that attest to the efficacy of psychotherapy.

* Correspondence address: Department of Psychology, Northwestern University, 2029 Sheridan Road, Evanston, Illinois 60208-2710, USA.

Research Foundations for Psychotherapy Practice. Edited by M. Aveline and D. A. Shapiro.

In fact, it seems that psychotherapy is the best documented medical intervention in history. No other medical intervention has anywhere near the empirical scientific support that psychotherapy enjoys. What are we to make of this? Basically, the conclusion that psychotherapy 'works' is akin to finding that antibiotics 'work'. We are left with the problem of determining which antibiotics are appropriate treatments for which kind of infections. Obviously, this is a daunting task, since there are a variety of treatments (psychotherapies, antibiotics) and a variety of illnesses (psychopathologies, infections). Parloff (1982) warned us of this need for specificity in his classic article, 'Bambi Meets Godzilla'. Our obsession with documenting the efficacy of psychotherapy has only recently abated to the extent necessary to mount empirical studies of such specificity.

In order for the findings of psychotherapy outcome research to be clinically relevant, patients, therapists, therapies and treatment settings must be representative of some specified populations, and outcome measures must be clinically relevant. Ultimately, of course, the individual clinician must decide the extent to which research findings are based on samples, procedures and measures that are applicable to his or her own clinical practice. In what follows, we shall present some general methodological considerations, some theoretical models for the guidance of outcome research and our own recent work.

EMPIRICAL SCIENCE

There have been two broad approaches to establishing the relationship between the domain of empirical data (facts) and the domain of theoretical ideas (explanations). One approach—the hypothetico-deductive method applied to formal theory—emphasizes the primacy of theory and adduces evidence to either support or refute theoretical hypotheses. This method has been described as the *confirmatory approach* (even though a theory can only be corroborated, not proved, and the best that we can do is to reject the null hypotheses). The other approach emphasizes the primacy of data and seeks to provide explanations for observed patterns of phenomena— this determination of patterns leads to concatenated theory (Kaplan, 1964). This approach has been described as the *exploratory (naturalistic) approach*.

Each of these approaches has advantages and disadvantages, but there is a natural tension between them. The confirmatory approach emphasizes internal validity (cf. Campbell & Stanley, 1963) at the expense of the generalizability of findings. It invokes the assumption of *ceteris palibus* in the search for verification of theoretical hypotheses. The confirmatory approach entails a sufficient condition methodology that seeks to demonstrate how things *could* happen.

The exploratory approach, on the other hand, emphasizes generalizability (external validity), often at the expense of failing to eliminate alternative plausible explanations for the results (i.e. at the expense of internal validity). It admits the influence of a large number of potentially confounding variables and seeks to minimize or explore that influence through the use of statistical technology. The exploratory approach seeks to establish how things *do* happen.

Following Fisher's (1925) introduction of the analysis of variance and its associated methodology, Underwood (1949) and others further explicated the value of controlled experiments for research. Underwood's influential treatise helped to establish the hypothetico-deductive method as the 'official' methodology for empirical inquiry in the social and behavioural sciences. The *sina qua non* of this methodology is the random assignment of subjects to conditions (levels of the independent variable). Such random assignment is intended to ensure that any observed differences in the dependent variable (across levels of the independent variable) can be attributed solely to the influence of the independent variable and not to the influence of pre-existing differences among the subjects. Hence, a causal relationship between an independent and a dependent variable can be corroborated. Statistical methods were developed, moreover, to ensure that any observed difference between conditions was not due to chance. These statistical methods provide standards for rejecting the null hypothesis of no real difference. With random assignment, it is thought, *chance* is the only rival explanation for observed differences. In the absence of random assignment, however, pre-existing differences between groups—subject variables, and other 'uncontrolled' confounds—become rival explanations.

This confirmatory methodology also has been adopted as the 'official' model for psychotherapy research. This approach potentially is most useful for addressing the sufficiency of an intervention: '*Can* this treatment produce this desirable effect?' However, three problems arise when this methodology is applied to patient populations. First, the process of random assignment in itself does not ensure generalizability because it does not correct the process through which patients enter treatment; patients who persevere through an assigned treatment are not a random sample of any population. Second, because of the multitude of potentially causally relevant (independent) variables, the sample size is never sufficient to ensure that random assignment will equate groups with regard to possible confounds (i.e. will satisfy a *ceteris paribus* clause). Thus, even when groups are randomly 'equated' and significant outcome differences between them are obtained, we cannot be sure that these differences were a result of the causal influence of the selected independent variables. Finally, in conducting research with patients it is virtually impossible to avoid missing data (attrition) because patients routinely fail to provide complete information at all data points and routinely fail to complete treatment regimens as defined by the investigator. Missing data, therefore, *always* compromises random assignment and, thus, the randomized experiment is reduced to the status of a poorly designed quasi-experiment.

With regard to data loss, once patients have been formally included in a study sample, various factors determine whether they remain in it and contribute complete data for the full duration of the data collection (cf. Howard *et al.*, 1986b; Howard *et al.*, 1991). Therapists may come to feel that the conditions to which their patients have been assigned are inappropriate, and may exercise their professional responsibility in a way that effectively removes these patients from the study. Patients, of course, may also decide to discontinue, for reasons relevant or irrelevant to their disorder. Patients may remain in treatment but effectively drop out of the study by failing to complete some or all of the measures requested of them during the course of treatment, at termination, or at follow-up.

No matter how carefully a study is designed, the unavailability of certain data is inevitable. Data will be lost at many points during a study—both before patients have met inclusion criteria and after they have been randomly assigned to treatment conditions. Methodologically, data attrition is of paramount importance because it undermines the equivalence presumably established among comparison groups by undoing the effects of randomization in the selection and assignment of cases. There have been a variety of methods put forth in an attempt to compensate for the effects of data attrition, but all of them can be shown to rest on the untenable assumptions that attritors and completers do not bias estimates of the relationships between independent and dependent variables and that they represent equivalent samples of the same patient population. In other words, these repair methods rely on the assumption of random attrition (see Howard *et al.* 1986b, for a more detailed treatment). If attrition is not random, there is no way to compensate for its influence on main effects and interactions.

The characteristics of any therapeutic phenomenon are certainly jointly, and probably interactively, determined by the particular patients, the particular therapists, the particular interventions actually involved, and by the particular organizational, sociocultural, geographic and historical contexts in which treatment occurs. Each study of a treatment reflects a limited sampling of patients, therapists, therapies and contexts. If the comparisons made between different therapies, different types of patients, different types of therapists or different contexts are to be internally valid and are to be an accurate representation of differences in the population (externally valid), then investigators must find ways to cope with deleterious effects of data attrition from any of these sources.

In fact, however, there are no certain remedies for data loss. It appears that we must accept the fact of attrition, and realize that we cannot know, much less repair its effects after it has occurred. We cannot depend on estimates of the counter-to-fact-conditional—what would have happened, if patients who rejected treatment would have accepted it. It is never very satisfactory to speculate about events that could never happen or to come to the conclusion that a treatment would have been effective if only all patients would accept it. That is a very big *if*.

In clinical research, what one is really seeking are optimal points in a space defined by all of the plausible independent (potentially causal) variables in terms of dependent variable outcomes. Optimum-seeking (e.g. Adby & Dempster, 1974) involves seeking the points in the independent variable space associated with satisfactory outcomes, and it is surely the most sensible approach to research designed to provide clinically relevant findings. It is the knowledge of which points in the independent variable space are associated with the best outcomes that is ultimately of clinical importance. Some sample methodology and a more technical explication is provided by Howard *et al.* (1986b).

In clinical research, we are rarely in the position to test formal theoretical hypotheses, in any case. No matter what the methodology, clinical research has to be judged ultimately on the basis of its informativeness. It is clearly self-defeating to espouse a methodology that we must always fail to properly implement and to be forced to move (apologetically) to *post hoc*, secondary analyses to make sense of our data. Instead, we recommend the adoption of more exploratory methodology and a greater emphasis on the generalizability and the constructive *replication* of findings. This entails more explicit attention to quasi-experimental methods and to the design of studies in a manner that allows us to evaluate the most plausible, inevitable threats to internal and external validly.

This exploratory, naturalistic approach is an outgrowth of the case study method, which has had a tremendous impact on the practice of psychotherapy. A sophisticated case-based approach, which is useful for building concatenated theory and is sufficient to address potential threats to internal validity, has been outlined by Kazdin (1981). This case method entails the use of objective data, continuous assessment, a predictable model of problem stability, diverse and heterogeneous samples of patients, and clear evidence of an effect which can be measured for its magnitude and used to modify treatment processes. A case-based method that realizes all of these features can address potential confounds of history (events occurring in time), maturation (processes of change within the individual), testing (repeated exposure to assessment), instrumentation (changes in scoring criteria in the course of treatment) and regression (reversion of the score to the mean).

Since our goal is to optimally assign patients to treatments, our first task is to select relevant patient characteristics.

A MODEL OF RELEVANT PATIENT PRESENTING CHARACTERISTICS

The DSM-III-R diagnostic system has not been very useful for assigning patients to treatments. In addition to the obvious problem in employing a more or less arbitrary (and seemingly ever changing) system, most clinicians are unwilling or unable to get the required training or to spend the time necessary to arrive at a reliable and valid diagnosis for a patient. For example, McNeilly (1991) found virtually no relationship between the diagnoses of clinical screeners (at intake) and diagnoses based on the Structured Clinical Interview for DSM-III-R (SCID)(Spitzer *et al.*, 1988). Therapists are usually confronted by a patient who is quite upset—anxious, frustrated, depressed—and must deal directly with this presentation. At this level, what do we need to assess?

Based on extensive literature reviews (Howard & Orlinsly, 1972; Orlinsky & Howard, 1978, 1986), we have developed a conception of some important psychological characteristics that have an impact on the patient's utilization of individual psychotherapy. This conception is comprised of three variables:

(1) psychopathology (presenting symptoms or syndromes);
(2) pathology-proneness (psychological vulnerabilities);
(3) life stress.

Psychopathology refers to the manifest psychiatric symptomatology of the patient. It is concerned with the intensity and types of distressing experiences and behaviours. DSM-III-R, Axis I diagnoses and symptomatic distress are examples of variables in this domain.

All persons are *pathology-prone* to some extent, in the sense that there are some life situations with which each of us would have difficulty coping. Some people have relatively stable psychological and behavioural vulner-abilities or deficits that make it difficult for them to cope with the challenges and stresses of a wider variety of life situations. Their proneness to psychopathology may stem from severe emotional conflicts, from maladaptive social attitudes, from dysfunctional cognitions, etc. DSM-III-R, Axis II diagnoses and dysfunctional cognitive styles are examples of variables in this domain.

In the context of a person's resources and vulnerabilities, *environmental stress* may overwhelm his or her social supports and coping capacities and, thus, trigger manifest psychopathology and the need for therapeutic intervention.

In assessing a patient's condition it is important to measure the kind and severity of symptoms, the kind and pervasiveness of vulnerabilities, and the likely recurrence of the environmental stressors.

THE DOSAGE AND PHASE MODELS

The dosage model of psychotherapeutic effectiveness (Howard *et al.*, 1986a) demonstrated a lawful linear relationship between the log of the number of sessions and the normalized provability of patient improvement. The dose–response function was shown to be different for different syndromes of pathology (e.g. Borderlines require a higher treatment dose than do Depressives). Subsequent work (e.g. Howard *et al.*, 1993b; Horowitz *et al.* 1988; Kopta *et al.*, in press) has provided evidence of the differential responsiveness to psychotherapy of various symptoms and syndromes.

The dosage model has also had some methodological impact in that it allows for a more refined understanding of the findings of studies. For example, Eysenck (1952) reported that the spontaneous remission rate for neurosis was 67%, almost identical to the improvement rates he found for psychotherapy. He concluded that psychotherapy was no more effective than spontaneous remission. A re-analysis of Eysenck's data based on the dosage model (McNeilly & Howard, 1991), however, showed that the impact of a few months of psychotherapy was equal to the impact of 2 years of all other forms of help available to an individual—a striking testimony to the efficacy of psychotherapy!

The dosage model gave rise to the following three-phase conception of psychotherapy (see Howard *et al.*, 1993b).

1. REMORALIZATION

Some patients are so beset by problems that they become demoralized and feel that they are at their 'wit's end'. This experience is so pervasive and of such an intensity that the patient's ability to mobilize personal coping resources is severely disrupted, and he or she begins to feel 'I am just barely holding on by my fingernails'. This state responds to a variety of interventions—medication, vacation, emotional support, etc.—and in psychotherapy would usually abate in the period between making an appointment with a therapist and a few sessions of (supportive or crisis-oriented) psychotherapy. For many patients, this reduction of distress will allow them to mobilize their own coping resources in a way that leads to resolution of their current life problem(s). Other patients will move on to a second phase of therapy.

2. REMEDIATION

A second phase of therapy is focused on remediation of the patient's symptoms, symptoms that have led the patient to feel so upset that he or she had to seek treatment. (Some patients will begin therapy in phase two, i.e. they will seek help before the emergence of immobilizing distress.) During this second phase, treatment is concerned with refocusing the patient's coping skills in a way that brings symptomatic relief. This phase usually lasts 3 or 4 months (about 16 sessions), but can be shorter or longer depending on the number and extent of current symptoms. At this point many patients will terminate treatment, but some will realize that the problem(s) that brought them to therapy have been encountered often in their lives (e.g. instability in employment, problematic interpersonal relationships) and are probably the result of long-standing patterns (habits, character) that are maladaptive and/or are hindering the achievement of life goals (e.g. finding a satisfying career, forming a long-lasting intimate relationship). These patients will move on to the third phase.

3. REHABILITATION

A third phase of treatment is probably what has traditionally been thought of as 'psychotherapy' in that it is focused on unlearning troublesome maladaptive long-standing patterns and establishing new ways of dealing with various aspects of life (prevention, rehabilitation). (Again, some patients may enter therapy at this point, having had no specific precipitating problem or acute distress.) This phase of therapy may last many months or years, depending on the accessibility and malleability of these maladaptive patterns.

To the extent that these phases are distinct, they imply different treatment goals and, thus, the selection of different outcome variables. This model also suggests that different interventions will be appropriate for different phases of therapy and that certain tasks may have to be accomplished before others can be undertaken. Our own work (Howard *et al.*, 1993b) has demonstrated that these phases are sequentially dependent: remoralization → remediation → rehabilitation. Also, there is the implication that different therapeutic processes may characterize each phase. Remoralization might be accomplished through the use of encouragement and empathic listening; assertiveness training may be useful for rehabilitation, etc. We next turn to a consideration of such process variables.

THE GENERIC MODEL OF PSYCHOTHERAPEUTIC PROCESS

Based on three major reviews of the psychotherapy literature (Howard &

Orlinsky, 1972; Orlinsky & Howard, 1978, 1986), we developed a generic model of psychotherapy (Orlinsky & Howard, 1987). This model distinguishes five major process components:

(1) the therapeutic contract;
(2) therapeutic interventions;
(3) the therapeutic bond;
(4) the participants' personal self-relatedness;
(5) therapeutic realizations.

The *therapeutic contract* specifies the terms under which therapy is to be carried out (schedule, fees, etc.), and subsumes an 'understanding' between the parties (i.e. patient and therapist) concerning the goals and methods of treatment. *Therapeutic interventions* made to implement the contract include both a 'diagnostic' phase (the formulation of patient's problem) and a complementary 'technical' phase (the techniques of treatment).

On the other hand, the *therapeutic bond* (and the emergent interpersonal issues that limit or compromise it) constitutes the ubiquitous and inevitable human relationship that forms between patient and therapist. We view the bond between patient and therapist in terms of three interrelated dimensions: their *working alliance*, defined as their reciprocal personal role-investments; their *empathic resonance*, determined by the expressive attunement of their communicational styles, and the congruence of their personal imageries and idioms; their *mutual affirmation*, as manifested in their respectful, yet at times also challenging, concern for each other's personal interests and wellbeing.

In addition to forming a relationship with one another, each participant inevitably stimulates and responds to him/herself as part of the overall interaction. *Personal self-relatedness* is the inclusive name we use to describe how an individual manages the reflexive or self-directed aspects of social encounter (self-responsiveness, self-cognition, self-control, self-esteem). In the context of therapeutic interaction, these reflexive dimensions determine the openness (or defensiveness) of the persons who face one another as patient and therapist and encompasses the issue of defensiveness.

Finally, *therapeutic realizations* are the yield of the therapeutic process (e.g. insight, catharsis, encouragement) occurring in the context of the treatment relationship, either during therapy sessions or in the intervals between sessions when the patient is thinking about therapy or the therapist.

The interrelationships postulated in this model have been subjected to empirical test (Grawe *et al.*, 1989; Kolden, 1991; Kolden & Howard, 1992) and have held up very well so far. The generic model points the way to the selection of outcome variables that are specific to therapeutic change processes, e.g. the formation of a therapeutic bond, the amelioration of patient defensiveness. Here, we have a clear example of the blurring of the distinction between process and outcome in that process variables become the target of interventions and, thus, become outcomes.

THE NORTHWESTERN/CHICAGO PSYCHOTHERAPY RESEARCH PROGRAM

The example that we shall present is a naturalistic (exploratory) study of psychotherapy supported by grants from the National Institute of Mental Health. This is the largest study of psychotherapy ever mounted. In this study we did not directly interfere with the treatment episode of any patient. We did not assign patients to therapists, we did not limit the number of sessions, we did not tell therapists how to conduct their sessions, etc. We had unknowable indirect impact, however, since patients and therapists knew that they were participating in research (and had consented to this participation) and had completed research questionnaires that inquired about the treatment and its effects. We have tried to make all of this treatment-syntonic and 'user-friendly' by designing and selecting procedures that have face validity (for patients, therapists, and the clinical research community) in terms of the goals and procedures of individual psychotherapy.

Most importantly, our selection of assessments was theory based. In accordance with the dose–response model, we assessed outcome regularly throughout the course of treatment. This approach has been reinforced by our findings that termination is an arbitrary (and often clinically irrelevant) phenomenon (Stacy, 1992) and that many early terminators quickly enter psychotherapy in another setting (Schwartz, 1991). The selection of outcome measures was based on the phase model of psychotherapy. As we have already noted, this phase model is based upon an hypothesized causal sequence of change: remoralization → remediation → rehabilitation. Remoralization was monitored by a measure of subjective wellbeing and showed a quick response to treatment. Remediation was monitored by a measure of psychiatric symptoms and responded somewhat more gradually, once remoralization was accomplished. Rehabilitation was monitored by a measure of life functioning and responded even more gradually, once remediation had been accomplished.

The major advantage of our approach is that, not only have we demonstrated the pattern of response to treatment of these variables, we also have developed a system for feedback to the therapist regarding the status and progress of each case (Howard *et al.*, 1992). This feedback can be used in supervision and case management, as well as in making treatment decisions. In this way, we have translated the clinical relevance of research from a general conceptual issue into a practical clinical application.

PATIENTS

As can be seen in Table 1.1, the typical patient in our sample was single (61%), female (68%), between the ages of 25 and 35 (52%), with at least some college education (90%). Sixty-eight per cent of the patients had had some previous treatment, 57% had had prior individual psychotherapy, with 38% of the total sample having had previous individual treatment with a duration of at least eight sessions. Patients were self-referred and were treated for a variety of mild to moderate psychological disorders. The patient sample was reasonably representative of the psychotherapy out-patient population (Vessey & Howard, 1993), although our patients were more likely to be single, are better educated, and somewhat younger. Via a clinical screening interview, all patients were determined to be appropriate for psychodynamically-oriented, intensive, individual therapy. Participation in the research project was voluntary), informed consent was obtained, and confidentiality of responses was assured.

With regard to treatment duration, the number of sessions of psycho-therapy ranged from 1 to over 400—the median number of sessions was 15.

Table 1.1. Demographic comparison of people who make a mental health visit

	National estimate ($N = 1429$)	Chicago clinic ($N = 540$)
Gender		
Male	34%	32%
Female	66%	68%
Education		
Grammar school or less	9%	0%
Some high school	11%	3%
High school graduate	33%	7%
Some college	23%	28%
College graduate	24%	62%
Marital status		
Married	50%	19%
Widowed	5%	2%
Separated or divorced	22%	18%
Never married	23%	61%
Age (years)		
18–20	4%	3%
21–30	22%	53%
31–40	33%	30%
41–50	20%	10%
51–60	11%	3%
61+	10%	1%

For purposes described below, the sample was divided into three duration of-treatment groups: 1–8 sessions ($N = 232$); 9–26 sessions ($N = 196$); and, more than 26 sessions ($N = 226$).

THERAPISTS AND SETTING

This study was conducted at Northwestern Memorial Hospital's Institute of Psychiatry. This is the teaching facility for the Department of Psychiatry and Behavioral Sciences at Northwestern University's School of Medicine. There were many outpatient programmes at the Institute: chemical dependence, eating disorders, sexual dysfunction, etc. The psychotherapy programme, which was our data collection site, was designed to serve the kinds of patients generally found in the private practice of psychotherapy. In general, therapy was once-weekly.

As noted, the orientation of the programme was psychodynamic; supervisors espoused this therapeutic approach, case presentations followed this model, and attempts were made to conceptualize each case from this perspective. For a subsample of therapists, responses to the Therapeutic Procedures Inventory-Revised (TPI-R) (Orlinsky et al., 1987) were examined. The TPI-R is a therapist questionnaire that assesses interventions used in therapy sessions. Thus, it provides quite specific descriptive information about the type of therapy that is characteristically delivered in a treatment setting. An analysis of 155 TPI-Rs indicated that the most frequently endorsed therapeutic interventions were (in order of frequency of endorsement):

(1) conveyance of non-judgemental acceptance;
(2) attention to personal reactions as a means of understanding the patient;
(3) facilitation of patient focusing on inner feelings and experiences;
(4) reflection of feelings;
(5) direction of attention to patterns or themes;
(6) examination of meaning of thoughts, behaviours or feelings;
(7) reframing of problem formulation.

There were 118 therapists in the Institute's psychotherapy programme. The majority of the therapists were in some stage of training—psychology practicum students, psychology interns and psychiatry residents— although most had had some prior experience. As can be seen in Table 1.2, about 68% of these therapists had seen more than 10 patients in individual psychotherapy. Over 70% of the therapists were psychologists, 16% were psychiatrists, and 8% were social workers. Over 90% were under 40 years of age; 64% were female and 51% had never been married. Over 90% of the therapists were currently in, or had had, personal therapy.

Table 1.2. Therapist demographics ($N = 118$)

Gender	
Male	36%
Female	64%
Marital status	
Married	40%
Widowed	2%
Separated or divorced	7%
Never married	51%
Age (years)	
21–30	55%
31–40	38%
41–50	6%
51+	1%
Experience	
Less than 1 year	2%
1–5 years	79%
6–10 years	11%
11+ years	8%
Number of patients seen	
1–5 patients	14%
6–10 patients	18%
11–20 patients	21%
21–40 patients	23%
40+ patients	24%

MEASUREMENT OF CHANGE

The determination of outcome was based on the change from pre-therapy status to status assessed after a specified number of sessions of psychotherapy. Outcome at termination (regardless of length of treatment) was also assessed because of its relevance to the evaluation of psychotherapeutic services. The scores for all of our assessment variables are expressed in T-scores (mean = 50, SD = 10) on the basis of our entire patient sample at intake ($N \cong 500$). In all cases a higher score indicates healthier status.

Speer (1992), borrowing from Nunnally (1967), proposed a method for correcting pre-treatment scores for the expected effect of regression (true score adjustment). This entails multiplying the reliability of a measure by the difference between an individual's score and the population mean, and adding the mean to this product. Thus, in our case, if a person scored 40 on a measure whose reliability was 0.80, we would subtract 50 (the mean) from 40 and multiply the difference (−10) by 0.80; we would then add 50. In this case the score of 40 would now be 42. This true score adjustment is applied to all pre-therapy measures.

Some patients at intake had a score near the ceiling of a measure (or had a score that was already in the 'normal range' for that measure), making improvement impossible to attain (or assess). Consequently, in order to evaluate improvement on a particular measure, we first eliminated patients whose pre-therapy true score adjusted scores were already in the normal range. This normal range criterion will be described in greater detail below.

For clinical purposes, we were interested in a dichotomous classification —improved vs. not improved. The notion of reliable improvement has to do with our confidence that a new score for a patient was significantly higher than a pre-therapy true score estimate. The criterion of reliable improvement (Jacobson & Truax, 1991) makes use of the concept 'standard error of measurement' (SEM), which is a function of measurement instrument reliability. The SEM is defined as the product of the observed standard deviation of that measure and the square root of (1 – reliability), and represents the standard deviation of a theoretical distribution of observed scores around a patient's true score. Ninety per cent of observed scores are expected to fall below that value that lies 1.25 SEMs above a patient's true score. Based on the SEM, it is possible to construct a 90% confidence interval with one upper bound around each patient's true score estimate. If a patient's subsequent score was above this cut-off, that is at least 1.25 SEMs above the intake true score estimate, we concluded that the subsequent score was indeed higher than the pre-therapy score and, thus, that the patient had improved.

In order to eliminate patients whose score on a particular variable was too high at intake, we determined a 'normal range'. For some measures we had data on samples of non-patients (normals). When such data are available, Jacobson and Truax (1991) have provided a system for assessing whether a score is more likely to belong to a patient or to a non-patient population. Assuming homogeneity of variance and normal distributions, they state this criterion as follows:

> (c) The level of functioning subsequent to therapy places that client closer to the mean of the functional population than it does to the mean of the dysfunctional population (p.13).

In the event that variances of functional and dysfunctional populations are unequal, Jacobson and Truax present an alternative method for calculating (c). Whether variances are equal or not, Jacobson and Truax's method is functionally the same as fitting two normal frequency functions to patient and non-patient data, respectively, and estimating that point at which a score is more likely to have been sampled from the non-patient population than the patient population.

DATA COLLECTION

Data were gathered through the use of questionnaires and rating scales.

Data points were spread from clinic registration through therapy to a 3-month follow-up. Therapists completed forms at intake, sessions 1, 3, 6, 16, 26, etc., and at termination. Patients completed forms at registration, intake screening, sessions 1, 2, 3, 4, 5, 6, 10, 15–17, 25–27, etc., and at the 3-month follow-up.

MEASURES AND RESULTS

For present purposes we have only presented results based on patient self-reports. These results are organized in relation to the phase model of therapy.

Subjective Wellbeing—a measure of remoralization

In keeping with the vast literature on wellbeing and quality of life, the content sampling for this scale includes both positive and negative affect (Diener, 1984; Watson and Tellegen, 1985), and health and contentment (Cowen, 1991; Viet and Ware, 1983). The four items include the dimensions of distress, energy, and health, emotional and psychological adjustment,

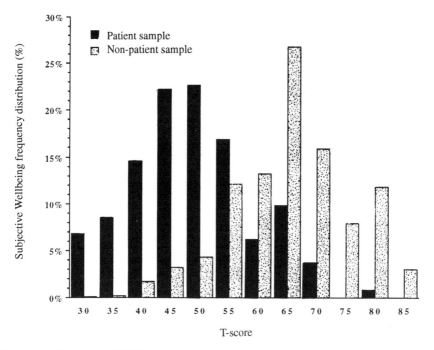

Figure 1.1 Subjective Wellbeing: frequency distribution of T-scores for a sample of 531 patients at intake and a sample of 979 non-patients

and current life satisfaction. Structured responses are provided for each item.

Based upon a sample of 197 patients prior to treatment, the internal consistency of the Subjective Wellbeing scale was 0.79. Based upon a sample of 93 patients over a 1-week interval (before beginning treatment), the test–retest reliability was 0.82.

In a sample of 108 non-patients (67% were female, mean age was 20 years), the Subjective Wellbeing scale correlated 0.79 with the 22-item General Wellbeing Scale (Dupuy, 1977), 0.51 with a 10-item measure of positive affect and -0.70 with a 10-item measure of negative affect (Watson and Tellegen, 1985), 0.73 with the total score of the SF-36 (Stewart *et al.*, 1988) and 0.76 with the five-item mental health index of the SF-36 (Ware and Sherbourne, 1992). The mean T-score for a heterogeneous sample of 979 non-patients was 64.9 with a standard deviation of 9.37 (t (1529) 29.49, $p < 0.0001$. It is clear that psychotherapy patients (mean = 50; SD = 10) have lower subjective wellbeing than do non-patients. Figure 1.1 shows the frequency distribution of T-scores for 531 patients at intake and 979 non-patients.

Figure 1.2 displays the percentage of patients who were below the

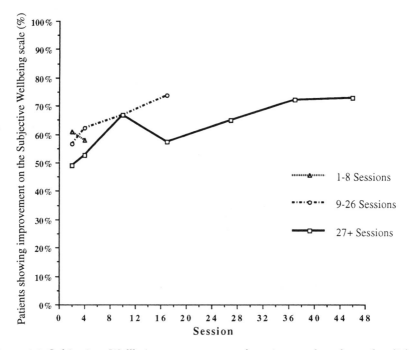

Figure 1.2 Subjective Wellbeing: percentage of patients who showed reliable improvement at selected sessions for three duration-of-treatment groups

normal range at intake and showed reliable improvement at session assessments. In order to evaluate the effect of attrition, patients were separated into three treatment-duration groups: 1–8 sessions, 9–26 sessions, more than 26 sessions. Figure 1.2 shows that there was little systematic effect of attrition i.e. with the possible exception of session 17, there is little evidence that patients are terminating therapy because of their subjective wellbeing status. Figure 1.3 shows the dose–response relationship of subjective wellbeing for the full sample of patients. As can be seen, 55 percent of patients reliably improved by session 2, with steady improvement to session 10, after which there was little additional gain. As hypothesized, demoralization appears to respond quickly to psychotherapy.

Current symptoms—a measure of remediation

Because of the growing interest in empirically-based diagnosis, we devised a new symptom checklist. From clinical diagnoses based on 140 Structured Clinical Interviews for the DSM-IIIR (SCID), we found that 74.3% of outpatients had at least one of the following six Axis I diagnoses: adjustment

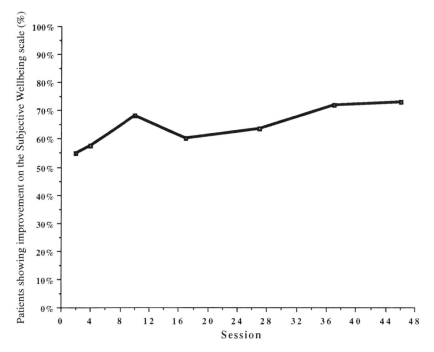

Figure 1.3 Subjective Wellbeing: percentage of patients who showed reliable improvement at selected sessions

disorder, anxiety, bipolar disorder, depression, obsessive–compulsive disorder, phobia. Of those patients who qualified for *any* DSM Axis I diagnoses, 92.0% had one of these six diagnoses.

Using the Diagnostic and Statistical Manual for Mental Disorders (DSM-III-R) (American Psychiatric Association, 1987), we listed the signs and symptoms for these six diagnoses, and also the diagnosis of substance abuse, and recast them as a 40-item patient self-report symptom checklist. There are at least three questions for each diagnosis; however, the higher the prevalence of the diagnosis the greater the number of questions pertaining to that diagnosis.

The Current Symptoms (CS) scale employs a five-point, fixed-response format. However, in keeping with DSM-III-R, the response categories reflect the *frequency* of experiencing each symptom in the past month.

Based on a sample of 160 patients at intake, the total score of the Current Symptoms scale had an internal consistency of 0.94. Test–retest correlations were computed over a 3–4 week period from registration to the second session of psychotherapy. The test–retest reliability for the total score was 0.85 ($N = 53$).

The total score on the CS scale correlated 0.91 ($N = 249$) with the total score of an abbreviated (47-item) version of the SCL-90R (Derogatis, 1977).

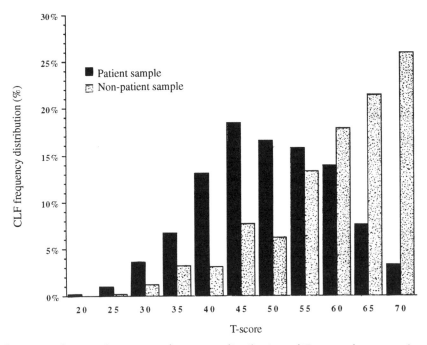

Figure 1.4 Current Symptoms: frequency distribution of T-scores for a sample of 530 patients at intake and a sample of 574 non-patients

We examined, in some detail, the concurrent validity of the Depression subscale of the CS. The Depression subscale correlated 0.68 ($N = 272$) with the Center for Epidemiologic Studies of Depression (CES-D scale) (Radloff, 1977) and 0.87 ($N = 44$) with the Beck Depression Inventory (Beck et al., 1961). Using the Structured Clinical Interview for Diagnosis (SCID) (Spitzer et al., 1988), 13 patients diagnosed with depression were compared to 15 patients with other Axis I disorders. SCID diagnosed Depressives' CS mean Depression score was nearly a full standard deviation higher than the mean for all patients. The mean Depression score for Depressed patients was significantly ($t(26) = 2.94$; $p < 0.01$) above that of patients with other DSM-III-R Axis I diagnoses. The mean T-score for a heterogeneous sample of 574 non-patients was 57.4 with a standard deviation of 7.93 ($t(1120) = 14.42$, $p < 0.0001$). Again, we see that psychotherapy patients are more symptomatic than are non-patients. Figure 1.4 shows the frequency distribution for 530 patients, at intake, and 574 non-patients. As can be seen, psychotherapy patients are more symptomatic than are non-patients. There is a fair degree of overlap of these distributions, however, indicating that non-patients report a relatively high level of psychiatric symptomatology.

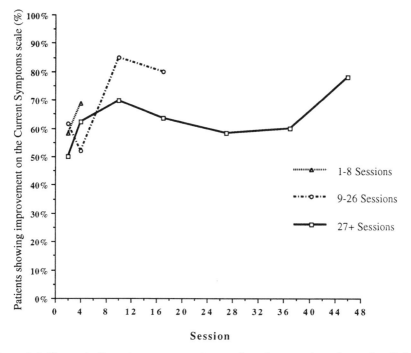

Figure 1.5 Current Symptoms: percentage of patients who showed reliable improvement at selected sessions for three duration-of-treatment groups

Figure 1.5 shows the dose–response relationship for symptoms for the three duration-of-treatment groups. Again, with the possible exception of sessions 10 and 17 for the 2–26 session group, improvement in symptoms did seem related to continuation in treatment. Figure 1.6 shows the dose–response relationship of psychiatric symptoms for the full sample. About 55% of patients showed reliable improvement by session 2, with a linear increase to about 80% by session 10. This was followed by a reappearance of symptoms and another period of remediation.

Current Life Functioning, Interpersonal Problems and Self-esteem—measures of rehabilitation

In the Current Life Functioning (CLF) scale, the patient is asked to report the degree to which his/her emotional and psychological problems are interfering with functioning in six life areas. The intent of this scale is to assess the extent of disability caused by the patient's emotional and psychological condition. A variety of inventories, including the Social Security disability guidelines, were examined and self-report items were generated to cover the range of disabilities. Factor analyses of these items

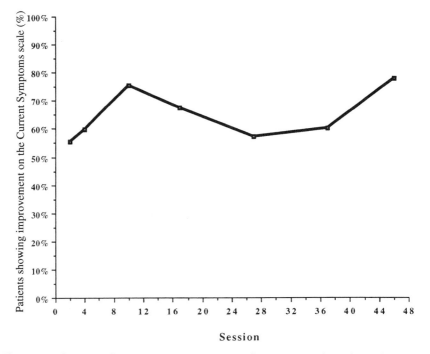

Figure 1.6 Current Symptoms: percentage of patients who showed reliable improvement at selected sessions

suggested six dimensions. The 24 items in this questionnaire are categorized into these six life areas so that there are at least three questions per area. The Family Functioning, Intimate Relationships and Social Relationships scales inquire about the patient's interactions with others, and carrying out his/her responsibilities to these people. The Health and Grooming scale addresses the patient's health habits and hygiene, and the Work Functioning scale assesses the patient's workplace interactions and ability to complete tasks. The Self-management scale assesses the patient's control over, conceptions of, and satisfaction with him/herself.

Based on a sample of 70 patients prior to treatment, the internal consistency (alpha) of the total score of the CLF was 0.93. The 3–4 week (from registration to the second therapy session) test–retest correlation was 0.76 ($N = 48$).

We have no data regarding the relationship of the CLF scale with other self-report measures of functioning. The mean T-score for a heterogeneous sample of 495 non-patients was 59.4 with a standard deviation of 10.16 ($t(1043) = 15.52$, $p < 0.0001$). Figure 1.7 shows the frequency distribution for 521 patients, at intake, and 495 non-patients. Again patients scored significantly lower than did non-patients.

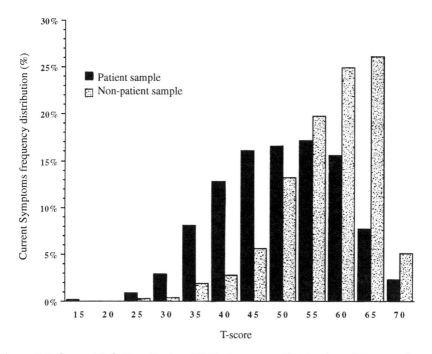

Figure 1.7 Current Life Functioning (CLF): frequency distribution of T-scores for a sample of 521 patients at intake and a sample of 495 non-patients

Figure 1.8 shows the dose–response relationship for life functioning for the three duration-of-treatment groups. In the 1–8 session group, 40% of patients showed reliable improvement by session 2 and 60% by session 4. The pattern of improvement for the other two groups were similar to one another. Figure 1.9 shows the dose–response relationship of life functioning for the full sample. About one-third of of the patients showed reliable improvement by session 2, with an increase to about 60% by session 10. This was followed by an erratic pattern of rehabilitation.

A 26-item version of the Inventory of Interpersonal Problems (IIP) (Horowitz et al., 1988) was included in patient questionnaire packets. The top loading three to five items for each of the six problem scales in the original IIP were selected for inclusion. Subsequently, the 26-item set was analysed to produce three interpersonally-homogeneous problem scales, labelled Control, Detached, and Self-effacing. Each of the 26 items was projected onto the interpersonal circumplex formed by the original version of the IIP (see Gurtman, 1993, for details on this method). Three groupings of items were apparent from this analysis, covering roughly two-thirds of the circumplex space. These became the basis for three scales, which were labelled in accordance with their predominant interpersonal content, as

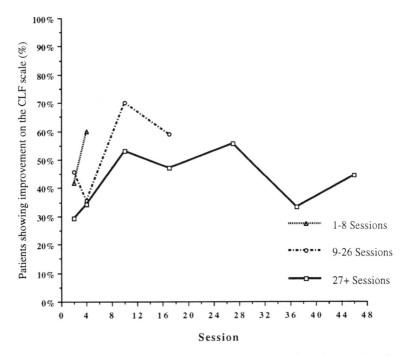

Figure 1.8 Current Life Functioning (CLF): Percentage of patients who showed reliable improvement at selected sessions for three duration-of-treatment groups

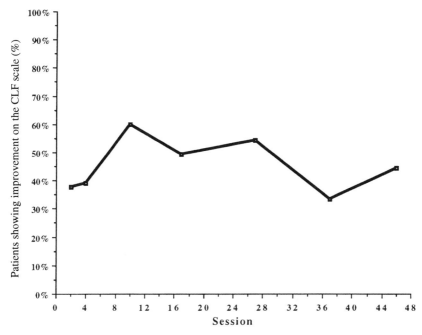

Figure 1.9 Current Life Functioning (CLF): Percentage of patients who showed reliable improvement at selected sessions

revealed by their respective circumplex projections (e.g. Kiesler, 1983). As confirmation of these groupings, a hierarchical cluster analysis (average linkage method) was performed on the item intercorrelations, first on non-patient data ($N = 1093$) and then, as cross-validation, on a patient sample at intake ($N = 472$). The results were virtually identical and consistent with the scale designations. Finally, the patient data were subjected to a principal components analysis. Three clear, orthogonal factors emerged with eigenvalues of 6.55, 2.40 and 2.20, respectively (the fourth and fifth factors had eigenvalues of 1.32 and 1.17, respectively). The three factors were then rotated to the Varimax criterion, and the respective factor scores were correlated with the three scales. The correspondence between the factors and the scale scores was nearly perfect, with Control correlating 0.95 with Factor II, Detached correlating 0.95 with Factor I, and Self-effacing correlating 0.93 with Factor III.

Tables 1.3 to 1.5 show the per cent endorsement of the items for the three interpersonal problem scales for 307 patients and 1093 non-patients. The response format had five points ranging from 1 ('not at all distressed') to 5 ('extremely distressed'). An item was considered to be endorsed if the patient responded 3 ('moderately distressed') or higher. As Table 1.3 shows, patients and non-patients were quite similar on the Control scale

Table 1.3. Interpersonal Problems: Control scale—per cent endorsement for 307 patients at intake and a sample of 1093 non-patients

Item	Endorsement (%)	
	Patients	Non-patients
Hard to accept another person's authority over me.	45.0	28.1
I tell personal things to other people too much.	37.1	25.0
I try to control people too much.	26.2	23.6
I argue with other people too much.	23.8	26.8
I am too aggressive toward other people.	23.5	23.1
I manipulate other people too much to get what I want.	21.2	20.2
I clown around too much.	18.2	35.0
Average item endorsement (%)	27.9	26.0
Alpha	0.77	0.75

Table 1.4. Interpersonal Problems: Detached scale—per cent endorsement for 307 patients at intake and a sample of 1093 non-patients

Item	Endorsement (%)	
	Patients	Non-patients
Hard to feel comfortable around people.	51.0	29.1
I keep other people at a distance too much.	49.7	35.2
Hard to open up and tell my feelings to another person.	49.3	46.8
Hard to feel close to other people.	48.0	21.5
Hard to make a long-term commitment to another person.	45.2	30.9
Hard to make friends.	41.1	29.1
Hard to have someone dependent on me.	36.2	34.5
Hard to be honest with other people.	33.8	16.2
Hard to get along with people.	23.2	12.1
Hard to be supportive of another person's goals in life.	15.6	10.6
Average item endorsement (%)	39.3	26.6
Alpha	0.84	0.82

(non-patient mean = 50.8, SD = 8.89; $t(1579 = 1.71$, $p < 0.05$), although patients had notably more difficulty accepting another person's authority. The results in Table 1.4 indicate that patients were more Detached (in interpersonal-relationships) than were non-patients (non-patient mean = 54.3, SD = 8.22; $t(1579) = 8.44$, $p < 0.0001$). Patients also were more Self-effacing (non-patient mean = 54.1, sd = −8.93; $t(1579) = 7.89$, $p < 0.0001$).

Table 1.5. Interpersonal Problems: Self-effacing scale—per cent endorsement for 307 patients at intake and a sample of 1039 non-patients

Item	Endorsement (%)	
	Patients	Non-patients
I am too sensitive to rejection.	78.8	58.7
I worry too much about disappointing other people.	72.2	54.9
I put other people's needs before my own too much.	65.0	42.0
Hard to let other people know what I want.	63.2	50.5
Hard to get out of a relationship I don't want to be in.	56.2	56.7
I let other people take advantage of me too much.	55.3	38.3
Hard to accept praise from another person.	43.2	38.8
Hard to stick to my own point of view and not be swayed.	42.7	28.8
Hard to disagree with other people.	41.9	28.2
Average item endorsement (%)	57.6	44.1
Alpha	0.82	0.79

Clinical improvement was assessed by the use of the original response categories. In general, a patient was deemed as improved if he or she originally endorsed the item (checked 3, 4, or 5) and subsequently changed at least two response categories (e.g. from 'extremely distressed' to 'moderately distressed, from 'moderately distressed' to 'not at all distressed'). Figure 1.10 shows the dose–response relationships of these three scales. In order to attenuate change due to decrease in overall distress, session 2 (rather than intake) was selected as the baseline. Nonetheless, about 45% of patients showed clinically significant change from session 2 to session 4 on the Control scale. with a linear increase in response after session 10. There was a linear increase in response for the Detached scale after session 17, while no dose–response relationship for the Self-effacing scale was apparent in the first 38 sessions.

Patients also rated the 10-item Rosenberg Self-esteem scale items on a six-point Likert scale (1 = totally agree, 6 = totally disagree). Rosenberg (1979) originally reported an alpha of 0.77. Byrne and Shavelson (1986) reported an alpha of 0.87 for a sample of 832 Canadian youth. Silter and Tippett (1965) reported a 2-week test–retest correlation of 0.85 for a sample of 28 college students, while Byrne (1983) reported a 7-month test–retest coefficient of 0.63 for a sample of 990 Canadian high-school students. Based on intake data from 336 patients in our sample, a principal components factor analysis of these items resulted in more than 50% of the

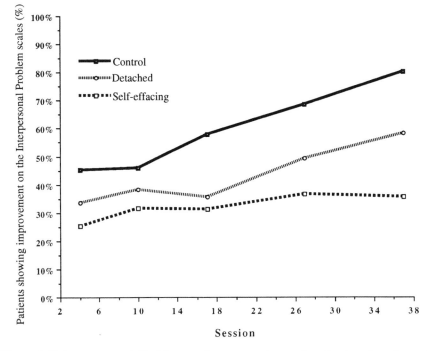

Figure 1.10 Interpersonal Problem scales: percentage of patients who showed reliable improvement from session 2 at selected sessions

total variance being accounted for by a single factor. Based on our patient sample, Cronbach's alpha for the 10-item scale was 0.89

Figure 1.11 shows the frequency distributions for the total scale score for a sample of 490 patients at intake and a sample of 298 non-patients. The non-patient sample had a mean T-score of 60.3 and a standard deviation of 6.51 ($t(786) = 15.82$, $p < 0.0001$). Patients at intake clearly reported lower self-esteem than did non-patients.

Figure 1.12 shows the dose–response relationship for self-esteem for the three duration-of-treatment groups. There was no suggestion of an attrition effect. Figure 1.13 shows the dose–relationship of self-esteem for the full sample. About one-quarter of the patients showed reliable improvement by session 2, with no subsequent increase until session 36. Here, we have the clearest indication of a variable that takes some time to respond to treatment.

The results, reported above, provide evidence for the differential responsiveness of psychological variables to individual psychotherapy. Since they were based on a naturalistic design, they should have consider-able generalizability to other samples of patients and therapists, though

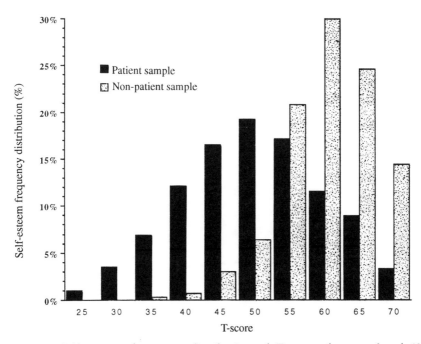

Figure 1.11 Self-esteem: frequency distribution of T-scores for sample of 490 patients at intake and a sample of 298 non-patients

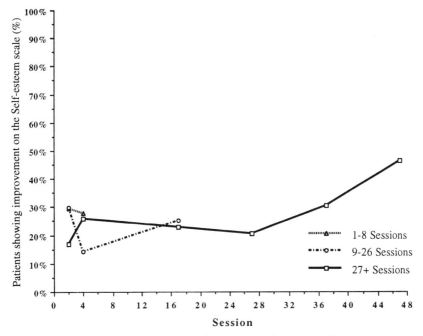

Figure 1.12 Self-esteem: percentage of patients who showed improvement at selected sessions for three duration-of-treatment groups

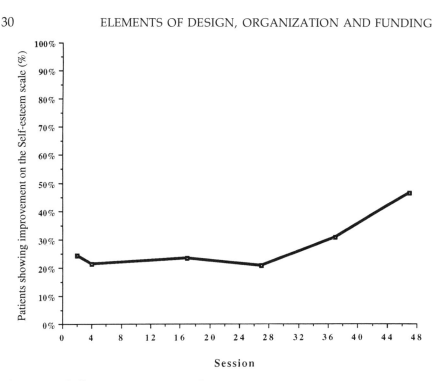

Figure 1.13 Self-esteem: percentage of patients who showed reliable improvement at selected sessions.

they may have limited application to other therapeutic orientations (e.g. cognitive behaviour therapy, pharmacological interventions).

SINGLE CASE APPLICATIONS

In order to be relevant to practice, research results must be interpretable in a way that allows application to the single case. Using the kinds of results presented in the previous section we have developed a system for assessing (tracking) a case in a manner that can inform therapists, supervisors and patients about the efficacy and direction of a psychotherapeutic treatment (Howard *et al.*, 1992). At its most general level, the system monitors each treatment using three measures: the Mental Health Index, the Clinical Assessment Index and the Therapeutic Bond. As noted above, all of these measures are normed to a large sample of psychotherapy patients at intake and are expressed as T-scores (with a mean of 50 and a standard deviation of 10).

MENTAL HEALTH INDEX

The Subjective Wellbeing, Current Symptoms, and Current Life Functioning scales, described above, are combined into a Mental Health Index that is based on patient self-reports. The Mental Health Index has an internal consistency of 0.87 ($N = 163$) and a 3–4 week test–retest stability of 0.82 ($N = 213$). The mean T-score for a heterogeneous sample of 493 non-patients was 61.1 with a standard deviation of 8.25.

CLINICAL ASSESSMENT INDEX

The clinician makes one overall rating of the patient along with ratings in six areas of the patient's life functioning.

The Global Assessment Scale (Endicott *et al.*, 1976) is a rating of 'the patient's lowest level of current functioning' using 'a hypothetical continuum of mental health-illness'. This scale, which is similar to Axis V of the DSM-IIIR, consists of ten 10-point intervals, which correspond to a description of a patient's general functioning, on a 1–100 rating continuum with 100 representing superior status. The clinician was asked to categorize the patient into one of these intervals by providing a specific numerical rating within the range. For example, if the clinician believed that the patient belonged in the lowest interval, 1–10, he/she must indicate where in that range the patient falls: 1, 4, 7, etc.

Several studies have assessed the reliability of this measure. Endicott *et al.* (1976) conducted five studies (on samples which were primarily comprised of inpatients) resulting in test–retest reliabilities ranging from 0.69 to 0.91. Clark and Friedman (1983) found the GAS test–retest reliabilities ranged from 0.74 to 0.78 decreasing as the length of time between assessments increased. In a study of chronic outpatients, interrater reliabilities for the GAS were obtained after four different training sessions involving either previously trained mental health professionals, untrained mental health professionals, or a mixture of trained and untrained clinicians. These interrater reliabilities ranged from 0.66 to 0.92 with greater reliabilities associated with the trained clinician groups (Dworkin *et al.*, 1990). Based on our sample, the test–retest correlation (2–3-week interval, rated by different clinicians at each time point) was 0.68.

The Life Functioning scales were devised to assess the patient's status in each of the six life areas described above: family functioning, intimate relationships, social relationships, health and grooming, work functioning, and self management. Each of the six domains was rated on a dimension ranging from 0 to 100 with increments of five marking the scale. Each scale has five descriptive anchors with each anchor occupying one-fifth of the dimension space. Anchors ranged from 'severe impairment, virtually unable to function' on the low end (0–20), to 'no impairment, high level of functioning' on the high end (80–100) of the functioning dimension. Five behavioural examples, each of 20–50 words were provided with each of the five levels to help standardize the labelled levels of functioning. Raters were asked to 'circle the number (from 0 to 100) that best applies to this patient's level of, for example, family functioning'.

The intercorrelations ($N = 1521$) of ratings of the six domains ranged from 0.36 to 0.67. The sum of the ratings for the six life areas had an internal consistency of 0.86 ($N = 1237$) and a test–retest (2–3-week interval, rated by different clinicians at each time point) correlation of 0.77 ($N = 81$). Corrected item–total correlations ranged from 0.55 to 0.66. This indicated the presence of an overall dimension of functioning, but also indicated that there was meaningful content heterogeneity across the domains. Each of the separate domains and the sum of the six domains of life functioning were correlated with the Global Assessment Scale (GAS). The summed domains correlated 0.74 ($N = 1459$) with the GAS; the separate domains correlated with the GAS in a range from 0.47 to 0.67. This further indicated convergence of the underlying dimension with another global measure of functioning.

The Clinical Assessment Index (CAI) is based on clinicians' ratings on the Global Assessment scale and the sum of Life Functioning scales. The CAI has an internal consistency of 0.84 ($N = 1459$) and a 2–3-week interval rated by different clinicians at the two time points) test–retest stability of 0.77 ($N = 81$).

THERAPEUTIC BOND

The system also included an assessment of the therapeutic relationship based on our conception of the therapeutic bond (Orlinsky & Howard, 1987). The bond has three theoretical components: working alliance, empathic resonance and mutual affirmation. Working alliance has to do with the effort the patient and the therapist put into implementing their respective roles. Empathic resonance relates to the patient's perception that the therapist understands him/her, and mutual affirmation pertains to an open, caring regard and trust between the patient and the therapist. Originally, a 50-item Therapeutic Bond Scale (Saunders et al., 1989) was developed to assess these three components. Based on item–total correlations, we selected the best four items for each of the three bond constructs. The sum of the 12 items correlated 0.81 with the 50-item Therapeutic Bond scale. Based on a sample of 1256 sessions, the internal consistency of this scale was 0.88. Based on a sample of 157 patients, the test–retest correlation (session 1 to session 3) was 0.62.

Using these three scores, the Mental Health Index, the Clinical Assessment Index and Therapeutic Bond, and the component scores from these indices, we can produce a report which depicts the course of therapy for a patient. The following are four illustrative cases from our study.

A SHORTER TERM SUCCESS

This patient was a single male in his late twenties. He was employed full-time, but felt that he was working at a dead-end job. He stated that he wanted to earn a college degree, but admitted to a lack of motivation to 'move ahead in life'. He presented for treatment at the suggestion of his girfriend, stating that he wanted to discover whether or not he really did have a problem. In addition, he wanted to increase his level of motivation and to reduce his impulsivity and tendency to express his anger in inappropriate ways.

The patient reported that his parents had a 'shaky ' relationship. He described his father as 'cool and stupid'. He stated that he 'adores' his mother and did not understand how she put up with his father. The patient had few friends, no close male friends, and felt that he had never had a satisfying relationship.

The patient considered his girlfriend to be a very stable and energetic person with a strong personality. He claimed that the problems with his girlfriend began about 2 months previously when she 'unloaded' on him. Preceding their difficulties, he was in an automobile accident in which he broke a leg. He felt that his girlfriend was not very supportive and told her that he saw her as distant and unfeeling. She responded by telling him that she was a strong person and he was weak, and that perhaps therapy would help strengthen him and reduce his negative thinking. Initially, he resented her suggestion that he enter psychotherapy, but he wanted the relationship to continue and stated that he cared about her and wanted the relationship to be more stable.

In the intake interview, the patient alternated between alert, task oriented thought processing to very slowed, depressed cognitive functioning that included some suicidal indication. He reported a pattern of bingeing with alcohol and marijuana to 'self-medicate' his depression. The intake diagnoses were: adjustment disorder; narcissistic personality disorder; alcohol abuse; cannabis abuse. His diagnosis based on the Structured Clinical Interview for DSM-III-R (SCID) (Spitzer et al., 1988) was adjustment disorder.

Therapy was conducted by a married female clinical psychology pre-doctoral intern. Her therapeutic orientation was eclectic psychodynamic with an emphasis on object relations and self-psychology. She had previously seen only 6–10 patients in individual psychotherapy.

Figure 1.14 shows the course of therapy for this patient. As can be seen, the therapeutic bond was stable and slightly higher than average. By the end of this 11-session treatment, from both the perspective of the therapist

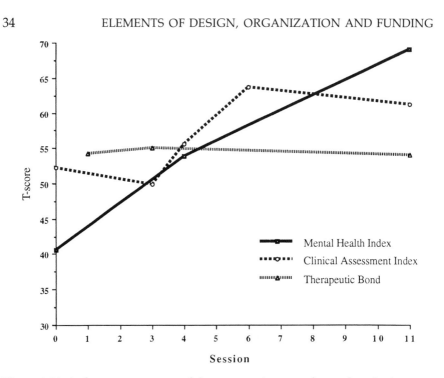

Figure 1.14 A shorter term successful treatment (see text for explanation)

(Clinical Assessment Index) and the patient (Mental Health Index), the patient was seen as functioning in the normal range (a T-score above 60) by session 7.

At discharge, the therapist wrote:

> [t] reatment was focused on helping the patient to look at the underlying feelings of the childhood experiences that he talked about...Direct empathic verbalization of such feelings on the therapist's part were met with defensiveness...An ego-supportive approach...proved to be more successful. That is, the patient was supported in his own exploration of his experiences and affects. This helped the patient to internalize a self-reflective stance... However, he also became aware...that his problems were more pervasive and deep than he had thought. He 'decided' not to continue exploring these at this point.

At termination, the therapist rated the patient as 'slightly improved', 'slightly distressed', and as getting along 'fairly well; has his ups and downs'.

A SHORTER TERM FAILURE

This patient was a young, divorced, female who had no children. She was well educated and appropriately employed. At intake, she stated she had been dating a man for over 2 years.

> We have an overall good understanding and communicative relationship. Lately, he has been pressuring me to make a commitment. I can't seem to decide what is right for me to do. I don't think I want to make that commitment right now, but I'm confused as to why I feel this way. I wonder that if after such a long relationship, if I don't want to make a commitment to this man, if I ever will. I also worry that past events (my divorce) are scaring me away. My current situation with a job with a heavy workload and an intense study schedule continue to make me feel like I don't have enough time for myself.

The patient is an oldest child. She described her father as 'not a real bright person, old fashioned attitudes, and lesser educated than my mom'. Her father had a drinking problem while the patient was growing up but had been sober for the last several years. She described her mother as:

> Very intelligent, level headed, loving if somewhat protective...she can be phony sometimes and try to shy away from what's really going on. She would rather nobody know what's going on with her. She was always there for me but I don't think I ever really felt like I was in touch with her.

In high school she was 'not popular' and did not feel close to anyone in her family. She recalled feeling depressed often. She had had suicidal thoughts but never had plan or intent. She saw college as a 'chance for a fresh start' and as a time when she was much happier.

The patient met her husband while they were both at college. They were married shortly after graduation. After less than a year of marriage her husband informed her that he was moving out. She stated that she knew there were problems in their relationship, 'but they were things I thought we could work out'. Shortly after the divorce, her ex-husband moved in with another woman. He has remarried and has a child. The patient admitted that 'I still have strong feelings for him'.

She described her current boyfriend as 'a sensitive person, nice almost to a fault, masculine looking, somewhat smaller physically than my idea of the most attractive man'. She says that he is easier to get along with than her ex-husband but she cannot help comparing the two.

The intake diagnosis was interpersonal problem. Based on the SCID, her diagnosis was adjustment disorder.

The therapist was a young, married, male clinical psychology pre-doctoral intern. He described his therapeutic orientation as 'interpersonal-psychodynamic'. He had previously seen over 40 patients.

According to the therapist, in the early sessions of psychotherapy, the patient resolved that she was not ready to get engaged and was unsure of her investment in the relationship with her boyfriend. Therapy focused primarily on working to understand the patient's general sense of unfulfil-ment which was manifest in her ambivalence both about her relationship and about her choice of a career. With regard to both of these, she complained of a recurrent sense of boredom and a recurrent desire to move on to a new job and a new relationship.

The therapist provided the following case formulation. The patient's ambivalence about both her relationship with her boyfriend and her career appeared to mirror an underlying struggle between obedience and defiance which was characteristic of her somewhat compulsive, conforming person-ality style. The patient described a pattern of parental overcontrol to which she adapted for the most part by compliant, good behaviour. This adapt-ation, however, took its toll when in adolescence her own internalized restrictions, in combination with those imposed by her parents, served to inhibit and restrain her strivings to achieve a sense of autonomy and self-competence. The patient felt liberated and excited about her life at college. Unfortunately, during college and throughout her adult life, her most intensive efforts to break free of the restrictions she had internalized have ultimately led to disappointment or loss. Most notable, her relationship with her first college boyfriend resulted in a rebuke from his mother, her marriage to a man with whom she was passionately in love ended in divorce after he suddenly left her, and the enjoyment she found working in a restaurant ended when her parents persuaded her to quit. In the wake of these events, she has attempted to compromise in recent years by choosing a 'risk averse' profession and staying involved in a relationship which is 'comfortable' but often frustrating and unfulfilling. These choices ultimately proved to be unsatisfactory, and have resulted in considerable ambivalence about her direction in life as well as a sense of being some-what ungrounded.

At discharge, the therapist stated that the patient '...responded well to psychodynamic, insight-oriented therapy which focused on under-standing her current ambivalence and sense of unfulfillment in terms of the adaptations she had made during childhood and adolescence'. The therapist stated that the patient chose to discontinue therapy (after 24 sessions) because she could see another therapist free of charge in another setting under her company's insurance programme.

Figure 1.15 shows the course of therapy for this patient. As can be seen, the therapeutic bond was somewhat unstable and very low. From both the

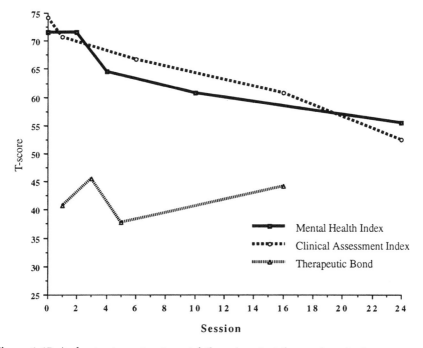

Figure 1.15 A shorter term treatment failure (see text for explanation)

perspective of the therapist (Clinical Assessment Index) and the patient (Mental Health Index), the patient's condition worsened over treatment.

However, at the time of discharge, the therapist reported that the patient had achieved substantial clarification of the nature of her dissatisfactions. The therapist reported that the patient made progress in dealing with surface problems but that, at a deeper level, she was likely to experience angry, defiant urges to assert herself and break free of the restrictions she has imposed upon herself. The therapist rated the patient as 'moderately improved', 'slightly distressed', and as getting along 'fairly well; has her ups and downs'.

A LONGER TERM SUCCESS

This patient was a middle-aged, divorced female. She was employed full-time in a professional capacity. The patient has three children, none at home. She was socially active, particularly with female friends.

The patient presented for therapy with the dominant complaints of depression, loneliness and a general disappointment about the course of her life. She wanted a better understanding of herself and an opportunity to learn new coping strategies to meet life stressors.

The patient reported that her parents had always been emotionally distant. She characterized her father as having a bad temper and her mother as critical and unaffectionate. She believed that she was the least favoured of her siblings. She described herself as a rebellious adolescent and felt that she never received positive attention from her family.

When the patient presented for treatment, she had been divorced for a long time. Since her divorce, the patient has worked successfully at several jobs and returned to school to pursue an advanced degree. However, she feels unsatisfied and stated that her life is meaningless and is going nowhere. With the recent loss of a friend, her last child away at school, and no real close friends in the area, she feels increasing loneliness.

The patient was experiencing several symptoms of depression, such as feelings of hopelessness, and overeating. She was emotionally constricted (except when drinking alcohol), had difficulty maintaining long-term relationships, had abused alcohol over an extended period of time, and had a history of sporadic abuse of marijuana and other illicit drugs. Her intake diagnoses were: dysthymia; alcohol dependence, in partial remission; cannabis abuse; nicotine dependence. Her diagnoses based on the Structured Clinical Interview for DSM-IIIR (SCID) (Spitzer *et al.*, 1988) was major depression.

The therapist was a young, married female. She was a third-year graduate student in clinical psychology participating in a practicum experience. Her therapeutic orientation was psychodynamic with an emphasis on the 'here-and-now' interaction and the utilization of transference interpretations. She had only seen 6–10 patients in individual psychotherapy.

Figure 1.16 shows the course of therapy for this patient. As can be seen, the therapeutic bond was reasonably stable and maintained at a very high level. From both the perspective of the therapist (Clinical Assessment Index) and the patient (Mental Health Index), the patient initially responded to treatment with a decrease in functioning. Following this initial decrease, the patient made steady progress and was seen as functioning in the normal range (a T-score of above 60) by session 60.

This course of therapy at the clinic was terminated after 135 sessions because the therapist was leaving to begin a pre-doctoral internship. At discharge, the therapist rated the patient as 'moderately improved' 'slightly distress', and as getting along 'quite well; has no important complaints'. However, the patient continued in treatment with the same therapist at the new location.

A LONGER TERM FAILURE

The patient was a single female in her late twenties. She was employed in a managerial position.

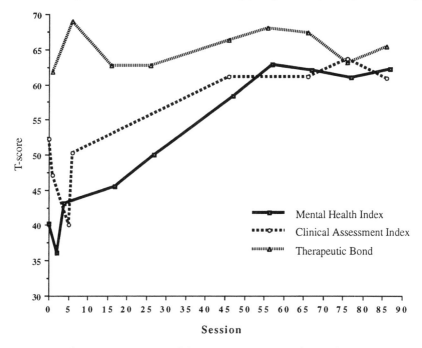

Figure 1.16 A longer term successful treatment (see text for explanation)

The patient had a history of unsatisfying relationships with peers, both male and female; she had never had an intimate relationship with anyone. She was reasonably successful in her vocation. However, the patient reported that she never developed avenues for enjoyment—no hobbies or recreational activities.

She grew up in a home with alcoholic parents. During her childhood, the patient was sexually abused, for several years, by her father. In other respects, the patient felt that she served as the family scapegoat. Even though the patient competed with her younger siblings for attention, she 'protected' them so that they would not experience the same abuse that she had experienced. As an adult, the patient continued to be hyper-vigilant for abuse in interpersonal relationships in a way that rendered her unable to approach others to help satisfy emotional needs. She experienced high levels of interpersonal tension and, consequently, extreme distress—anxiety, sadness, anger and frustration.

The patient had recently suffered a life-threatening injury. In addition, she was mourning the recent death of her father which followed a prolonged illness, while working through her anger at him for his abuse of her. The patient was engaged in multiple heterosexual relationships that were conflictual in nature. She was drinking alcohol to an abusive extent, and had a tendency to overeat in a pattern of bingeing. Though she viewed her alcohol consumption as problematic, she drank to excess every day.

The patient reported crying easily, experienced mood swings, and generally felt 'stressed out'. She was consistently unhappy and had a tendency to act out in self-destructive ways (i.e. with alcohol, in relation-ships, and binge eating). Her intake diagnoses were: borderline person-ality disorder; post-traumatic stress disorder; adjustment disorder with atypical features; alcohol dependency; nicotine dependency; asthma and allergies. She did not meet criteria for any Axis I SCID diagnosis, but the clinical interviewer ruled her very high on borderline personality disorder (Axis II) characteristics.

The therapist was a young, single female. She was an advanced graduate student in clinical psychology participating in a clinical practicum experience. Her therapeutic orientation was psychodynamic with an emphasis on object relations. She had previously seen 11–20 patients in individual psychotherapy.

Figure 1.17 shows the course of therapy for this patient. As can be seen the therapeutic bond was very unstable, illustrating a borderline relational style. From both the perspective of the therapist (Clinical Assessment Index) and the patient (Mental Health Index), the patient showed a steady decline in functioning over the course of this treatment.

The therapist characterized the first year of treatment (one session a week) as '...supportive with some interpretive work accomplished' and the second year (two sessions a week) as '...marked by much progress and movement toward insight'. Over the course of this treatment, the patient missed 31 scheduled appointments and made considerable use of the clinic's crisis hot-line. Therapy at the clinic was terminated after 121 sessions because the therapist was leaving to begin a pre-doctoral internship. At discharge, the therapist rated the patient as 'moderately improved', 'extremely distressed' and as getting along 'quite poorly; can barely manage to deal with things'. The patient continued in treatment with the same therapist at the new location.

A REVIEW OF SOME METHODOLOGICAL ISSUES

The four cases presented here illustrate the use of naturalistic research in building concatenated theories that are useful to practicing psycho-therapists. At the base of this approach is a sophisticated case methodology which, when individual cases are aggregated, can address many com-peting explanations of the casual efficacy of psychotherapy processes. Moreover, the observable data can be highly informative to the practicing clinician and has the potential of altering the psychotherapist's behaviour in the delivery of an individual treatment. Using the four prototypic cases just presented, we shall discuss each of the methodological attributes in

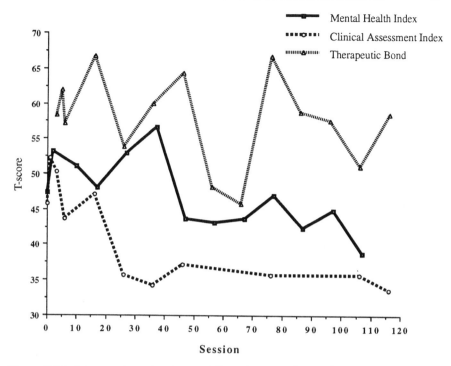

Figure 1.17 A longer term treatment failure (see text for explanation)

turn and identify critical information that could potentially influence the delivery of psychotherapy.

How objective are the data?

A comparison of the patient self-reported MHI and the clinician-rated CAI provides one index of the value of the monitored progress in treatment. Conventional wisdom holds that convergence of data across measurement modes reflects strength of data. However, it is interesting that among the four cases presented here, two showed such convergence at intake and two did not. Surprisingly, the two treatment failures showed convergence of MHI and CAI at intake, one at high levels and the other at slightly below-average. In the two successful cases, the CAI was higher at intake than the MHI, and this relationship was maintained through the early course of treatment. We have shown in previous work on rates of improvement as a function of treatment duration that clinicians tend to rate patients as more improved early in treatment than patients rate themselves, but that the direction of the relationship reverses later in treatment (Howard *et al.*, 1986). Complete agreement may not be a cause for alarm, but widely discrepant ratings would alert the therapist to the possibility that the basis for beginning psychotherapy will have to be re-evaluated.

Is there clear evidence of a treatment effect (improvement)?

Our prior work has shown that reliable improvement requires a change of approximately half a standard deviation (Howard *et al.*, 1993). Three of the four cases show evidence of improvement. For the two treatment successes the intake-to-termination change was about two-and-a-half standard deviations. For the third case, improvement could be noted at several points, but the overall intake-to-termination change was unimproved. At termination, the two successful cases clearly were more like non-patients than like patients on both MHI and CAI scores.

Does continuous measurement show stability?

For the above case, the number of self-report data points ranged from 3 to 13, and the number of clinician-rated points ranged from 5 to 11. General patterns or trends are somewhat unstable early in treatment, but become more predictable in three of the cases. In the fourth case, a sawtooth pattern is evident in the self-report data, and less confidence can be placed in the temporal stability of the patient's condition. Reliable improvement was evident on at least two occasions for each of the successful cases in addition to the stable general pattern of improvement. Although reliable improvement was evident on at least two occasions in the long-term failure case, the overall pattern of change was unstable. For the fourth case, the pattern of non-improvement was consistent, but no reliable change was observed.

Has an effective therapeutic alliance been established?

Both of the successful therapies involved a stable therapeutic bond that was maintained at a slightly higher than average level. Conversely, the two treatment failures involved therapeutic bonds which were erratic or lower than average. The level and stability of the therapeutic bond within the first half-dozen sessions appears to be predictive of later developments in treatment, given the same conditions of therapist, patient and therapeutic approach. Objective information about the quality of the therapeutic alliance lends evidence that the therapy process is on the right track, that the therapist–patient match is not yet working and is in need of adjustment, or that the patient is incapable of establishing an effective working relationship with this particular therapist at this particular time. At any rate, continuation of an erratic or poor therapeutic relationship bodes ill for the outcome of the treatment.

Are the patient's dysfunctional areas being addressed? Conversely, are the patient's functional areas respected?

The shorter term failure case is an excellent example of misreading the functional status of a patient. The therapist conceptualized the patient's work and intimacy functioning as a common failure to make commitments that had origins in earlier family experiences. The patient showed little distress before therapy began, but lost confidence in her self-appraisal as the therapist pursued a passive, non-commitment theme. In contrast, the therapist in the shorter term success case adjusted her therapeutic interventions on the basis of her understanding of the patient's responsiveness to her initial approach. The longer term success case presented with work, family and friends intact. Her main dysfunction was being the kind of person she wished to be. The good therapeutic bond that developed evidently provided her with an environment in which to successfully address this goal. The long-term failure case presented with some functional and some dysfunctional areas. Although employed as a manager, she was dissatisfied with her friendships, had not established effective intimate relationships, and had few satisfying leisure activities. The therapeutic bond in this case was never firmly established.

CONCLUSION

How can we design relevant psychotherapy research? The time-honoured approach is the randomized clinical trial. For this to be informative to a practitioner or policy maker, the study sample must represent some identifiable patient population and main effects must be unambiguous. With regard to the first, attrition and subject recruitment criteria provide unavoidable barriers to generalization. With regard to the second, overlap of outcome distributions (i.e. some control subjects have better outcomes than some experimental subjects) (Howard *et al.*, 1994) and within-cell variation that exceeds measurement error (Lyons & Howard, 1991) compromise the application of findings. Since we are very unlikely to ever mount a clinical trial with a representative patient population, can rarely avoid attrition, outcome distributions will overlap, and there will be reliable within cell variation, this approach seems fruitless in terms of clinical relevance.

Another approach would be to standardize a treatment and determine the type of patient who responds well to this treatment (e.g. Howard *et al.*, 1993b). This exploratory approach would entail a self-correcting learning model that continuously incorporates the response to treatment of new patients in order to clearly determine the relevant patient group. The main problem with this approach is that it is unlikely that we could ever standardize a treatment in such a way that it could be delivered in the same way by the same therapist across patients, by different therapists, or in different settings.

The approach that we have taken (as illustrated in the preceding sections) is to develop a theoretical framework, the dosage, generic, and phase models, operationalize and test the concepts, and develop a system for feedback to the clinician about the course of a single treatment. In this system, the clinician (therapist, supervisor or case manager) can view the treatment at several levels:

1. the overall course of the Mental Health Index, the Clinical Assessment Index, and the Therapeutic Bond;
2. the status of the component scores (e.g. subjective wellbeing, life functioning);
3. the subscores within each component (e.g. depression symptoms, work functioning)
4. critical single items (e.g. amount of distress, trouble concentrating).

We can also add modules (e.g. self-esteem) that are seen as important in a particular clinical context. Our experience, so far, is that therapists welcome this feedback and are able to use it in a manner that enhances the treatment of a special case.

Convincing clinicians that psychotherapy research has a practical value is a daunting challenge. Numerous authors have discussed the disjuncture between research and practice (cf. Cohen *et al.*, 1986; Strupp, 1989; Whiston & Sexton, 1993). Strupp, in addressing his own empirical work, perhaps said it best, 'Although I have greatly profited from the investigation of others, nothing is as convincing as one's own experience' (Strupp, 1989, p. 717). The key to impacting clinicians' self-experience is to provide information relevant to the current case in treatment.

ACKNOWLEDGEMENTS

This work was partially supported by grants RO1 MH42901 and KO5 MH00924 from the National Institute of Mental Health. We are grateful for the clinical work of Cori Hillmann, the statistical work of Bruce Briscoe, and the helpful comments of Merton Krause, Elizabeth Bankoff and the Northwestern/Chicago psychotherapy research group.

REFERENCES

Adby, P.R. & Dempster, M.A.H. (1974). *Introduction to Optimization Methods*. London: Chapman & Hall.

American Psychiatric Association. (1987). *Diagnostic and Statistical Manual of Mental Disorders*, 3rd edn, revised. Washington, D.C.: American Psychiatric Association.

Beck, A.T., Ward, C.H., Mendelson, M., Mock, J. & Erbaugh, J. (1961). An inventory measuring depression. *Archives of General Psychiatry* 4, 561–571.

Byrne, B.M. (1983). Investigating measures of self-concept. *Measurement and Evaluation in Guidance*, 16, 115–126.

Byrne, B.M. & Shavelson. R.J. (1986). On the structure of adolescent self-concept. *Journal of Educational Psychology* 78, 474–481.

Campbell, D.T. & Stanley, J.C. (1963). *Experimental and Quasi-experimental Designs for Research*. Chicago: Rand McNally & Company.

Clark, A. & Friedman, M.J. (1983). Nine standardized scales for evaluating treatment outcome in a mental health clinic. *Journal of Clinical Psychology* 39(6), 939–950.

Cohen, L.H., Sargent, M.M. & Sechrest, L.B. (1986). Use of psychotherapy research by professional psychologists. *American Psychologist*, 41, 198–206.

Cowen, E.L. (1991). In pursuit of wellness. *American Psychologist* 46, 404–408.

Diener, E. (1984). Subjective well-being. *Psychological Bulletin* 95(3), 542–575.

Derogatis, L.R. (1977). *SCL-90: Administration and Procedures Manual-I for the R(evised) Version*. Baltimore: Clinical Psychometrics Research.

Dupuy, H.J. (1977). *A Current Validational Study of the NCHS General Well-being Schedule* (DHEW Publication No. HRA 78-1347). Hyattsville, MD: National Center for Health Statistics, U.S. Department of Health, Education and Welfare.

Dworkin, R.J., Friedman, L.C., Telschow, R.L., Grant, K.D., Moffic, H.S. & Sloan, V.J. (1990). The longitudinal use of the Global Assessment Scale in multiple-rater situations. *Community Mental Health Journal* 26(4), 335–344.

Endicott, J., Spitzer, R.L., Fleiss, J.L. & Cohen, J. (1976). The Global Assessment Scale: procedure for measuring overall severity of psychiatric disturbance. *Archives of General Psychiatry* 33, 766–771.

Eysenck, H.J. (1952) The effects of psychotherapy: an evaluation. *Journal of Consulting Psychology* 16, 319–324.

Fisher, R.A. (1925). *Statistical Methods for Research Workers*. Edinburgh: Oliver & Boyd.

Grawe, K., Ambuhl, H. & Drew-Foppa, S. (1989) *Research on Orlinsky and Howard's Generic Model of Psychotherapy*. Symposium presented at the meetings of the Society for Psychotherapy Research, Toronto, Canada, June, 1989.

Gurtman, M. B. (1993). Constructing personality tests to meet a structural criterion: application of the interpersonal circumplex. *Journal of Personality* 61, 237–263.

Horowitz, L.M., Rosenberg, S.E., Baer, B.A., Ureño. G. & Villasenor, V.S. (1988). Inventory of interpersonal problems: psychometric properties and clinical applications. *Journal of Consulting and Clinical Psychology* 56, 885–892.

Howard, K.I., Brill, P.L., Lueger, R.J., O'Mahoney, M.T. & Grissom, G.R. (1992). *Integra Outpatient Tracking Assessment: Psychometric Properties*. Philadelphia: Integra, Inc.

Howard, K.I., Cox, W.M. & Saunders, S.M. (1991). Attrition in substance abuse comparative treatment research: the illusion of randomization. In *Psychotherapy and Counseling in the Treatment of Drug Abuse*. (L.S.Onken & J.D.Blaine, eds) Washington, D.C.: National Institute of Drug Abuse, Research Monograph Series 104, 66–79.

Howard, K.I., Kopta, S.M., Krause, M.S. & Orlinsky, D.E. (1986). The dose–effect relationship in psychotherapy. *American Psychologist* 41, 159–164.

Howard, K.I., Krause, M.S. & Lyons, J. (1993a). When clinical trials fail: a guide for disaggregation. In *Psychotherapy and Counseling in the Treatment of Drug Abuse* (L.S. Onken & J.D. Blaine, eds) Washington, D.C.: National Institute of Drug Abuse.

Howard, K.I., Krause, M.S. & Orlinsky, D.E. (1986b). The attrition dilemma: towards a new strategy for psychotherapy research. *Journal of Consulting and Clinical Psychology* **54**, 106–110.

Howard, K.I., Krause, M.S. & Vessey, J.T. (1994). Analysis of clinical trial data: the problem of outcome overlap. *Psychotherapy*.

Howard, K.I., Lueger, R., Maling, M. & Martinovich, Z. (1993b). A phase model of psychotherapy: causal mediation of outcome. *Journal of Consulting and Clinical Psychology* **61**, 678–685.

Howard, K.I. & Orlinsky, D.E. (1972). Psychotherapeutic process. *Annual Review of Psychology*. **23**, 615–668.

Jacobson, N.S. & Truax, P. (1991). Clinical significance: a statistical approach to defining meaningful change in psychotherapy research. *Journal of Consulting and Clinical Psychology* **59**, 12–19.

Kaplan, A. (1964). *The Conduct of Inquiry*. New York: Intext.

Kazdin, A.E. (1981). Drawing valid inferences from case studies. *Journal of Consulting and Clinical Psychology*, **49**, 183–192.

Kiesler, D.J. (1983). The 1982 interpersonal circle: a taxonomy for complementarity in human transactions. *Psychological Review* **90**, 185–214.

Kolden, G.G. (1991). The generic model of psychotherapy: an empirical investigation of patterns of process and outcome relationships. *Psychotherapy Research* **1**, 62–73.

Kolden, G.G. & Howard, K.I. (1992). An empirical test of the generic model of psychotherapy. *Journal of Psychotherapy Research and Practice* **1**, 225–236.

Kopta, S.M., Howard, K.I., Lowry, J. & Beutler, L. (In press). The psychotherapy dosage model and clinical significance: a comparison of treatment response rates over time for psychological symptoms. *Journal of Clinical and Consulting Psychology*.

Lyons, J. & Howard, K.I. (1991). Main effects analysis in clinical research: statistical guidelines for disaggregating treatment groups. *Journal of Consulting and Clinical Psychology* **59**, 745–748.

McNeilly, C.L. (1991) *The Relationship Between Structured Diagnoses, Screening Diagnoses, and Presenting Problems of Outpatients*. Unpublished doctoral dissertation, Northwestern University.

McNeilly, C.L. & Howard, K.I. (1991). The effects of psychotherapy: a reevaluation based on dosage. *Psychotherapy Research* **1**, 74–78.

Nunnally, J.C. (1967). *Psychometric Theory*. New York: McGraw-Hill.

Orlinsky, D.E. & Howard, K.I. (1978). The relation of process to outcome in psychotherapy. In *Handbook of psychotherapy and behavior change 2nd edn* (S.L. Garfield & A.E. Bergin, eds), New York: John Wiley.

Orlinsky, D.E. & Howard, K.I. (1986). Process and outcome in psychotherapy. In *Handbook of psychotherapy and behavior change 3rd edn* (S.L. Garfield & A.E. Bergin, eds), New York: John Wiley.

Orlinsky, D.E. & Howard, K.I. (1987). A generic model of psychotherapy. *Journal of Integrative and Eclectic Psychotherapy* **6**, 6–27.

Orlinsky, D.E., Lundy, M., Howard, K.I., Davidson, C.V. & O'Mahoney, M.T. (1987). *Therapeutic Procedures Inventory-Revised*. Chicago, IL: Northwestern University.

Parloff, M.B. (1982). Psychotherapy research evidence and reimbursement decisions: Bambi meets Godzilla. *American Journal of Psychiatry* **139**, 718–727.

Radloff, L.S. (1977). The CES-D scale: a self-report depression scale for research in the general population. *Applied Psychological Measurement* **1**, 358–401.

Rosenberg, M. (1979). *Conceiving the Self*. New York: Basic Books. Inc.

Saunders, S.M., Howard, K.I. & Orlinsky, D.E. (1989). The therapeutic bond scales: psychometric characteristics and relationship to treatment effectiveness. *Psychological Assessment: A Journal of Clinical and Consulting Psychology* **1**, 323–330.

Shapiro, D.A. & Shapiro, D. (1982). Meta-analysis of comparative therapy outcome studies: a replication and refinement. *Psychological Bulletin* **92**, 581–604.

Schwartz, D. (1991). *Early Termination from Individual Psychotherapy: A Follow-up Study of the Earliest-leaving Patients (and Prospective Patients)*. Northwestern University.

Silber, E. & Tippet, J. (1965). Self-esteem: clinical assessment and measurement validation. *Psychological Reports* **16**, 1017–1071.

Smith, M.L., Glass, G.V., & Miller, T.I. (1980). The Benefit of Psychotherapy. Baltimore: The Johns Hopkins Press.

Speer, D.C. (1992). Clinically significant change: Jacobson and Truax (1991) Revisited. *Journal of Consulting and Clinical Psychology* **60**, 402–408.

Spitzer, R.L., Williams, J.B.W., Gibbon, M., & First, M.B. (1988). *Structured Clinical Interview for DSM-III-R*. Washington, D.C.: American Psychiatric Press, Inc.

Stacy, M. (1992) *Reasons why Patients Terminate Therapy: Patient and Therapist Perspectives*. Paper presented at the annual meeting of the Society for Psychotherapy Research, Berkeley, California.

Stewart, A.L., Hays, R.D. & Ware, J.E. (1988). *The MOS Short-form General Health Survey Medical Care*, **26**, 724–735.

Strupp, H.H. (1989). Psychotherapy: can the practitioner learn from the researcher? *American Psychologist*, **44**, 717–724.

Underwood, B.J. (1949). *Experimental Psychology*. New York: Appleton-Century-Crofts.

Vessey, J.T. & Howard, K.I. (1993). Who seeks psychotherapy? *Psychotherapy* **30**, 546–553.

Viet, C.T. & Ware, Jr., J.E. (1983). The structure of psychological distress and well-being in general populations. *Journal of Consulting and Clinical Psychology* **51**, 730–742.

Ware, Jr., J.E. & Sherbourne, C.D. (1992). The MOS 36-item short form health survey (SF-36): I. Conceptual framework and item selection. *Medical Care* **30**, 473–483.

Watson, D. & Tellegen, A. (1985). Toward a consensual structure of mood. *Psychological Bulletin* **98**, 219–235.

Whitson, S.C. & Sexton, T.L. (1993). An overview of psychotherapy outcome research: implications for practice. *Professional Psychology: Research and Practices*, **24**, 43–51.

2 Therapy Process Research and Clinical Practice: Practical Strategies

ROBERT ELLIOTT
University of Toledo, Ohio, USA

ABSTRACT The first task of this chapter is to present suggestions for carrying out informative, useful and scientifically-valid research on the process of therapy in the real world of clinical practice. A second task is to offer suggestions for creating a better 'working marriage' between researchers and therapists. These tasks are pursued through a series of discussions: political, conceptual, methodological, practical, and organizational. I begin my presentation by reviewing arguments for the interdependence of therapy research and practice and suggesting collaboration between researchers and therapists is a political necessity. This is followed by a conceptual framework that defines the domain or menu of possibilities for therapy process research. Next, I outline an 'appropriate methodologies' strategy of letting research questions guide choice of methodological alternatives in psychotherapy process research. A diverse array of practical research questions and options for data collection in clinical settings is then offered. Different organizational models for researcher–therapist collaboration are then considered, concluding with a set of recommendations which also serve to summarize the presentation.

INTRODUCTION

This chapter is based on a set of four assumptions which I find useful for the practice of therapy process research.

1. Therapy researchers are trying to improve the practice of therapy, i.e. therapy research is an area of applied research.
2. Therapy process and outcome are best studied in conjunction with one another.
3. Psychotherapy is a highly complex set of processes; research should reflect this complexity, or risk triviality.

Correspondence address: Department of Psychology, University of Toledo, Toledo, Ohio 43606, USA.

Research Foundations for Psychotherapy Practice. Edited by M. Aveline and D. A. Shapiro.
Copyright © 1995, Mental Health Foundation and Individual Contributors.
Published 1995 by John Wiley & Sons Ltd

4. Methodological pluralism is a very useful philosophical position from which to approach therapy research; i.e. a range of different methods are needed to understand therapy and appropriate for answering different research questions.

Not all therapy researchers will agree with these assumptions; others will undoubtedly disagree with at least some of the points I will make. Nevertheless, I believe that taking these assumptions seriously provides the basis for carrying out more clinically interesting and relevant therapy process research. In this chapter, I hope to illustrate the value of this position, organized under five headings: political, conceptual, methodological, practical, and organizational.

A POLITICAL PERSPECTIVE: RESEARCHERS AND THERAPISTS NEED EACH OTHER

The gap between psychotherapy research and practice is a frequent theme in the psychotherapy research literature and one on which my colleagues and I have often harped (Barker *et al.*, in press; Elliott, 1983; Morrow-Bradley & Elliott, 1986). It has been traditional for researchers to blame therapists for not utilizing research findings and for being biased, irresponsible, or antiscientific (Giles, 1993). Although not so vocal, when they are asked, therapists will readily blame researchers for not researching common or clinically-relevant treatments or populations and for being generally scientistic, boring, and irrelevant (Morrow-Bradley & Elliott, 1986).

Elsewhere, I have argued that a useful approach to the relationship between therapy researchers and practitioners might be to reframe it as a troubled marriage, in which the two partners need to accept the fact that each lives in a different reality and to stop trying to change the other (Elliott & Morrow-Bradley, 1994). It seems to me that while therapists and researchers do live in different worlds, they are nevertheless 'married' to each other—'for better or for worse'. In other words, therapists and researchers are mutually dependent on one another because of a number of political considerations.

First, psychotherapy practice is the source of most psychotherapy research ideas (Stiles, 1992). Many or most of the key developments in psychotherapy (e.g. transference, therapeutic alliance, the client-centred facilitative conditions, systematic desensitization, cognitive therapy) were developed through careful qualitative/clinical observation by practicing therapists who also happened to be researchers.

These developments have typically undergone a subsequent 'translation' into researchable questions, generally in quantitative terms. Sometimes the clinical innovator went on to carry out research on his or her clinical models (e.g. Luborsky's, 1976, research on therapeutic alliance; Laura Rice's work on Problematic Reaction Points, Rice & Sapiera, 1984). However, in many instances, professional researchers have taken up the clinical insights as 'consumers' of the ideas, sometimes in order to support them, but at other times to disprove them (Stiles, 1992). The important point here is that it is easy to forget that the original source of these ideas was the practice of psychotherapy, and it is highly unlikely that they would have developed outside of this context.

Second, psychotherapy practice is a primary justification for psychotherapy research. It is true that some psychotherapy researchers identify themselves as basic researchers who just happen to use psychotherapy as a convenient site for studying basic psychological processes such as the organization of talk (e.g. Labov & Fanshel, 1977) or verbal and non-verbal modes of human information processing (e.g. Dahl et al., 1988). However, most therapy researchers describe themselves as applied researchers attempting in one way or another to improve the effectiveness or efficiency of therapy (e.g. Elliott, 1983; Garfield & Bergin, 1986, Rice & Greenberg, 1984). Thus, most therapy researchers justify their interest in their research area by reference to its potential practical benefits, or at least pay lip service to improving practice in their grant proposals.

Third, therapy research can also justify practice. In the United States, therapists are experiencing increasing restrictions on whom they can see and for how long. Moreover, some (e.g. Giles, 1993) are attempting to use existing psychotherapy research to justify restrictions on the type of therapy which will be reimbursed; i.e. short-term cognitive-behavioural treatments.

However, psychotherapy research does not need to be a bludgeon for beating down psychodynamic and experiential therapies. Instead, recent meta-analytic reviews of outcome research on these treatments can be called upon to support their efficacy (e.g. Crits-Christoph, 1992; Greenberg, et al., 1994), thus empowering therapists who follow these approaches. In addition, we need more research examining the benefits of long-term treatments for serious or chronic disorders, such as chronic or recurrent depression, eating and substance abuse disorders, and personality disorders, especially borderline personality disorder. If health care officials could be shown that there is a cost benefit, for example, in the form of fewer hospitalizations or significant improvements in occupational/social functioning in persons with these diagnoses, they may be more willing to spend money on longer-term outpatient treatments for them.

Fourth, research can help therapists do a better job. Although practicing therapists report finding little of value in psychotherapy research, this does not mean that research has had no influence on the practice of therapy or that there is no potential for useful input from research. It seems likely that research has benefitted practice in indirect ways. For example, therapy research may help therapists develop an attitude of thoughtful reflection about practice. In addition, therapy research provides 'sensitizing concepts' (Strauss & Corbin, 1990), which can sharpen clinical observation skills. For example, for contemporary dynamic therapists, Luborsky's Core Conflictual Relationship Theme research method (Luborsky & Crits-Christoph, 1990) is readily adapted as a helpful framework for therapists to use in listening to their clients' accounts of their interactions with other people, and suggests the value of interpretations of recurrent interpersonal conflict themes (Luborsky, 1984). Finally, the clearest example of how research can facilitate practice is the proliferation of research-based treatment manuals which has occurred over the past 15 years, primarily dating from the Beck *et al.* (1979) classic treatment manual for cognitive therapy of depression.

Thus, it appears that research and practice exist in dialogue with each other. While this dialogue has not always been productive (as when recriminations are hurled), it has the potential to be mutually rewarding. In fact, for the *realpolitik* reasons I have offered, it is now necessary for the researcher–therapist dialogue to move to a higher plane in order to help both researchers and therapists to face the challenge of increased competition for diminished resources. I will return near the end of my presentation to a discussion of specific ways in which this can be done, after a bridging presentation of how clinically-oriented prospective researchers can go about selecting their topics, research questions, general methodological approaches, and specific data collection methods.

A CONCEPTUAL PERSPECTIVE: WHAT IS THERAPY PROCESS? A FIVE-DIMENSIONAL MODEL

In beginning to think about researching the process of psychotherapy, it is useful to have an overall understanding of what 'therapy process' refers to. Etymologically, the word 'process' originally meant 'a going forward'. This origin is preserved in the main contemporary meaning of the word, which is 'A continuous and regular action or succession of actions, taking place or carried on in a definite manner, and leading to the accomplishment of some result' (Oxford English Dictionary, 1971). Some of the elements implied in this and similar definitions are actions or events; sequence and continuity; order or purpose; and effect or outcome.

A comprehensive definition of what comprises therapy process is given in the Five Dimensional Model of therapy process (Elliott, 1991). This model, derived primarily from earlier work by Kiesler (1973), Russell and Stiles (1979), and Schaffer (1982), offers a 'menu' of therapy process, describing the different options which are available. The five dimensions in this conceptual model (see Table 2.1) are:

1. Perspective of Observation (who observes the process: client, therapist, or researcher).
2. Person/Focus (who is observed: client, therapist, or dyad).
3. Aspect of Process (what communication feature is observed: content, action, style, or quality).
4. Unit Level (the hierarchical order or scale at which observation occurs, e.g sentence/idea, speaking turn, episode, session, or treatment.

Table 2.1. Five dimensions of therapy process

---

1. *Perspective of Observation*: what is the point of view of the person providing the data on the therapy process?
   Client
   Therapist
   Researcher
2. *Person/ Focus*: which element of the therapeutic system is studied?
   Client(s)
   Therapist(s)
   Dyad/system (emergent properties of system; e.g. alliance)
3. *Aspects of Process*: what kind of communication variable is studied?
   Content: what is said or meant (kinds of propositions, themes)
   Action: what is done by what is said (speech act, intention, task, response mode)
   Style/state: how it is done or said, and what that reveals about the speaker's feelings or attitudes (paralinguistic and non-verbal behavior, vocal quality, mood, feelings about the other)
   Quality: how well it is done or said (therapist skillfulness, client working)
4. *Unit Level*: at what level or 'resolution' is the therapy process studied? Selected useful units:
   Speaking turn (interaction unit): a response by one speaker, surrounded by utterances of the other speaker
   Episode (topic/task unit): a series of speaking turns organized by a common task or topic
   Session (occasion unit): a time-limited situation in which client and therapist meet to work on the therapeutic tasks
   Treatment (relationship unit): the entire course of a treatment relation
5. *Sequential Phase*: what temporal orientation or purpose in studying a unit of therapy process (i.e. towards what happened before, during and after the unit)?
   Context: what has led up to a unit of process (e.g. previous episodes, sessions)
   Process: the process variable (perspective/person/aspect) which is targeted for study at a given level (unit)
   Effects: the sequelae of a unit of process (e.g. immediately following turns, treatment outcome)

Adapted by permission from Elliott (1991).

5. Sequential Phase (the temporal orientation or purpose of observation: to understand either context, process, or impact).

The process and outcome positions on the Sequential Phase dimension define the major domains of current therapy research: outcome (effects) research, process research, and change process research, the latter focusing on the connections between therapy process and outcome (Greenberg, 1986).

Specifying one's interests on each of these five dimensions seems to be essential for measuring and describing the therapy process. Thus, in planning research on the process of psychotherapy, one begins by making a series of decisions about:

1. who observes (Perspective);
2. who is observed (Person);
3. what is observed (Aspect);
4. at what resolution (Unit Level);
5. for what purpose (Sequential Phase).

A further implication is that, where possible, the researcher will want to capture more of the richness of therapy process by measuring it from more than one position on each dimension; in other words, more than one perspective, person, aspect, unit level or phase (Elliott & Anderson, 1994).

A METHODOLOGICAL PERSPECTIVE: LETTING RESEARCH QUESTIONS GUIDE METHODS

The social sciences are currently in a period of creative ferment as old theoretical and research paradigms are called into question and new alternatives are considered and debated. One result of this is a movement towards pluralistic, integrative approaches to theory and research in psychology and psychotherapy. A wider range of research methods is becoming acceptable, and the signs are that this process will continue and perhaps accelerate in psychotherapy process research.

As I see it, this emerging methodological pluralism is part of an overall philosophy of 'appropriate methodologies' (by analogy to the current catch-phrase 'appropriate technologies'). In general, this means that the narrow range of traditionally acceptable research methods (e.g. control group designs, available quantitative process measures, ANOVA) should not determine the research questions one can study, or the settings in which one can do research. Instead, the type of research question and the realities of the practice situation being investigated should dictate the investigative methods used.

Table 2.2. Genres of psychotherapy process research

Type of question	Appropriate methods
1. *Definition* What is the nature of a particular therapeutic phenomenon? What defines or constitutes it?	Phenomenological research; grounded theory (some types); ethogenic research; structuralism
2. *Description* What kinds of events or aspects exist in therapy? What features, types, or patterns do these events or aspects have?	Naturalistic qualitative research; grounded theory (some types); ethnography: quantitative content, cluster, interaction analysis
3. *Interpretation* What is the meaning of a therapeutic event or process? Why did it happen? How did it develop?	Interpretative research; narrative case study research; comprehensive process analysis; task analysis
4. *Critique/Action* What is wrong with how things are now? How could it be made better?	Feminist research; participant action research
5. *Quantification* How frequent is a type of event? To what degree or intensity is a property generally present in therapy? What is typical?	Descriptive quantitative methods: surveys; rating scales; category systems; descriptive statistics
6. *Comparison* Does a type of therapy, event or phase of therapy have more of something than some other type, event or phase? Which therapy is better for this disorder?	Quantitative experimental and quasi-experimental designs; inferential statistics; meta-analysis of comparisons
7. *Relationship* Which aspects of therapy vary together? What types of event typically precede or follow another? What predicts therapy process and outcome?	Bivariate and multivariate correlational methods; sequential analysis; prediction research; path analysis; meta-analyses of relational research
8. *Method Quality* How well (reliably, validly) can an aspect or event be measured by means of a particular process or outcome measure?	Psychometric or measure development research
9. *Deconstruction* What implicit assumptions are made in this research? Whose interests are served or ignored?	Conceptual analysis; self-reflection; systematic analysis and critique of typical practice; collage; blurred genres

In the appropriate methodologies strategy, the process researcher begins with his or her topic of interest, perhaps defined in terms of the Five Dimensional Model presented in the previous section (e.g. clients' perceptions of therapist empathic quality in within-session episodes preceding therapeutic breakthroughs). Then, on the basis of personal interests or the relevant theoretical or research literature, the researcher next identifies a single, primary research question, a question for which he or she is interested in finding answers.

The key assumptions of the appropriate methodologies strategy are that there are a relatively small number of different types of question in scientific research, and that different methodological approaches lend themselves to answering specific types of question. Traditionally, the most widely accepted or 'scientific' methods have legitimated a narrow subset of research questions, forcing researchers to abandon or drastically revise research questions which don't fit this mold. Together, the research questions and the methods that lend themselves to answering those questions form a 'genre' of research (in Kuhn's, 1962, terms this would be called a 'paradigm') within a wider field of study such as psychotherapy research.

What then are the main types of research question in psychotherapy process research? By my analysis, there are roughly nine of these, as presented in Table 2.2. Table 2.3 offers examples of each of these questions for the three typical therapy research domains of outcome, process, and change process research.

QUESTIONS WHICH LEND THEMSELVES TO QUALITATIVE RESEARCH

The first four kinds of question are most appropriate for qualitative research, because they do not require descriptions of quantity. These are questions of definition, description, interpretation, and critique/action.

Definition

Definitional questions ask about the nature or essence of some aspect of therapy; i.e. what makes it what it is and not something else. For example, What do we mean by insight in therapy? How does it differ from awareness? (e.g Elliott *et al.*, in press). Such questions seem to me to be best addressed by qualitative phenomenological research methods, especially the Duquesne method (Giorgi, 1975; Wertz, 1985).

Description

Descriptive research questions seek answers about features, types, or patterns within therapy process. In particular, they ask about the existence

Table 2.3. Sample research questions of different types for three domains of therapy

Domains of therapy process	Sample research questions
1. *Treatment outcome*	What can I learn from this client about what it means to be improved in therapy? (*Definitional*) In what ways has this client changed? (*Descriptive*) What is the story of this client's improvement? (What led to what? What was the process of change?) (*Interpretive*) If the client is dissatisfied or desires further change, what could be done to bring this about? (*Critique/Action*) How much, if any, improvement did this client show? (*Quantificational*) Which of these two clients, seen in two different forms of this treatment, has improved more? (*Comparitive*) To what extent do these two different types of outcome vary together? (*Relational*) To what extent can judges agree on changes in this client's dynamic outcome criteria? (*Methodological*) What are the cultural and sociopolitical implications of the way in which outcome has been measured for this client (e.g. focusing on pathology vs. health) (*Deconstructive*)
2. *Therapy processes*	What is the nature of a particular experience (e.g. feeling misunderstood) in this therapy? (*Definitional*) What different kinds of things (or experiences) happened in this therapy? (*Descriptive*) Why did this event occur at moment in therapy? How did this event, process or theme unfold in therapy? (*Interpretive*) What therapeutic difficulties occurred in this treatment? How could they have been dealt with? (*Critique/Action*) What did the client and therapist typically do or experience in therapy? (*Quantificational*) Did client and therapist do or experience different things in early vs. late sessions of therapy? (*Comparitive*) How much do client and therapist agree with each other about what happened? (*Relational*) How well can judges agree on this client's core conflictual relationship themes? (*Methodological*) What kinds of assumptions have we made about what is important to know about therapy process? (*Deconstructive*)
3. *Change processes* (i.e. processes connected to effects)	What does it mean to this client to be helped in therapy? (*Definitional*) What aspects or events in therapy did the client experience as helpful? (*Descriptive*) Why did this client improve? Why did certain aspects of therapy help or hinder the client? (*Interpretive*) What aspects or events in this therapy are/were hindering for this client? How could they be overcome? (*Critique/Action*) Which helpful events or aspects of therapy are described most frequently by this client? (*Quantificational*) Do clients who resolve in-session tasks show better posttreatment improvement? (*Comparitive*) Which therapy processes appear to predict posttreatment improvement in these cases? (*Relational*) How consistent over time are clients' reports of what was most helpful in therapy? (*Methodological*) To what extent does this research assume that the therapist is the 'prime mover' in therapy? (*Deconstructive*)

and diversity of particular therapeutic phenomena, for example, 'Can empathy ever be harmful?' 'What different kinds of positive change occur after this therapy?' These questions seem most simply answered by naturalistic qualitative research methods (e.g. Patton, 1990; Lincoln & Guba, 1985), and grounded theory (Strauss & Corbin, 1990), a systematic approach to qualitative data analysis which tends to appeal to researchers who are trying to move beyond their quantitative training. It is worth noting that quantitative methods are commonly pressed into service for descriptive purposes (e.g. content, cluster or interaction analysis). However, such methods provide more information than is needed to answer questions of existence and diversity and may in fact overlook important variations because of their use of pre-set categories.

Interpretation

The third type of question which lends itself to qualitative investigation is interpretation. Interpretive questions seek explanation or understanding for events or phenomena in therapy. They attempt to create under-standings of things which are not clear, such as why an event happened, how it came about, or what it means. For example, the researcher may be faced with a puzzling or interesting event in therapy, such as instances in which a therapist totally misunderstands a client, and may want to try to come to an understanding of the various factors which seem to have brought this event about. Interpretive questions lend themselves to hermeneutic (Packer & Addison, 1989) or narrative (Rosenwald & Ochberg, 1992) methods of research, including traditional narrative case studies. Two examples of systematic interpretive therapy process research are comprehensive process analysis (Elliott, 1989) and the early steps of task analysis (Greenberg, 1984). The emphasis here is developing under-standings rather than verifying them.

Critique/action

Critique/action questions ask what is wrong with a situation, with an interest in determining how it can be changed. These questions attempt to go beyond definition, description and interpretation by seeking to empower or liberate individuals in oppressive or difficult circumstances. Such questions may address general conditions, such as demonstrating the distorting effects of male communication style on female therapy clients. However, the question may instead focus on a particular difficulty, such as an impasse in treatment. Such questions are most consistent with qualitative research which is guided by a liberating theory, such as feminism or Marxism. In order to be consistent with their goal of emancipation, such research seeks to create an equalitarian relationship between researcher and informant. Although very little psychotherapy research of this type has been carried out to date (see Hutchison, 1986), this genre is important in education (Belenky *et al.*, 1986; Lather, 1991) and industry (Whyte, 1991).

QUESTIONS WHICH LEND THEMSELVES TO A QUANTITATIVE RESEARCH

In contrast to the first four types of question, which seem most consistent with qualitative research (or which at least do not require quantification to be answered), the next four questions require quantitative data collection and analysis in order to be answered. These are questions of quantification, comparison, relationship, and method quality.

Quantification

Quantitative questions ask how often or to what degree or intensity something happens. Given an adequate definition of an event or aspect of therapy, such as mutative therapist interpretations, and some indication of its existence, it is useful to try to find out how often these interpretations occur and how long, important or powerful they are. Questions of this type cannot be answered without the usual paraphernalia of quantitative process research, including category systems, rating scales and descriptive statistics.

Comparison

Comparative questions seek to compare quantities; e.g. Does cognitive-behavioural therapy work better than interpersonal- psychodynamic therapy in treating depressed clients? How do resolved and unresolved misunderstandings differ from one another? These questions lend themselves to quantitative experimental or quasi-experimental designs, as well as inferential statistics.

Relationship

It is also useful to distinguish relational questions, which ask about aspects of therapy which vary with one another. Relational questions also include covariation across time, as well as complex relationships among sets of variables. Examples include, 'What are the basic dimensions of therapy process?' and, 'Does client pre-treatment functioning moderate the relationship between therapist technique and outcome?' Such questions lend themselves to quantitative bivariate and multivariate correlational, sequential and predictive methods, including path analysis.

Method quality

A closely related type of research question concerns method quality, e.g. With what degree (quantity) of reliability or validity can a feature or event within therapy be measured by means of a particular measure? Such questions include issues of consistency or generalizability across observers or judges, across items or measures of the same thing, across times of observation, and across more or less controlled observation situations (Wiggins, 1973). For example, How consistent are clients' ratings of therapeutic alliance over the course of therapy? Such questions are most readily addressed through psychometric or measure development research, and often play a supportive role in answering quantification, comparison, and, relationship questions.

DECONSTRUCTION

A final type of research question is meta-methodological, in the sense that it addresses the research method itself. Research motivated by deconstructive questions scrutinizes the implicit assumptions, values, or limitations in existing research. Deconstruction may include consideration of whose sociopolitical and scientific interests are well- or ill-served by the research. An example might be an analysis into the guiding metaphors in process research and the assumptions about client–therapist relationships contained in those metaphors (cf. Olds, 1992). Also, research that deliberately combines different genres and questions can be used to reveal the varying interests and assumptions of different forms of therapy research. (I know of no psychotherapy research which does that consciously, but Lather, 1991, provides an example of educational research in the deconstructionist 'collage' genre.)

The appropriate methodologies approach is a strategy of systematic methodological pluralism. This strategy indicates the need for a wider range of methods than is currently used in psychotherapy process research. However, this does not amount to an abandonment of standards for evaluating therapy research. This is because each genre has its own unique standards of excellence, which differ from those of other genres (e.g. Elliott, 1993). The main point is that consistency between research questions and methods is desirable.

A PRACTICAL PERSPECTIVE: OPTIONS FOR DATA COLLECTION IN CLINICAL SETTINGS

Assuming that the clinically-based process researcher has identified a key research question (and perhaps some secondary questions) and, on that basis, has selected an appropriate methodological approach, he or she is then ready to plan the actual data collection procedures. From the point of view of beginning research in a real-world clinical setting, a useful strategy is to start small and simple and work up to large and complex Thus one can begin by focusing on a single treatment case or even a single significant therapy event within a case.

A practical initial strategy here is to improve the quality of information obtainable from one's own treatment cases, in order to enrich and strengthen the conclusions which may be drawn from them. This form of research, the systematic case study, utilizes systematic qualitative or quantitative methods in order to reduce many of the problems with traditional case studies, including reliance on memory, anecdotal data collection, and narrative smoothing (Kazdin, 1981; Spence, 1986). In general, systematic case studies are careful investigations of particular treatment cases, using a variety of observational (non-experimental) research methods (see Elliott, 1983, for a review; also Jones, 1993).

Although any of the kinds of therapy process defined by the Five Dimensional Model can be the subject of investigation, the three domains of outcome, process and change process are most often addressed in systematic case studies. Consistent with the argument in the previous section, each of these domains can be the subject of any of the nine types of research question (see Table 2.3).

A number of research measures and procedures, involving varying degrees of time and effort, may be utilized in trying to answer these questions. As noted earlier, it is useful to begin small and then to enlarge the data collection to the limits of one's interests and setting. Table 2.4 contains lists of three sets of suggestions for carrying out systematic case studies on one's own clients (see also Barker et al., in press). Within each set, the options are ordered from least to most time-consuming or intrusive, so that one may begin with a minimum requirement and work up to more elaborate procedures. Because I believe that it is valuable to assess outcome as well as process, I include a list of options for that domain as well. The recommendation that methods be consistent with one's research questions should be kept in mind in considering these options.

EVALUATING TREATMENT OUTCOMES

The idea here is to do better than anecdotal impressions. Kazdin (1981) lists five principles for improving the credibility of assessments of degree (quantity) of change in case studies:

1. use systematic, quantitative data collection;
2. use multiple assessments of change over time;
3. if possible, look for change in previously chronic or stable problems;

Table 2.4. Practical suggestions for carrying out systematic case studies

(each set of suggestions is ordered from least to most time-consuming or intrusive)

Treatment effects
1. Measure the client on one well-chosen change measure before and after therapy (e.g Beck Depression Inventory with a cognitively-depressed client) (*Quantificational*)
2. Add an individualized change measure (target complaints, personal questionnaire) (*Quantificational*)
3. Use three or more standard measures of change/outcome (*Quantificational*)
4. Ask the client: 'What has changed since therapy started?' (Better: have a third person interview the client for you, e.g. change interview, Elliott *et al.*, 1990.) Also ask: 'What do you think has brought about these changes?' (*Interpretive*)
5. Add more assessments:
   (a) initial screening/intake;
   (b) midtreatment (or every 8–10 sessions);
   (c) follow-up (e.g. 1 year post-treatment) (*Quantificational*)
6. Add weekly change measures (*Quantificational*)
7. Add cases, creating a clinical replication series (Barlow *et al.*, 1984; a form of multiple baseline design)

Therapeutic processes
1. Tape sessions; take detailed process notes from tape or transcribe illustrative passages (*Descriptive*)
2. Administer periodic (e.g. three sessions) relationship measures, such as the Barrett Lennard Relationship Inventory, Penn Helping Alliance Questionnaire, or Working Alliance Inventory (all reviewed in Greenberg & Pinsof, 1986), (*Quantificational*)
3. Give standard self-report session measures to client and therapist, e.g. Session Evaluation Questionnaire (Stiles, 1980); Session Impacts Questionnaire (Elliott & Wexler, 1994); Therapy Session Reports (see Orlinsky & Howard, 1986) (*Quantificational*)
4. Use therapist- (or supervisor-) completed treatment-specific measures after each session to assess therapist adherence with treatment model or client progress, e.g. Therapist Action Questionnaire and Patient Action Questionnaire (Hoyt, 1980), Cognitive Therapy Scale (Beck *et al.*, 1979); Experiential Task Completion scales (Greenberg *et al.*, 1993; NIMH treatment adherence scale (DeRubeis *et al.*, 1982) (*Quantificational*)
5. Use your detailed process notes and the questionnaire data to write a narrative description of the case describing how the therapy evolved over time (*Interpretive*) (Better: have someone check your analysis against the original recorded sessions or process notes.)

Change processes
1. Ask the client: use helpful factors or significant event client self-report measures, including the Helpful Aspects of Therapy (HAT) form (Llewelyn *et al.*, 1988); or a posttreatment questionnaire or interview (e.g. Llewelyn & Hume, 1979). (*Descriptive*) Ask client to explain what made the factors or events helpful (*Interpretive*) (Also: ask the therapist to do the same)
2. Select a notable treatment success or failure (or both) and analyse (or compare) them (e.g. Strupp, 1980); or compare the most and least helpful or effective session (based on client or therapist ratings or ratings on a weekly change measure) (e.g. Parry *et al.*, 1986) (*Comparitive*)
3. Select an interesting or important event in therapy; transcribe it and analyse it thoroughly, using either task analysis (Rice & Greenberg, 1984) or comprehensive process analysis (Elliott, 1989) (*Interpretive*)

4. if possible, look for immediate or marked effects following the onset or key point of the intervention;
5. use multiple cases.

A combination of these features substantially improves the researcher's ability to infer that the treatment was related to the outcome.

Thus, client change can be measured by beginning with one well-chosen change measure administered before and after therapy (e.g. the BDI for a depressed client); to this can be added an individualized change measure such as the Personal Questionnaire (a set of target complaints which are rated on a regular basis throughout therapy; Phillips, 1986). Next, one can add three or more standard outcome measures, perhaps including a general symptom measure (e.g. the SCL-90; Derogatis, 1983), a measure of social or interpersonal functioning (e.g. the Interpersonal Problems Inventory; Horowitz *et al.*, 1988), and a treatment-relevant measure (e.g. Dysfunctional Attitudes scale: Weissman & Beck, 1978). At this point, one has reasonable approach to measuring quantity of change. Change can be further quantified by adding more assessments (e.g. mid-treatment, follow-up, weekly). In addition, if one is interested in other research questions that lend themselves to qualitative investigation, then interviewing the client about his or her experience and understanding of change can prove to be very illuminating (e.g. Elliott *et al.*, 1990).

EVALUATING THERAPY PROCESS

There are a number of practical, easily-administered methods for assessing what happens in sessions and the experiences clients have in them. A useful compendium of these can be found in Greenberg and Pinsof (1986). For example, therapy process can be assessed simply by taking good process notes or periodic relationship measures (e.g. the Penn Helping Alliance Questionnaire; Alexander & Luborsky, 1986), to which can be added weekly client self-report measures such as the Therapy Session Report (Orlinsky & Howard, 1986). Therapist- or supervisor-completed treatment adherence measures can also be utilized. Finally, one can use all these materials to write a narrative case study, which can additionally be checked by someone else to address issues of bias, or integrated with outcome data.

EVALUATING CHANGE PROCESSES

Beyond assessing therapy outcome and processes, change processes (effective ingredients) can be evaluated, for example by asking clients to quantify helpful factors on a post-treatment questionnaire (e.g. Sloane *et al.*, 1977). A more descriptive approach is to ask clients to write descriptions of what was most helpful after each session (e.g. on the Helpful Aspects of Therapy form; Llewelyn *et al.*, 1988). A more time-consuming comparative alternative is to identify two comparable cases with better vs. worse outcomes, and compare the processes (e.g. Strupp, 1980). Finally, one may adopt an interpretive approach by carrying out a task analysis of a repeating type of event (e.g. Rice & Greenberg, 1984) or by doing a comprehensive process analysis of a single interesting significant event (e.g. Elliott & Shapiro, 1992).

Parry *et al.* (1986) provides an excellent example of a systematic case study illustrating a number of these possibilities. The ultimate strength of systematic case studies is their potential for linking clinical practice and research in non-behavioural treatments. They also provide an initial entry into psychotherapy research and an opportunity to try out measures and data collection procedures, before plunging into larger, more involved studies. Finally, the availability of useful measures and data collection procedures can provide a useful stimulus to thinking about research questions, even though I am arguing that it is not wise to let available methods fix one's research questions.

AN ORGANIZATIONAL PERSPECTIVE: MODELS OF RESEARCHER-PRACTITIONER COLLABORATION

At the beginning of this chapter, I argued for more productive collaboration between therapy researchers and practitioners. In this final section I will review a number of different organizational models for conducting real-world psychotherapy process research, including both non-collaborative and collaborative models.

NON-COLLABORATIVE MODELS

Traditionally, there has been relatively little collaboration between practicing therapists and therapy researchers. Instead, researchers have had to become therapists, or else therapists have had to become researchers.

Most often, academically-based researchers have mounted studies in a research setting, usually a university or medical center. This typically occurs in a research or training clinic within a larger academic or medical organization. However, in reviewing research on treatments for children, Weisz *et al.* (1992) have argued that these research clinics differ greatly from typical practice situations, in which broadly-trained, loosely-supervised therapists carry large case loads of referred clients. In contrast, studies in research clinics typically involves specially-trained, closely-supervised therapists with small case loads of specially-recruited clients.

The ecological validity of such research is highly questionable (cf. Shapiro & Shapiro, 1983).

On the other hand, I have been focusing on the alternate model of therapists becoming researchers in their own settings, often through the means of the systematic case studies. As I see it, this approach has advantages to the research clinic model, particularly in terms of external validity. However, clinical practice situations are often not geared to research and practicing therapists often lack resources which make it easier to sustain the long process of designing research, carrying it out, and analysing and writing up the results. For example, schedules are often crowded, paperwork is constantly increasing (in the USA, private practice therapists are estimated to spend a quarter of their time dealing with insurance companies), sources of research expertise may be far away and computing facilities may not be available. Under these conditions, it is easy to become discouraged and give up.

COLLABORATIVE MODELS

Instead of these non-collaborative models of therapy research being conducted by researchers or therapists in isolation from each other, I am proposing the development of research organizational models in which researchers and therapists collaborate on topics of mutual interest, pooling their scientific and clinical skills and energies. Below, I describe three such models, giving examples of each.

Facilitator model

Here, the researcher's role is to help practicing therapists figure out what they want to learn about and to help them study it. Even though the therapists do not have a specific research agenda, the researcher does not impose one on them, but instead helps the therapists find a topic, research questions and methods. It is also highly appropriate (and fun also!) for the researcher to participate in the process of carrying out the research.

This model is illustrated by research on therapist difficulties carried out with a group of practicing therapists (headed by Marcia and John Davis) in the British National Health Service who were also affiliated with the University of Warwick. When I was on sabbatical at the University of Sheffield in 1984, I was offered the opportunity to carry out research with this group on a research topic to be determined. After I presented my list of favourite research topics, I asked this team of therapists what they were interested in studying. Their answer was 'none of the above'.

Instead, as we talked, they decided they wanted to study a topic not on my list at all: 'therapist difficulties', meaning situations in therapy in which they ran into problems of various sorts. From the point of view of these practicing therapists, moments of difficulty were the times when they felt in greatest need of help from research; these were times when they most wanted to turn to research. For starters, they wanted to know what these difficulties were and if other therapists also had trouble with the same situations (a descriptive question), and they also wanted to know what other therapists did about these difficulties (critical/action-oriented question).

Starting with this interest, we began by studying ourselves, each writing down descriptions of a number of difficulties we had encountered. Using a grounded theory approach (e.g. Strauss & Corbin, 1990), we then developed a taxonomy of therapist difficulties. This taxonomy included the following common therapist difficulties: incompetent, damaging, puzzled, threatened, out of rapport, personal issues, painful reality/ethical dilemma, stuck, and thwarted (Davis *et al.*, 1987). We next carried out a survey of therapist difficulties in a broad sample of psychotherapists working in the NHS, subsequently developing a similar taxonomy of coping strategies used by therapists for dealing with difficulties (Schröder *et al.*, 1987). We then used our taxonomies to quantify rates of different difficulties and coping strategies. Currently, quantitative questionnaire versions of the difficulties and coping strategies taxonomies are being used as part of a large-scale international study on the personal and professional development of therapists over time (Orlinsky *et al.*, 1991).

Consultant model

A second paradigm of researcher–practitioner collaboration is the consultant model, in which practicing therapists begin with a specific research interest and then ask researchers to help them design and carry out this research. The researcher's role here is to help the therapists to clarify their specific research questions and, using the strategy of appropriate methodologies, to offer alternative approaches to answering the research questions. As with the facilitator model, in the consultant model the researcher may become involved in actually carrying out the research, but this is probably less likely to happen than with the facilitator model, because the therapist team begins from a position of greater autonomy.

The consultant model is exemplified by a study currently being conducted by Cher Morrow-Bradley, a colleague of mine who has a private practice specializing in the treatment of clients diagnosed with borderline personality disorder. She asked me to consult with her on how to research the treatment of borderline clients in real-world private practice and agency settings. Through conversations, she and her team developed an interest in using clients' post-session self-reports to develop a data-based theory of treatment for borderline personality disorder. Data collection is accomplished by having the client fill out a Helpful Aspects of Therapy questionnaire

(Llewelyn *et al.*, 1988) after each session. Clients also complete a locally-generated quantitative post-session questionnaire asking them to report their feelings about themselves, their therapist, and various aspects of their treatment, including the therapeutic relationship. This team is particularly interested in extracting the sequential information from client's descriptions of most helpful events. For example, early data has suggested a common helpful sequence of client anger or emotional expression followed by therapist tolerance or support.

In consulting with this team, I have most often offered suggestions about possible measures and data analysis. When they found themselves uncomfortable with my demonstration of a standard qualitative grounded theory analysis, I offered a more traditional content-analytic approach as a compromise. I find that I am much less involved in the actual conduct of this study than was the case with the Warwick team, described above. Instead, the team asks me to come in whenever they are having difficulties or reach a new stage in the research. Because of my growing interest in the interpersonal dynamics of teams of researchers who do consensual rating, I have also offered more guidance on the relationship aspects conducting this sort of research. Finally, I have found this team's energy and enthusiasm, as well as their ups and downs, to be engaging and stimulating.

Dialogical model

A third collaborative model is dialogical in nature. Here, therapists and researchers study the same thing, using their divergent perspectives to illuminate it. Unlike the consultant model, both therapist and researcher are equally involved in the research, but bring different perspectives and interests to it, differences which are exploited to enrich the analysis.

An example of this dialogical model can be found in a study I carried out with David Shapiro (Elliott & Shapiro, 1992), who was wearing his therapist's hat at the time. This research has gone through a couple of stages. First, we developed a more efficient recall procedure (Brief Structured Recall, Elliott & Shapiro, 1988) and carried it out on two cases, in which David was the therapist. Our first interest was methodological: Could a more efficient recall procedure be developed? Next, we wanted simply to describe the different kinds of significant events identified in this way and how frequent they were (descriptive and quantificational).

Finally, we carried out an interpretive research genre study (Elliott & Shapiro, 1992), in which client and therapist data gathered in the brief recall procedure were used to provide participant-based interpretations of the meaning, effects and context of a single significant change event. As the researcher, I analysed the event, using the comprehensive process analysis method (Elliott, 1989). Following this, I organized and carefully integrated client, therapist and researcher analyses into a single version. I then gave this integrated analysis to the therapist for correction. Finally, after therapist and researcher had agreed on the analysis, the therapist wrote a commentary on the process. A portion of this commentary is worth quoting here:

> From the point of view of the interface between research and clinical practice, this analysis of contrasting perspectives can be highly educational for the therapist. The data allow for the corroboration and revision of therapist views.... In effect, taking part in this research was for the therapist like receiving supervision from the client and research colleagues. The therapist was required to make explicit his rationales and purposes, and to have these checked out against the experiences of the client and the sense made of the process by observers. Like all good supervision, this process is often challenging and stressful, but can be very useful for continued learning, particularly if the validity of the therapist perspective is recognized and respected.

Thus, based on my own experiences, I am excited about the possibilities for research collaboration between therapists and researchers, and see them as important avenues for fostering better working relationships and more clinically interesting research.

SUMMARY RECOMMENDATIONS: PRACTICAL STRATEGIES

I conclude with a summary of the main points of this chapter, in the form of a set of recommendations.

CONSIDER A RANGE OF POSSIBILITIES IN CHOOSING A KIND OF PROCESS TO STUDY

It is valuable to consider the full range of possible aspects of therapy process in order to identify an aspect or aspects of therapy process that you are interested in. The five dimensional model of therapy process, presented earlier, can serve this function, as can Orlinsky and Howard's (1987) generic model of therapy. In addition, consider the possibility of investigating more than a single perspective, person, aspect, unit level or phase of therapy process.

APPROPRIATE METHODOLOGIES

It is useful to follow the Appropriate Methodologies strategy of letting your research question guide your choice of a general research method/genre.

It is better at least to start by formulating a key research question or two that you are interested in, before considering the research methods available. You can always compromise later, and may instead be inspired to invent a method for answering your questions! In addition, consistency between research questions and methods is desirable; do not attempt to force one into the other. However, in keeping with the 'collage' approach sometimes used in deconstructive research (e.g. Lather, 1991), it is alright to combine different process research genres in a single study (referred to as 'triangulation' by educational researchers; cf. Patton, 1990); this allows them to inform and 'deconstruct' each other.

COMMON MEASURE/DESIGN SELECTION OPTIONS

Consider a range of data collection options for your research, from minimally intrusive and time-consuming to as extensive as your questions suggest and your setting allows

COLLABORATIVE RESEARCH TEAMS

Rather than going it alone as a therapist-researcher or researcher-therapist, consider researcher–practitioner collaboration. However, if you are a practicing therapist working with researcher as consultant or facilitator, don't allow yourself to be pushed around or to have a foreign research agendas or favoured methods forced on you. Instead, find a researcher who is interested in listening to you and working collaboratively with you, as a facilitator, consultant or dialogue partner. That way it will be more fun for everyone!

REFERENCES

Alexander, L.B. & Luborsky, L. (1986). The Penn helping alliance scales. In *The Psychotherapeutic Process: a Research Handbook* (L. Greenberg & W. Pinsof, eds). New York: Guilford, pp. 325–366.

Barker, C., Pistrang, N. & Elliott, R. (In press). *Research Methods in Clinical and Counseling Pychology*. Chichester, England: John Wiley & Sons.

Barlow, D.H., Hayes, S.C. & Nelson, R.O. (eds) (1984). *The Scientist Practitioner: Research and Accountability in Clinical and Educational Settings*. New York: Pergamon.

Beck, A.T., Rush, A.J., Shaw, B.F. & Emery, G. (1979). *Cognitive Therapy of Depression*. New York: Guilford.

Belenky, M.F., Clinchy, B.M., Goldberger, N.R. & Tarule, J.M. (1986). *Women's Ways of Knowing: the Development of Self, Voice, and Mind*. New York: Basic Books.

Crits-Christoph, P. (1992) The efficacy of brief dynamic psychotherapy: a meta-analysis. *American Journal of Psychiatry* **149**, 151–158.

Dahl, H., Kaechele H. & Thomae, H. (eds) (1988). *Psychoanalytic Process Research Strategies*. New York: Springer-Verlag.

Davis, J.D., Elliott, R., Davis, M.L., Binns, M., Francis, V.M., Kelman, J. & Schroeder, T. (1987). Development of a taxonomy of therapist difficulties: initial report. *British Journal of Medical Psychology* **60**, 109–119.

Derogatis, L.R. (1983). *SCL-90-R Administration, Scoring and Procedures Manual—II*. Towson, MD: Clinical Psychometric Research.

DeRubeis, R.J., Hollon, S.D., Evans, M.D. & Bemis, K.M. (1982). Can psychotherapies for depression be discriminated? A systematic investigation of cognitive therapy and interpersonal therapy. *Journal of Consulting and Clinical Psychology* **50**, 744–756.

Elliott, R. (1983). Fitting process research to the practicing psychotherapist. *Psychotherapy: Theory, Research & Practice* **20**, 47–55.

Elliott, R. (1989). Comprehensive Process Analysis: understanding the change process in significant therapy events. In *Entering the circle: Hermeneutic Investigation in Psychology* (M. Packer & R.B. Addison, eds). Albany, NY: SUNY Press, pp. 165-184.

Elliott, R. (1991). Five dimensions of therapy process. *Psychotherapy Research* **1**, 92–103.

Elliott, R. (1992). *Modes of Explanation in Psychotherapy Research*. Unpublished manuscript, University of Toledo.

Elliott, R. (1993). *Proposed Guidelines for Evaluating Qualitative Clinical Research*. Unpublished manuscript, University of Toledo.

Elliott, R. & Anderson, C. (1994). Simplicity and complexity in psychotherapy research. In *Psychotherapy Research: Assessing and Redirecting the Tradition*. (R.L. Russell, ed.). Plenum Press.

Elliott, R., Clark, C., Wexler, M., Kemeny, V., Brinkerhoff, J. & Mack, C. (1990). The impact of experiential therapy of depression: initial results. In *Client-centered and Experiential Psychotherapy Towards the Nineties* (G. Lietaer, J. Rombauts & R. Van Balen, eds). Leuven, Belgium: Leuven University Press, pp. 549–577.

Elliott, R. & Morrow-Bradley, C. (1994). Developing a working marriage between psychotherapists and psychotherapy researchers: identifying shared purposes. In *Research Findings and Clinical Practice: Bridging the Chasm* (P.F. Talley, H.H. Strupp & S.F. Butler, eds). New York: Basic Books.

Elliott, R. & Shapiro, D.A. (1988). Brief structured recall: a more efficient method for identifying and describing significant therapy events. *British Journal of Medical Psychology* **61**, 141–153.

Elliott, R. & Shapiro, D.A. (1992). Clients and therapists as analysts of significant events. In *Psychotherapy Process Research* (S.G. Toukmanian & D.L. Rennie, eds). Newberry Park, CA: Sage, pp. 163–186.

Elliott, R., Shapiro, D. A., Firth-Cozens, J., Stiles, W.B., Hardy, G., Llewelyn, S.P. & Margison, F. (in press). Comprehensive process analysis of insight events in cognitive-behavioural and psychodynamic interpersonal psychotherapies. *Journal of Counseling Psychology*.

Elliott, R. & Wexler, M.M. (1994). Measuring the Impact of Treatment Sessions: The Session Impacts Scale. *Journal of Counseling Psychology* **41**, 166–174.

Garfield, S.L., Bergin, A.E., (Eds.). (1986). Handbook of Psychotherapy and Behavior Change, (3rd edn). New York: Wiley.

Giles, T.R. (ed.). (1993). *Handbook of Effective Psychotherapy*. New York: Plenum Press.

Giorgi, A. (1975). An application of phenomenological method in psychology. In *Duquesne Studies in Phenomenological Psychology*: vol. 2 (A. Giorgi, C. Fischer & E. Murray, eds). Pittsburgh, PA: Duquesne University Press.

Greenberg, L.S. (1984). Task analysis: the general approach. In *Patterns of Change* (L. Rice & L. Greenberg, eds). New York: Guilford Press, pp.124–148.

Greenberg, L.S. (1986). Change process research. *Journal of Consulting and Clinical Psychology*, **54**, 4–9.

Greenberg, L.S., Elliott, R. & Lietaer, G. (1994). Research on humanistic and experiential psychotherapies. In *Handbook of Psychotherapy and Behavior Change*, 4th edn (A.E. Bergin & S.L. Garfield, eds). New York: Wiley.

Greenberg, L.S. & Pinsof, W.M. (1986). *The Psychotherapeutic Process: a Research Handbook*. New York: Guilford.

Greenberg, L.S., Rice, C.N. & Elliott, R. (1993) *Facilitating emotional change: the moment-by-moment process*. New York: Guilford.

Horowitz, L.M., Rosenberg, S.E., Baer, B.A., Ureño, G. & Villaseñor, V.S. (1988). Inventory of interpersonal problems: psychometric properties and clinical applications. *Journal of Consulting and Clinical Psychology* **56**, 885–892.

Hoyt, M.E. (1980). Therapist and patient actions in "good" psychotherapy sessions. *Archives of General Psychiatry*, **37**, 159–161.

Hutchinson, C.H. & McDaniel, S.A. (1986). The social reconstruction of sexual assault by women victims: a comparison of therapeutic experiences. *Canadian Journal of Community Mental Health*, **5**, 17–36.

Jones, E.E. (1993). Special section: single-case research in psychotherapy. *Journal of Consulting and Clinical Psychology*, **61**, 371–430.

Kazdin, A.E. (1981). Drawing valid inferences from case studies. *Journal of Consulting and Clinical Psychology* **49**, 183–192.

Kiesler, D.J. (1973). *The Process of Psychotherapy*. Chicago: Aldine.

Kuhn, T.S. (1962). *The Structure of Scientific Revolutions*. Chicago: University of Chicago Press.

Labov, W. & Fanshel, D. (1977). *Therapeutic Discourse*. New York: Academic Press.

Lather, P. (1991). *Getting Smart: Feminist Research and Pedagogy With/in the Postmodern*. New York: Routledge.

Lincoln, Y. & Guba, E.G. (1985). *Naturalistic Inquiry*. Beverly Hills, CA: Sage.

Llewelyn, S.P., Elliott, R., Shapiro, D.A., Firth, J. & Hardy, G. (1988). Client perceptions of significant events in prescriptive and exploratory periods of individual therapy. *British Journal of Clinical Psychology* **27**, 105–114.

Luborsky, L. (1976). Helping alliances in psychotherapy. In *Successful Psychotherapy* (J.L. Claghorn, ed.). New York: Brunner/Mazel, pp. 92–116.

Luborsky, L. (1984). *Principles of Psychoanalytic Psychotherapy: a Manual for Supportive–Expressive Treatment*. New York: Basic Books.

Luborsky, L. & Crits-Christoph, P. (1990). *Understanding Transference: The CCRT Method*. New York: Basic Books.

Morrow-Bradley, C. & Elliott, R. (1986). The utilization of psychotherapy research by practicing psychotherapists. *American Psychologist* **41**, 188–197.

Olds, L.E. (1992). *Metaphors of Interrelatedness*. Albany, NY: SUNY Press.

Orlinsky, D.E. & Howard, K.I. (1986). The psychological interior of psychotherapy: explorations with the therapy session reports. In *The Psychotherapeutic Process: a Research Handbook* (L. Greenberg & W. Pinsof, eds). New York: Guilford, pp. 477–501.

Orlinsky, D.E. & Howard, K.I. (1987). A generic model of psychotherapy. *Journal of Integrative and Eclectic Psychotherapy* **6**, 6–27.

Orlinsky, D., Aapro, N., Backfield, J., Dazord, A., Geller, J., Rhodes, R. & The Collaborative Research Network (Oct. 1991). *The Development of a Psychotherapist's Core Questionnaire: a New Research Instrument and its Rationale*. Workshop presented at meeting of North American Society for Psychotherapy Research, Panama City, FL.

Oxford English Dictionary (Compact Edition). (1971). New York: Oxford University Press.

Packer, M.J. & Addison, R.B. (eds). (1989). *Entering the Circle: Hermeneutic Investigation in Psychology*. Albany, NY: SUNY Press.

Parry, G., Shapiro, D.A. & Firth, J. (1986) The case of the anxious executive: a study from the research clinic. *British Journal of Medical Psychology* **59**, 221–233.

Patton, M.Q. (1990). *Qualitative evaluation and research methods*, 2nd edn. Beverly Hills, CA: Sage.

Phillips, J.P.N. (1986). Shapiro personal questionnaire and generalized personal questionnaire techniques: a repeated measures individualized outcome measurement. In *The Psychotherapeutic Process: a Research Handbook*. (Greenberg, L.S. & Pinsof, W.M., eds). New York: Guilford, pp. 557–590.

Rice, L.N. & Greenberg, L. (eds) (1984). *Patterns of change*. New York: Guilford Press.

Rice, L. N. & Sapiera, E.P. (1984). Task analysis and the resolution of problematic reactions. In *Patterns of Change* (L.N. Rice & L.S. Greenberg, eds). New York: Guilford, pp. 29–66.

Rosenwald, G.C. & Ochberg, R.L. (1992). *Storied Lives: The Cultural Politics of Self-understanding*. New Haven, CT: Yale University Press.

Russell, R.L. & Stiles, W.B. (1979). Categories for classifying language in psychotherapy. *Psychological Bulletin* **86**, 404–419.

Schaffer, N.D. (1982). Multidimensional measures of therapist behaviour as predictors of outcome. *Psychological Bulletin* **3**, 670–681.

Schröder, T., Binns, M., Davis, J.D., Elliott, R., Francis, V.M. & Kelman, J.E. (June, 1987). *A Taxonomy of Therapist Coping Strategies*. Paper presented at meetings of Society for Psychotherapy Research, Ulm, West Germany.

Shapiro, D.A. & Shapiro, D. (1983). Comparative therapy outcome research: methodological implications of meta-analysis. *Journal of Consulting and Clinical Psychology* **51**, 42–53.

Sloane, R.B., Staples, F.R., Whipple, K. & Cristol, A.H. (1977). Patients' attitudes toward behavior therapy and psychotherapy. *American Journal of Psychiatry*. **134**, 134–137.

Spence, D.P. (1986). Narrative smoothing and clinical wisdom. In *Narrative Psychology: The Storied Nature of Human Conduct* (T.R. Sarbin, ed.). New York: Praeger, pp. 211–232.

Stiles, W.B. (1980). Measurement of the impact of psychotherapy sessions. *Journal of Consulting and Clinical Psychology* **48**, 176–185.

Stiles, W.B. (1992). Producers and consumers of psychotherapy research ideas. *Journal of Psychotherapy Practice & Research* **1**, 305–307.

Strauss, A.L. & Corbin, J. (1990). *Basics of Qualitative Research: Grounded Theory Procedures and Techniques*. Beverly Hills, CA: Sage.

Strupp, H.H. (1980). Success and failure in time-limited psychotherapy. *Archives of General Psychiatry* **37**, 595–603; 708–717; 831–841, 947–954.

Weissman, A. & Beck, A.T. (1978). *Development and Validation of the Dysfunctional Attitude Scale*. Paper presented at the annual convention of the Association for Advancement of Behavior Therapy, Chicago.

Weisz, J.R., Weiss, B. & Donenberg, G.R. (1992). The lab versus the clinic: effects of child and adolescent psychotherapy. *American Psychologist* **47**, 1578–1585.

Wertz, F.J. (1985). Methods and findings in the study of a complex life event: being criminally victimized. In *Phenomenology and Psychological Research* (A. Giorgi, ed.). Pittsburgh: Duquesne University Press.

Whyte, W.F. (ed.). (1991). *Participatory Action Research*. Newbury Park, CA: Sage.

Wiggins, J.S. (1973). *Personality and Prediction: Principles of Personality Assessment*. Reading, MA: Addison-Wesley.

3 Aptitude × Treatment Interaction (ATI) Research: Sharpening the Focus, Widening the Lens

VARDA SHOHAM* AND MICHAEL ROHRBAUGH
University of Arizona, USA

ABSTRACT   Aptitude × treatment interaction (ATI) research is concerned with how 'different folks benefit from different therapeutic strokes'. 'Aptitude' refers to any individual-difference variable that may moderate effects of a treatment, and 'treatment' stands for any type of intervention, from a whole package to a specific, even single intervention. The main appeal of ATI psychotherapy research has been to provide empirical guidelines for matching clients to treatments. This promise, however, remains largely unfulfilled. Here we offer four proposals to improve this situation: (1) widen the aptitude (A) side of the paradigm to include relational (R) as well as person (P) moderator variables; (2) investigate change processes by distinguishing moderator variables assessed before therapy from moderator/mediator variables assessed during therapy; (3) distinguish treatment models or packages (big Ts) from specific treatment interventions (little ts), concentrating on the latter; and (4) use bottom-up, intra-individual change-curve methodologies to examine systematic individual differences in response to treatment.

> *Stultifying consequences for psychotherapy research can be foreseen*
> *if scientific curiosity is blunted by acceptance of the Dodo view.*
> *(Rachman & Wilson, 1980, p. 257)*

Despite the well-known Dodo bird verdict on comparative psychotherapy outcome research—that 'everybody has won and all must have prizes' (Luborsky, *et al.*, 1975)—most would agree it is too soon to abandon investigations of why specific treatments work and for whom they are most and least helpful. The finding that different treatments tend in general to have equivalent average outcomes overshadows the enormous intratherapy 'error' variance typically found in these studies (Beckham, 1990). Equivalent averages often represent a considerable range of clients, including

* Correspondence address: Department of Psychology, University of Arizona, Tucson, AZ 85718, USA.

Research Foundations for Psychotherapy Practice. Edited by M. Aveline and D. A. Shapiro.
Copyright © 1995, Mental Health Foundation and Individual Contributors.
Published 1995 by John Wiley & Sons Ltd

some who benefit a great deal from a particular treatment, others who remain unchanged, and still others who deteriorate. If outcome depends on the match (or mismatch) between specific characteristics of clients and the treatments they receive, it is reasonable to suspect that the null findings from comparative outcome studies may obscure systematic individual differences in response to specific treatments. The promise of Aptitude × Treatment Interaction (ATI) research (Cronbach & Snow, 1977; Shoham-Salomon, 1991)—by some accounts largely unfulfilled—has been to illuminate, or unpack, these relationships by showing how 'different folks benefit from different therapeutic strokes'.

Consider two recent findings. In a study of obese diabetics with obese spouses, Wing et al. (1991) assigned patients to two types of behavioural treatments. In one, patients were treated alone while their spouses attended assessment sessions only; in the other, patients were treated conjointly with their spouses. On the average, clinical outcomes for the two groups did not differ, either at termination or at 1-year follow-up. But does this mean that the treatments were equivalent (i.e. that including the spouse made no difference)? With further analysis, Wing et al. (1991) found a significant sex × treatment (statistical) interaction: Women benefited more from being treated with their spouses but men benefited more from being treated alone. Had the results not been broken down by gender, the researchers might have concluded, mistakenly, that couple-based treatment has no particular advantage for obese diabetics.

The second finding is from Ludwick-Rosenthal and Neufeld (1993), who studied two types of preparatory treatment for patients undergoing first-time cardiac catheterization. The treatments involved providing either high or low levels of information about the procedure, and the patients in the study were stratified according to whether their preferred coping style involved seeking information or avoiding it. Here, too, a statistical analysis of behavioural outcomes revealed a clear coping-style × treatment inter-action: patients experienced less anxiety and coped with cardiac catheteriz-ation more effectively when the level of preparatory information matched their coping style (i.e. when information seekers received more infor-mation and information avoiders received less).

Should obese diabetics be treated alone or with their spouses? Should patients undergoing an invasive medical procedure receive high or low preparatory information? The two findings just cited suggest that answers to these simple, yet important clinical questions depend on characteristics of the patients—and it is precisely such 'it depends' questions that ATI research presumes to address.

In the ATI paradigm, the 'A' (aptitude) stands for any individual-difference variable or client-characteristic that may moderate the effects of a treatment (T) on an outcome (O). The 'T' may represent any type of inter-vention, from a whole ('pure' or 'integrated') therapy package to a specific, even single intervention. The 'I' (interaction) term is used in a statistical (not social) sense, referring to the moderating effect of A on the relationship between T and O. As Smith and Sechrest (1991) point out, demonstrating an ATI requires a minimum of four data points: it is not sufficient to show that treatments X and Y have different outcomes for clients with a single characteristic (e.g. anxiety, extroversion or a stable marriage) *or* that some client characteristic predicts response to only a single treatment. Thus, an ATI finding usually presumes differential effects, moderated by a client variable, of at least *two* treatments* (but see the Snow, 1991, discussion of non-standard ATI designs). Smith and Sechrest also note that some empirically-based 'treatment of choice' recommendations have been established, at least in the behaviour therapy literature, without the benefit of ATI research. One of these concerns the well-documented efficacy of exposure therapy for certain anxiety-based disorders (Barlow, 1988; Rachman & Hodgson, 1980). Indeed, there seems to be an implicit consensus that this treatment would not be helpful for other conditions; for example, '...it is doubtful that simply exposing depressed patients to depressing stimuli will make them any better' (Smith & Sechrest, 1991, p. 235). Such treatments of choice *assume* that ATI exists without directly testing it.

As noted above, the main appeal of ATI research is pragmatic, since it may lead to empirically based matching of clients to treatments, thus improving therapeutic efficacy (Beutler, 1991). Both clinicians and clients seem to 'know' that different folks benefit from different strokes: the idea feels right. Indeed, clinicians often try to match clients to therapies on an intuitive basis—and even the strongest Dodo-bird advocates would be unlikely to advise clients to seek any form of therapy regardless of who they are (Rachman & Wilson, 1980). In any case, there can be little doubt that ATI research addresses clinically relevant questions: Who does and does not benefit from therapy X? Would the clients who do not benefit from therapy X benefit from therapy Y? Would clients who do well under therapy X benefit less if assigned to therapy Y? Are therapies for well-matched clients more effective than therapies for poorly matched or randomly assigned clients? Do therapies for well-matched clients yield a higher effect size than the average effect of psychotherapy?

Apart from the pragmatics of patient–treatment matching, the ATI paradigm has heuristic value for psychotherapy process research in that it

---

* To appreciate why ATI requires at least two treatments, consider the Jacobson *et al.* (1986) report that behavioural marital therapy is more beneficial for symmetrical, 'egalitarian' couples than for those in a marriage organized according to traditional gender roles. Even if this finding is robust (which it may not be), we would need to know that the moderator variable—egalitarian vs. traditional relationship—operates differently for some *other* form of marriage therapy. Otherwise there is no basis for inferring that an ATI exists or that BMT is differentially indicated.

can enrich our knowledge of *how* a particular intervention works (Shoham-Salomon & Hannah, 1991). As discussed below, heuristic applications focus on how aptitudes (e.g. client reactance) relevant to hypothesized change processes (e.g. compliance vs. defiance) moderate the effect of specific interventions (e.g. paradoxical vs. non-paradoxical directives) in order to mediate outcome (Shoham-Salomon, *et al.*, 1989; see also Borkovec & Mathews, 1988).

ATI designs also have certain methodological advantages (in addition to the prospect of explaining more outcome variance). Having a control group, for example, becomes less essential. Unlike comparative clinical-trial designs where suitable control conditions are crucial to demonstrating treatment effects, it is often possible to answer ATI questions without getting bogged down in the ambiguities of wait-list, measurement-only, or placebo control groups. These issues sometimes do have relevance in ATI designs (when the efficacy of the treatments being compared has not previously been established), but they are clearly secondary to the main question, 'who benefits more from which therapy?'

In light of the pragmatic, heuristic and methodological appeals of ATI psychotherapy research, the paucity of studies actually demonstrating solid ATI findings is striking. It has been more than a quarter-century since Keisler (1966) challenged our 'uniformity myths' and Paul (1967) recast the goal of comparative outcome research with his specificity question (What treatment, by whom, for what problem, etc.). Since then further calls for replicated 'prescriptions for optimal combinations' (Goldstein & Stein, 1976) have yielded only meager results (Lambert *et al.*, 1986; Stiles *et al.*, 1986). Indeed, the corpus of ATI findings in the literature is surprisingly sparse (Dance & Neufeld, 1988), and many of the findings appear tentative, *post hoc*, and difficult to replicate (Beckham, 1990; Smith & Sechrest, 1991).

Recent commentaries on the current state of ATI research (and how to resuscitate it) appear in a special section of the April, 1991, issue of the *Journal of Comparative and Clinical Psychology* and elsewhere (Dance and Neufeld, 1988). Each of these papers addresses important methodological issues and offers specific recommendations for improving ATI psychotherapy research. Interestingly, one of the strongest points of agreement was that ATI research should be driven by plausible hypotheses closely tied to clinical theory.

Since writing about the ATI paradigm seems to come easier than designing and carrying out actual ATI studies (e.g. the first author is guilty of publishing at least one review paper for each of her three reports of actual ATI findings), and since previous commentaries demonstrate substantial overlap, we will focus here on a limited set of issues and offer yet another set of prescriptions for sharpening (and widening) the lens of ATI psychotherapy research. We organize the essay by proposing remedies for four 'problems'—in a nutshell, that:

1. aptitudes are too restricted in scope;
2. aptitudes are not sufficiently assessed during therapy;
3. treatments are often too global;
4. outcomes are too aggregated.

The proposals are based partly on empirical evidence, partly on good sense, and partly on our own theoretical biases.

PROPOSAL 1: Widen the aptitude (A) side of the ATI paradigm to include relational (R) as well as person (P) moderator variables

The first and most challenging task of ATI research is to identify theoretically meaningful moderator variables pervasive enough to cut psychotherapy at a natural joint (Shoham-Salomon & Hannah, 1991). Ideally, a moderator should be central to clients' lives, yet differentially predict therapeutic outcome. To date, the vast majority of moderator variables in psychotherapy ATI research have been limited to characteristics of individual patient (e.g. coping styles, type or severity of disorder). Yet when these individual-patient variables are correlated with outcome, either directly or in interaction with treatments, they rarely account for much of the outcome variance (Beckham, 1990; Dance & Neufeld, 1988).

Given this apparently finite capacity of person moderator variables to explain ATI outcomes, where else (beyond the person) might we look? One possibility, familiar to family therapists, is that the patient's current intimate relationships may influence how he or she responds to different treatments. It seems self-evident that, as patients change, their spouses and significant others may act (or react) in ways that facilitate or hinder therapeutic progress. Furthermore, a growing body of research shows correlations between patterns of family interaction and many forms of psychopathology treated by psychotherapists (Jacob, 1987; Jordan et al., 1992) and demonstrates that family factors influence a patients' readiness to change and probability of relapse (McCrady et al., 1986; Murphy et al., 1982).

Since the theories of change underpinning most psychotherapies focus almost exclusively on characteristics of the thoughts, behaviour, or psychodynamics of the individual patient, it is not surprising that these are the main foci of measurement in psychotherapy research. Only rarely do researchers assess attributes of the (individual) patient's spouse or marriage and family relationships, † and when they do, these relational variables are usually measured from only the patient's view. In any case, if relational moderators do interact with treatments to influence outcome, they do so whether we measure them or not. ‡ While the importance we are attaching to relational moderators is admittedly speculative, research by Barlow and others suggests that at least one such variable—the quality of the patient's marriage—may be crucial in the treatment of anxiety disorders. Barlow *et al.*, (1984) reported an ATI involving pre-treatment marital satisfaction and whether or not an agoraphobic's husband was included in her behaviour therapy. They found that married agoraphobic women experiencing low marital satisfaction responded better to treatments that included their husbands. By contrast, the spouse's participation did not make much difference when marital satisfaction was high. Other studies of behaviour therapy for agoraphobia (e.g. Milton & Hafner, 1979; Bland & Hallam, 1981; Monteiro, *et al.*, 1985), though not ATI designs, provide supporting evidence of an inverse relationship between the quality of the pretreatment marriage and the likelihood for improvement (cf. Rohrbaugh & Shean, 1988). In general, patients who exhibit better marital and sexual adjustment seem to respond better to behavioural treatment and better retain therapeutic gains than those involved in poor marriages. Bland and Hallam (1981) suggest that spouses in distressed marriage are often either unable or unmotivated to support the patient during treatment; Barlow (1988) adds that the patients in these marriages may not be able to accept such support even if it is offered. One way or another, assessing characteristics of a patient's marriage or family relationships may help to explain additional outcome variance in ATI studies.

Apart from the quality of a marital relationship, there may be other potentially important relational moderators that can be readily assessed. One is whether or not the individual client is actually involved in a committed relationship with a partner. Another is whether or not the client comes to therapy around the time of an important family life-cycle transition (e.g. within 6 months of birth, marriage, children leaving home, divorce, etc.). While researchers often record this type of information to describe their clinical samples, they rarely analyse it as a moderator of

---

† Psychotherapy researchers have understandably invested a great deal of effort in developing and improving measures of the *therapeutic* relationship, yet as Coyne and Liddle (1992) point out, '...the psychotherapeutic relationship is only one among many in the patient's life and usually not the most important one' (p. 48)

‡ We believe that relational moderators are no less relevant to *individual* therapy than to marriage and family therapy. Because relational moderators tend to be less accessible, both conceptually and technically, to the individual psychotherapy researcher, their exclusion is less reversible; unmeasured relational variables will usually be lost to subsequent analyses. On the other hand, if a marital or family researcher suspects that a moderator variable may be contributing to outcome, he or she is often in a position to re-analyse taped marital or family sessions in a *posthoc*, exploratory manner.

treatment process and outcome. We would not be surprised, however, if a meta-analytic review someday reveals higher success rates for cognitive-behavioural than psychodynamic therapies with relationally 'committed' clients, and perhaps an opposite pattern for single clients. Nor would we be surprised to learn that supportive and problem-solving interventions tend to be more effective around the time of a family life-cycle transition, whereas psychodynamic therapies are relatively more effective when the patient's family life is not in transition.

Other potentially relevant couple-level moderators might involve inter-action styles and patterns of communication. For example, one promising dimension is whether a couple's dominant pattern of transaction tends to be symmetrical or complementary (Watzlawick *et al.*, (1967). In a symmetrical pattern the partners exchange similar behaviours, as in an egalitarian marriage; in a complementary pattern they exchange opposite behaviours, as in a marriage based on traditional sex roles, or one characterized by one spouse pursuing while the other distances, or one overfunctioning while the other underfunctions. Aspects of relational symmetry/complementarity have been assessed with both self-report (e.g. Christensen, 1987, 1988) and observational rating scales (e.g. Rogers & Bagarozzi, 1983), though we know of no applications of such measures in ATI psychotherapy research. Still, there is reason to hypothesize that clients involved in symmetrical relationships may benefit more from therapies emphasizing direct negotiation of behaviour exchange and communication skills (Jacobson *et al.*, 1986), while clients in rigidly complementary systems may do better with strategic interventions designed to challenge problematic transactions less directly (Goldman & Greenberg, 1992).

The literature on schizophrenia and depression calls attention to family-level variables such as 'expressed emotion' (EE), affective style, family burden, family cohesion, and family support that predict expression of symptoms and treatment outcomes—and some of these may be useful moderator variables as well (Hooley, 1986). It is well established, for example, that high levels of family criticism, hostility, and emotional over-involvement predict relapse and rehospitalization for schizophrenia patients (Goldstein & Strachan, 1987; Leff & Vaughn, 1985), and psycho-educational interventions designed to reduce EE have become the psycho-social treatment of choice for this disorder. It is also clear, however, that families of schizophrenics vary on the EE dimension and that psycho-educational family treatments are not a panacea. On the average, the effects of these interventions 'appear to be modest and of uncertain durability' (Bellack & Meuser, 1993, p. 317). For this reason, some investi-gators have called for ATI designs that will improve our ability to match different schizophrenia patients to different treatments (Bellack and Meuser, 1993)—and considering EE itself as a moderator variable would be a good place to start. For example, patients from high EE families may respond best to treatments that avoid confrontation and use normalizing and supportive interventions, whereas patients from low EE families may be amenable to more direct cognitive-behavioural interventions and skill training.

ATI research might also draw on family theoretical frameworks grounded in empirical research. One such framework is Olson's (1986) circumplex model, which offers well-developed client self-report and clinician-observation scales to locate families on the orthogonal dimensions of cohesion (enmeshed vs. disengaged) and adaptability (rigid vs. chaotic). It is not difficult to imagine how these different relational contexts may have differential treatment implications for clients who inhabit them. Clients from enmeshed families, for example, may be most likely to benefit from treatments that emphasize intra- or interpersonal boundary definition, while those from disengaged families may be better served by techniques oriented to establishing or participating more effectively in close relationships. Similar predictions might be made about differential indications of structured and unstructured (e.g. directive vs. non-directive) therapies for clients from rigid as opposed to chaotic families.

The possibility of studying relational and person moderators *concurrently* is illustrated by a current project comparing family-systems vs. cognitive-behavioural treatments for different alcoholism subtypes (Beutler *et al.*, 1993). The main hypothesis is that 'internalizing' alcoholics, whose drinking tends to be steady and interwoven with family dynamics, will benefit more from family-systems treatment than from symptom-focused cognitive and behavioural treatment, while the opposite should be true for 'externalizing' alcoholics who lack impulse control and tend to drink episodically. The rationale for the person moderator variable is based largely on Beutler's work (e.g. Beutler, 1983; Beutler *et al.*, 1991a; Beutler *et al.*, 1991b) distinguishing patients with 'internalizing' and 'externalizing' coping styles. However, we are also interested in a *relational* moderator construct having to do with the types of marriages internalizing and externalizing alcoholics have (which Beutler's previous work did not address). This relational focus follows not only from models of the 'alcoholic family' (e.g. Steinglass *et al.*, 1987), but also from intriguing evidence that (1) spouses of steady (but not episodic) drinkers often report *decreased* marital satisfaction when drinking stops (Dunn *et al.*, 1987), and (2) these couples actually show improved problem-solving when intoxicated (Jacob & Leonard, 1988) We are now attempting to distinguish 'relationship syntonic' and 'relationship dystonic' alcohol problems based on client reports and ratings of videotaped interactions. In addition to examining such measures of symptom–system 'fit' as a potential ATI moderator construct, we hypothesize that relational moderators will map onto person (coping style, drinking pattern) moderators. An especially salutary finding would be that some multi-faceted, composite moderator explains more ATI outcome variance than each (person or relational) moderator explains alone.

Whereas some relational and personal moderators may augment and complement each other to cover a larger proportion of the outcome variance, others may cancel each other out—and the latter situation is problematic for ATI research. For example, a client may have some person characteristic that should enable her to benefit from insight therapy, but at the same time she may be in a relationship that hinders her positive response to increased awareness of psychodynamic issues. Without assessing the relationship, we risk diluting potential interactions and missing effects worth detecting.

In summary, we propose giving more attention to relational moderators because we believe the noise factor in ATI research may have more melody to it than our one-channel (person-focused) recording devices can capture. (The fact that family life can be quite noisy does not diminish its potential for reducing 'noise' in the methodological sense.)

PROPOSAL 2: Distinguish dispositional moderator variables assessed *before* therapy from moderator/mediator variables assessed *during* therapy (big Ps and Rs vs. little ps and rs). The latter have a more fluid course and may themselves be responsive to earlier therapeutic interaction

A common misunderstanding is that 'aptitudes' in the ATI paradigm refer only to stable characteristics of the client and assumes a 'trait model' of personality (Dance & Neufeld, 1988). As originally conceived, the 'A' domain of moderator/aptitudes is much broader, including aptitudes as 'process' (Cronbach & Snow, 1977; Snow, 1980). In a more recent paper Snow (1991) emphasizes that:

> ... modifiability and continuity of aptitude differences represent questions for research... most ATI research has not yet examined aptitude change as a function of treatment, or the interrelation of traitlike and statelike properties of aptitude... . These seem to be especially important questions for further ATI research in psychotherapy (p. 206).

Despite skepticism that diagnostic categories will ever be useful differential predictors of psychotherapy outcome (Beutler, 1991), most ATI studies do focus on presumably stable patient characteristics such as personality traits and coping styles (see Dance & Neufeld, 1988, for a review). The stable moderator, measured before therapy, is hypothesized to interact with type of treatment to influence therapeutic outcome measured at termination or follow-up. This strategy makes good sense if one is concerned solely with matching patients to treatments, since matching will usually be based on what is known about clients before treatment begins. The picture becomes more complicated when treatment, by design or happenstance, produces change in the moderator variable itself. This is especially likely when a moderator variable is conceptually related to the problem or dynamics in a case. Thus, in cognitive therapy for depression, a therapist would probably focus directly on aspects of the patient's 'internalizing' coping style (i.e. repetitive blaming of self) in an attempt to modify 'irrational' attribution processes (Beck *et al.*, 1979). If this had occurred in an ATI study, the *moderator* variable of coping style could change face as it becomes a *mediator* of outcome.

Fortunately, the prospect of patient–treatment matching is not the only purpose or virtue of ATI research. Shoham-Salomon and Hannah (1991) have pointed to the 'heuristic' value of ATI designs in shedding light on change processes and how particular interventions work. An example comes from research by Shoham and colleagues suggesting that certain paradoxical directives (e.g. 'try to engage in your symptom') are effective because they arouse defiance in reactance-prone clients (Shoham-Salomon *et al.*, 1989). As shown in Figure 3.1(a), high levels of client reactance, measured (or induced) at the beginning of therapy, predicted the effectiveness of paradoxical interventions but not self-control interventions. Furthermore, non-reactant clients exposed to paradoxical interventions reported increased self-efficacy (for controlling their symptom) but showed an opposite pattern following self-control interventions (Figure 3.1(b)). We do not think these results have many implications for matching patients to therapies, in part because reactance-potential may not be particularly stable. The ATI findings simply tell us something about the underlying nature of paradoxical interventions and the way they produce their effects: non-reactant clients who do not experience symptom relief still perceive an increase in self-efficacy, whereas reactant clients exhibit symptomatic relief with little cognitive mediation. A subsequent study comparing various paradoxical and non-paradoxical treatments for insomnia (Shoham & Bootzin, 1991) provides for measuring the activation (and suppression) of reactance throughout treatment, so that its role in moderating and mediating outcome can be more thoroughly assessed.

Another example of heuristic concerns superseding pragmatic ones in ATI research comes from organizational psychology. Fiedler *et al.* (1979) found that, under stressful work conditions, amount of experience on the job is the best predictor of performance; under relaxed work conditions, however, intelligence is the best predictor. Such a finding obviously does

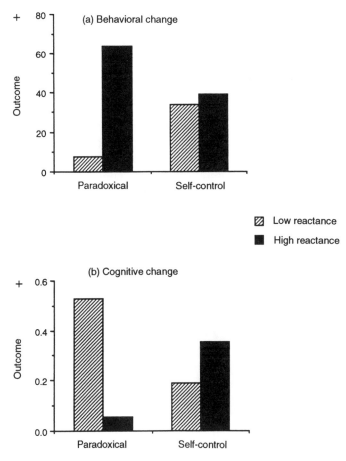

Figure 3.1 (a) Behavioural outcome as a function of reactance level and treatment condition (adapted from Shoham-Salomon *et al.*, 1989). (b) Cognitive outcome as a function of reactance level and treatment condition (adapted from Shoham-Salomon *et al*, 1989).

not imply that experienced but less intelligent employees should be assigned to stressful offices and smart but inexperienced ones to pleasant offices. Rather, the value of this research is in illuminating how stressful and relaxed work conditions produce the same effects through different routes: the former seems to activate survival skills and job smartness, while the latter activates intelligent problem-solving.

The idea of looking at aptitudes as *processes* rather than as fixed entities deserves more attention in the field of psychotherapy research. When a particular treatment or intervention 'benefits' individuals high on some relevant aptitude (e.g. anxiety), it is possible that the treatment does so by activating or suppressing that aptitude. Once activated, an aptitude can affect and be affected by treatments in a reciprocal (albeit differential) way. Thus, moderators, when shaped by the treatment, may turn into mediators and further interact with the treatment. Not surprisingly, studies that examine moderator/mediator relationships are rare (Shadish, 1992). It is easier to focus on pre-treatment moderators; continuous, ongoing mediators are less accessible, their course is more fluid, and they require repeated measurements and more complicated statistics. Nevertheless, a more fine-grained analysis of the continuous interaction between specific treatments and moderating/mediating patient characteristics should help us better understand therapeutic change processes and design more specific interventions to activate them.

As for proposal 2, we find it useful in thinking about ATI research to distinguish moderators measured only *before* therapy (we call these big As, or big Ps and Rs) from patient characteristics measured *during* therapy (little ps and rs), which in theory could serve as both moderators of treatment effects and mediators of outcome. Whether or not an upper-case variable is 'truly' a stable disposition of a person (P) or relationship (R), the fact that it is measured *before* therapy makes it a prospective marker for patient–treatment matching. Similarly, the fact that a potential moderator variable is measured *during* therapy—even if it appears dispositional or was also assessed before therapy—opens the possibility of it reflecting an ongoing, fluid change process and doubling as both a moderator and mediator of outcome. Thus, although big-letter moderators tend to be more analogous to traits and the little letters to states, the distinction is somewhat more complicated.

PROPOSAL 3: Distinguish treatment models or packages (big Ts) from specific treatment interventions and techniques (little ts) and concentrate on the latter

What exactly do client variables interact *with*? The 'T' in psychotherapy research is usually a large and multifaceted package that contains a variety of therapeutic components, some may augment each other, others may be redundant with each other, and still others may cancel out each other's effects. Do we expect certain patients characteristics to interact with the whole treatment package as if it was a single independent variable? With a specific phase of therapy? With a specific intervention or technique that is considered 'the heart' of the particular treatment? In an insightful comment about ATI research, Shulman (1981) wondered why we should expect to find generalizable interactions when we measure aptitudes with a micrometer and treatments with a divining rod.

The bigger the treatment package, the higher the risk of including (or inadvertently implementing) change mechanisms that may cancel each other's effects. This risk is somewhat reduced when

1. the theory underlying the design clearly states *what* in the T (i.e. its main change mechanisms) should interact with *what* in the A (i.e. the underlying theory of personality, psychopathology, or family dynamics, and other theoretically meaningful problem-maintaining patients' processes);
2. a treatment manual is carefully crafted to keep the change-mechanism in focus and to minimize irrelevant or contradictory therapeutic elements;
3. adherence to the principles outlined in the manual is continuously monitored.

Although meeting these conditions reduces the risk of implementing inconsistent therapeutic elements in a big-treatment ('big T') package, it is unrealistic to exepect that the risk would be completely eliminated.

There have been some notable 'big T' applications of ATI paradigm (e.g. Beutler, 1983; Beutler *et al.*, 1991a), but in our view the most promising and robust results have come from 'little t' designs where the 't' is a clearly operationalized intervention, the 'A' (usually a 'P') is a well-defined and reliably measured client characteristic, and there exists a theoretically sound reason to expect an interaction between the treatment and the aptitude. Not surprisingly, the best illustrations of this type of research come from behaviour therapy. A series of studies by Ost and colleagues, for example, examined moderating effects of behavioural vs. physiological response patterns on different interventions for different types of phobia. Hypothesized matches between moderators and treatments were based on Rachman's three-factor theory of anxiety, and conditions that matched treatments to outcome yielded consistently better outcomes than mismatched conditions. In one study, social phobics who were mainly behavioural reactors benefited most from social skills training, whereas physiological reactors benefited most from relaxation treatment (Öst *et al.*, 1981). In later studies, both agoraphobics and claustrophobics who were behavioural reactors responded best to exposure treatment, while the physiological reactors again did better with relaxation (Öst *et al.*, 1984). Tailoring specific interventions to patients' proneness to physiological arousal also proved fruitful in a recent study by Foa (1992): here, muscle relaxation techniques were more effective than muscle tensing techniques for anxiety-disorder patients with high proneness to physiological arousal, while the reverse was true for patients with low physiological proneness. Finally, in an investigation of interventions designed to reduce fear, Vallis and Bucher (1986) found that subjects who used visual imagery to cope benefited more from covert modeling than self-instructional training, while the latter was more efficacious for patients whose coping strategies were mainly verbal.

We will mention three other little-t examples of ATI findings to represent other disorders and therapeutic orientations. In their work with families of conduct-disordered adolescents, Patterson and Forgath (1985) observed that unco-operative parents showed relatively greater benefit than co-operative parents from 'facilitate-support' interventions compared to 'teach-confront' techniques. When exposed to the latter, unco-operative parents did far worse than the co-operative parents. Patterson and his colleagues used these findings to further sharpen their intervention techniques to fit different patterns of patients' resistance. The second example is from the McKnight *et al.* (1984) test of a compensatory treatment model for depression. These investigators found that depressed patients with social-skill deficits benefited from social skills training, whereas patients with irrational beliefs responded better to cognitive restructuring techniques that focused on attributing the depression to irrational beliefs. Finally, Beutler and colleagues (e.g. Daldrup *et al.*, 1988) contrasted emotion-focused interventions (e.g. the empty-chair technique for dealing with unfinished business) with cognitive-restructuring techniques for over-controlled and under-controlled depressed patients. As their work illustrates, little-t ATI designs can be used to compare the differential effects of specific (yet central) interventions derived from any treatment model amenable to a brief or analogue format.

Little-t designs have an additional, rather practical advantage over big-T designs: these well-focused studies are appealing in terms of *sample size* requirements. The cost of psychotherapy research often results in small samples that lack the power to detect the pursued differences. Researchers are justifiably reluctant to design ATI studies if they imply doubling the sample size. ATI findings that are pursued as a *posthoc* 'fishing expedition' often fall short of the significance level or fail to reach significance level when replicated. Indeed, in the absence of focused patient and treatment variables, or a good theoretical match between them, sample sizes for exploratory ATI designs have to be large. On the other hand, when the treatment is a well-defined technique and the hypothesized client-by-treatment interaction is theory derived, sample size considerations change dramatically. This is because enlarging sample size is not the only (or even best) way to reduce error variance and increase power. An alternative is to ensure specificity, clarity, and uniformity of treatments (Dance & Neufeld, 1988).

The following example from a recent research proposal illustrates how sample size was determined for a planned comparison between paradoxical and non-paradoxical (little-t) interventions for low vs. high reactance-prone clients (Shoham & Bootzin, 1991, pp. 38, 39):

> ... two specific questions of power are addressed. The first pertains to the magnitude of the overall effect sizes of paradoxical and non-paradoxical

interventions that are *worth detecting*. In Shoham-Salomon & Rosenthal's (1987) meta-analysis, the positively connoted paradoxical interventions (of the types used in the proposed study) were found to exhibit an effect size of $r = 0.52$.§ This effect size, in terms of Rosenthal's r means a true probability (expressed as a percentage) of 76% for the treated group to improve, when the untreated group has a 24% chance of improving. The average effect size of therapy, as found by Smith, Glass, and Miller (1980) and translated to Rosenthal's (1984) r is 0.42 (Shoham-Salomon & Rosenthal, 1987). Therefore, if the *true* average effect size or the paradoxical and the non-paradoxical interventions, across type of clients, is below 0.40, this effect may not be worth detecting. But if the effect size found by Shoham-Salomon and Rosenthal (1987) is a true effect size of paradoxical intervention, the planned design needs to have the power to detect such an effect. According to Chassan (1967), the probability of obtaining significant results (at the 0.05 level) when the true effect size is 0.50 is better than 0.70 when there are 10 clients in each [treatment] group. This probability increases to 0.98 with 25 clients in a group. In fact, even if the true effect size of the paradoxical or non-paradoxical treatments is the average $r = 0.40$, with 25 clients in a group the probability of detecting such an effect is close to 0.90, well above the cutting point recommended by Cohen (1977). Setting the number of clients per treatment on 20, will still enable us to detect a true effect of $r = 0.40$ with a probability of above 0.80 and an effect of 0.50 with a probability close to 0.90.

The second question deals with power to detect the interaction between the type of treatment and the level of the moderator variable (clients' reactance), if such interaction indeed exists. Since this question pertains to a theoretically derived match between a patient characteristic and the treatment, well matched clients should benefit much more from the treatments than the mismatched (or less well matched) clients. A worth-detecting effect size for the contrast of: (a) highly-reactant-clients-under-paradoxical-interventions and low-reactant-clients-under-nonparadoxical-interventions; *versus* (b) highly-reactant-clients-under-nonparadoxical-interventions and low-reactant-clients-under-paradoxical-interventions, should be higher than an overall treatment effect. If a true effect worth detecting is $r = 0.50$, then a sample size of 20 in each of these two combinations [(a) and (b)] should give us enough power (above 0.90) to detect such an effect. As mentioned in the previous power analysis, even a true interaction effect of $r = 0.40$ will be detected by such sample size (probability above 0.80).

Two features of these power calculations deserve emphasis: first, instead of pursuing effect sizes that *can* be detected, they pursue effect sizes that are *worth* detecting. This question invites researchers to establish a cut-off point (based on previous findings) below which they would not try to detect an effect. If, for example, the true effect is lower than the average effect size of psychotherapy, it might not make sense to enlarge the sample size to the point that such effect would be detected. The second point is that the sample size reached by the above power calculations could be

---

§ The transformation of effect size (ES) used by Smith *et al.* (1980) to a Rosenthal's r follows the formula: $r = ES/\sqrt{ES^2 + 2dfn}$ (Rosenthal & Rubin, 1986). Thus the overall effect of psychotherapy (ES = 0.89) is transformed into $r = 0.42$.

relatively small (20 patients in a treatment group) due to the fact that the theory-derived ATI hypothesis focused on well-defined intervention techniques (little ts) for which there is substantial accumulated knowledge.* In addition, it should be noted that main-effects were pursued in the above example in order to establish the superiority of the treated groups over the untreated control group. Where the ATI design does not involve a control group, power calculations can be done directly for the hypothesized interaction.

PROPOSAL 4: Use bottom-up methodologies (e.g. intra-individual growth curves) to examine systematic individual differences in response to treatment

We began this chapter by suggesting that unexplained outcome variance is detrimental to comparative outcome designs: a large variance around the means may mask differential effects for different clients. While ATI designs address this problem, they still leave much outcome variance unexplained. We then proposed including relational moderators and studying specific (little-t) treatment techniques as possible approaches to increasing the power of these designs. Both of these 'top-down' strategies make sense as long as ATI research focuses primarily on comparing averages or regression lines. Although four averages in a comparative outcome study are better than two, the ATI researcher may still face the unfortunate situation where the variance within each of the four groups is larger than the variance between them.

Change-curve methodology provides an alternative, 'bottom-up' approach to problems of over-aggregation. Change curves, also known as growth curves and widely used in experimental and developmental psychology, permit close examination of outcome data in small scale (small n) studies where repeated observations are feasible (Rogosa *et al.*, 1982; Rogosa & Willett, 1985). This methodology allows the researcher to treat outcome as a continuous, rather than discrete variable. Instead of breaking the aggregate into smaller and smaller units, it begins with the smaller unit of analysis, repeated observations of the individual case, and uses this to

---

* This argument stands in apparent contradiction with proposal 1. The recommendation to entertain hypotheses regarding multivariate moderators requires larger sample sizes (which is the case in our study of treatment × alcoholism subtypes, Beutler *et al.*, in press. As noted by Snow (1991), there is a '...trade-off between improving power and exploring complexity' (p. 212). The exploration of complex moderators may be better done as a first, separate study, where the sample is larger but the duration is shorter. An ATI treatment study with well-defined treatments and moderators (and increased power to detect effects in spite of small samples) can then follow.

build-up the aggregates. Moreover, instead of imposing a linear function on client change data, the methodology uses the form of best-fitting individual patients' change-curves (linear, quadratic, cubic) as its building blocks. Fitting change curves to the data in this way allows us to sift through the variability of multiple observations without averaging them or prematurely collapsing them into aggregates (Shoham-Salomon *et al.*, 1990).

Consider the following example from Shoham and Bootzin's (1991) pilot study of paradoxical and non-paradoxical one-session interventions with insomnia patients. Before treatment, the two patients represented in Figure 3.2 showed the same average baseline sleep-onset latency of about 60 minutes. Both then received an intervention that prescribed the symptom (' ...try to stay awake as long as you can... '), and 2 weeks later both were falling asleep in about 15 minutes. Guided only by these before and after latencies, one might easily conclude that the effects of the intervention for patients A and B were equivalent. Yet their change curves reveal striking differences in their immediate response to treatment, as well as in the shape of the curve and its current and projected stability.

Figure 3.2 shows the change curves that best fit each patient's daily sleep latencies over the 2 weeks following the paradoxical intervention. On the first few nights following the 'stay awake' directive, patient A promptly fell asleep while patient B, apparently complying with the prescription, stayed awake even longer than she had before. In addition, while patient A's latencies show a rising and falling (curvilinear) pattern, patient B's curve shows a steady trend towards change, reaching an asymptote. A clinician looking at these two curves would probably predict better prognosis for patient B, whose curve seems to have already stabilized. A researcher might be interested in the 'goodness of fit' between the best-fitting curve and the raw data (sleep latencies) it represents. In change-curve methodology one quantifies this fit by calculating an R^2 to represent an individual subject's variability around his or her own best-fitting (e.g. linear, quadratic) curve. In the above example, R^2 was lower for patient B's change curve than for patient A's, suggesting that B's pattern of change was less erratic than A's.

Interestingly, change-curve methodology can lead to the very same questions addressed by the ATI paradigm (Shoham-Salomon *et al.*, 1990). If we begin by examining how the shape or topography of change may vary with aptitudes and treatments, the ATI question can be reformulated in terms of a search for systematic individual differences in response to particular treatments. After each individual's raw outcome data have been plotted over time and modeled to a best-fitting curve, an investigator can group individual curves into change patterns or types, according to curve-shape (e.g. linear, cubic, quadratic), pace of change (reflected in the slope), variability (reflected in R^2) or some other parameters, depending on the

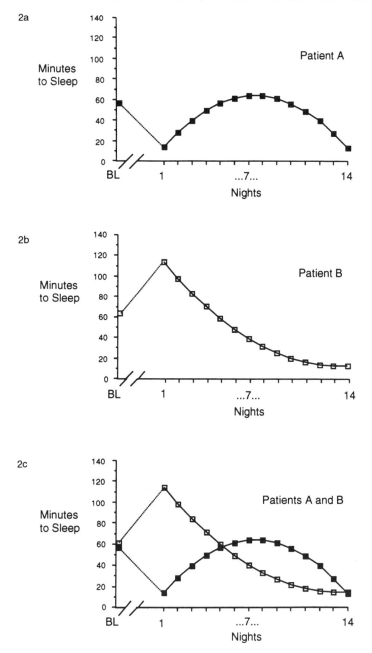

Figure 3.2 Change curves for two pilot patients receiving paradoxical intervention for insomnia. The first (BL) data point shows the mean sleep latency for 14 baseline nights; the remaining data points represent the best-fitting curve for sleep latencies on the 14 nights immediately following a paradoxical intervention.

specific data and questions at hand. One is then in a position to relate change types to meaningful inter-individual differences by asking three basic ATI questions:

1. Do change types correlate with moderator and mediator variables?
2. Do change types correlate with types of treatments?
3. Do these correlational patterns covary?

Thus, differential outcome questions can be rephrased: Do therapies differ from each other in the typical change-curve they produce? Is there a systematic difference in change-curves between the compared groups? Do systematic differences covary with theoretically meaningful moderating and mediating variables? In other words, do grouped curves reflect systematic individual differences in response to treatments? With this last question we have come full-circle to address the ATI questions through a bottom-up, rather than a top-down methodology, retaining the richness of individual differences rather than forcing our own functions on aggregated data.*

CONCLUSION

In the spirit of the conference in which this paper was originally presented, our intention is neither to bury ATI research or to praise it. In recommending remedies to a limited set of problems in this area, we proposed that ATI researchers might

1. explain more outcome variance by examining relational as well as person moderator variables;
2. produce more heuristic findings by distinguishing moderators measured before therapy from moderator/mediators measured during therapy;
3. distinguish treatment models from treatment techniques (big Ts from little ts), since the latter hold more promise for meaningful ATIs;
4. employ bottom-up, intra-individual change-curve methodologies to examine systematic individual differences in response to treatment.

Whether any of these proposals will pan out remains to be seen.

---

*In addition to detecting subtleties of change while retaining the richness of individual data, change-curve methodology has a practical appeal. Though demanding, the task of obtaining repeated measurement points is often less taxing than enlarging sample sizes. Simple repeated observation do not necessarily require excessive additional assessment hours. They can be done by quick checklists or even by introducing some standard questions into certain sessions and recording the data off the tapes. The sample size required for the analyses can be as small as a few cases or as large as a full-fledged ATI design, depending on the type of questions asked and the types of change obtained.

In the light of growing sentiment that the 'horse race' of comparative outcome research has run its course, one might reasonably wonder if traditional ATI questions should be stabled as well. We think not. Although the science of psychotherapy has so far offered embarrassingly few answers to the clinician's pressing question—What should I do for whom?—there is reason to continue the quest. With a wider range of As, more focused ts, and some disaggregated Os, research on this question may yet fulfil its potential.

ACKNOWLEDGEMENTS

This work was partially supported by NIMH grant No. 47451 and NIAAA grant No. 08970. We wish to thank Richard Bootzin, Douglas Tataryn, Lee Sechrest, and Deborah Bice-Broussard for their contributions to parts of this manuscript.

REFERENCES

Barlow, D.H. (1988). *Anxiety and its Disorders: The Nature and Treatment of Anxiety and Panic*. New York: Guilford.

Barlow, D.H., O'Brien, F.T. & Last, C.G. (1984). Couples treatment of agoraphobia: initial outcome. In *Advances in clinical behavior therapy*, (K.D. Craig & R.J. McMahon, eds). New York: Brunner/Mazel.

Beck, A.T., Rush, A.J., Shaw, B.F & Emery, G. (1979). *Cognitive Therapy of Depression*. New York: Guildford Press.

Beckham, E.E. (1990). Psychotherapy of depression research at a crossroads: directions for the 1990s. *Clinical Psychologyv Review* **10**, 207–228.

Bellack, A.S. & Meuser, K.T. (1993). Psychosocial treatment for schizophrenia. *Schizophrenia Bulletin* **19**, 317–336.

Beutler, L.E. (1983). *Eclectic Psychotherapy: a Systematic Approach*. New York: Pergamon.

Beutler, L.E (1991). Have all won and must all have prizes? Revisiting Luborsky *et al.*'s verdict. *Journal of Consulting and Clinical Psychology* **59**, 226–232.

Beutler, L.E., Engle, D., Mohr, D.C., Daldrup, R.J., Bergan, J., Meredith, K. & Merry, W. (1991a). Predictors of differential response to cognitive, experiential, and self directed psychotherapeutic procedures. *Journal of Consulting and Clinical Psychology* **59**, 333–340.

Beutler, L.E., Mohr, D.C., Grawe, K., Engle, D. & MacDonald, R. (1991b). Looking for differential effects: cross-cultural predictors of differential psychotherapy efficacy. *Journal of Psychotherapy Integration* **1**, 121–142.

Beutler, L.E., Patterson, K.M., Jacob, T., Shoham, V., Yost, E. & Rohrbaugh, M.J. (1993). Matching treatment to alcoholic subtypes. *Psychotherapy* **30**, 463–472.

Bland, K. & Hallam, R.S. (1981). Relationship between response to exposure and marital satisfaction in agoraphobics. *Behaviour Research and Therapy* **19**, 335–338.

Borkovec, T.D. & Mathews, A.M. (1988). Treatment of nonphobic anxiety disorders: a comparison of nondirective, cognitive, and coping desensitization therapy. *Journal of Consulting and Clinical Psychology* **56**, 877–884.

Chassan, J.B. (1967). *Research Design in Clinical Psychology and Psychiatry*. New York: Appleton-Century-Crofts.

Christensen, A. (1987). Detection of conflict patterns in couples. In *Understanding Major Mental Disorders: The Contribution of Family Interaction Research* (K. Hahlweg & M.J. Goldstein, eds). New York: Family Process Press

Christensen, A. (1988). Dysfunctional interaction patterns in couples. In *Perspectives on Marital Interaction*. (P. Noller & M.A. Fitzpatrick, eds). Clevedon, England: Multilingual Matters, pp. 31–52.

Cohen, J. (1977). *Statistical Power Analysis for the Behavioral Sciences*. New York: Academic Press.

Coyne, J.C. & Liddle, H.A. (1992). The future of systems therapy: shedding myths and facing opportunities. *Psychotherapy* **29**, 44–50.

Cronbach, L.J. & Snow, R.E. (1977). *Aptitudes and Instructional Methods: a Handbook for Research on Interactions*. New York: Irvington.

Dance, K.A. & Neufeld, R.W.J. (1988). Aptitude-treatment interaction research in the clinical setting: a review of attempts to dispel the 'patient uniformity' myth. *Psychological Bulletin* **104**, 192–213.

Daldrup, R.J., Beutler, L.E., Engle, D. & Greenberg, L.S. (1988). *Focused Expressive Psychotherapy: Freeing the Overcontrolled Patient*. New York: Guilford Press.

Dunn, N.J., Jacob, T., Hummon, N. & Seilhamer, R.A. (1987). Marital stability in alcoholic-spouse relationships as a function of drinking pattern and location. *Journal of Abnormal Psychology* **96**, 99–107.

Fiedler, F.E., Potter, E.H., Zais, M.M. & Knowlton, W.A. (1979). Organizational stress and the use and misuse of managerial intelligence and experience. *Applied Psychology* **64**, 635–647.

Fox, E.B. (1992, August). *What we Know and What we do not Know about Cognitive and Behavioral Treatments of Anxiety Disorders*. Paper presented at the 100th Annual Convention of the American Psychological Association, Washington DC.

Goldman, A. & Greenberg, L. (1992). Comparison of integrated systemic and emotionally focused approaches to couples therapy. *Journal of Consulting and Clinical Psychology* **60**, 962–969.

Goldstein, A.P. & Stein, N. (1976). *Prescriptive Psychotherapies*. New York: Pergamon.

Goldstein, M. J. & Strachan, A.M. (1987). The family and schizophrenia. In *Family Interaction and Psychopathology: Theories, Methods, and Findings* (T. Jacob, ed.). New York: Plenum, pp. 481–508.

Hooley, J.M. (1986). Expressed emotion and depression: interactions between patients and high versus low EE spouses. *Journal of Abnormal Psychology* **95**, 237–246.

Jacob, T. (ed.). (1987). *Family Interaction and Psychopathology: Theory, Methods and Findings*. New York: Plenum Press.

Jacob, T. & Leonard, K.E. (1988). Alcoholic-spouse interaction as a function of alcoholism subtype and alcohol consumption interaction. *Journal of Abnormal Psychology* **97**, 231–237.

Jacobson, N.S., Follette, W.C. & Pagel, M. (1986). Predicting who will benefit from behavioral marital therapy. *Journal of Consulting and Clinical Psychology* **54**, 518–522.

Jordan, B.K., Marmar, C.R., Fairbank, J.A., Schlenger, W.E., Kulla, R.A., Hough, R.L. & Weiss D.S. (1992). Problems in families of male Vietnam veterans with posttraumatic stress disorder. *Journal of Consulting and Clinical Psychology* **60**, 916–926.

Keisler, D.J. (1966). Some myths of psychotherapy research and the search for a paradigm. *Psychological Bulletin* **65**, 110–136.

Lambert, M.J., Shapiro, D.A. & Bergin, A.E. (1986). The effectiveness of psycho-therapy. In *Handbook of Psychotherapy and Behavior Change*, 3rd edn (S.L. Garfield & A.E. Bergin, eds). New York: John Wiley, pp. 157–212.

Leff, J. & Vaughn, C. (1985). *Expressed Emotion in Families*. New York: Guilford Press.

Luborsky, L., Singer, B. & Luborsky, L. (1975). Comparative studies of psycho-therapies: is it true that 'everyone has won and all must have prizes'? *Archives of General Psychiatry* **32**, 995–1008.

Ludwick-Rosenthal, R. & Neufeld, R.W.J. (1993). Preparation for undergoing an invasive medical procedure: interacting effects of information and coping style. *Journal of Consulting and Clinical Psychology* **61**, 156–164.

McCrady, B.S., Noel, N.E., Abrams, D.B., Stout, R.L., Nelson, H.F. & Hey, W.M. (1986). Comparative effectiveness of three types of spouse involvement in out-patient behavioral alcoholism treatment. *Journal of Studies in Alcoholism* **47**, 459–467.

McKnight, D.L., Nelson, R.O., Hayes, S.C & Jarret, R.B. (1984). Importance of treating individually assessed response classes in the amelioration of depression. *Behavior Therapy* **15**, 315–335.

Milton, F. & Hafner, J. (1979). The outcome of behavior therapy for agoraphobia in relation to marital adjustment. *Archives of General Psychiatry* **36**, 807–811.

Monteiro, W., Marks, I.M. & Ramm, E. (1985). Marital adjustment and treatment outcome in agoraphobia. *British Journal of Psychiatry* **146**, 383–390.

Murphy, J.K., Williamson, D.A., Buxton, A.E., Moody, S.C., Absher, N. & Warner, M. (1982). Long-term effects of spouse involvement upon weight loss and maintenance. *Behavior Therapy* **13**, 681–693.

Olson, D.H. (1986). Circumplex model VII: validation studies and FACES III. *Family Process* **25**, 337–351.

Öst, L., Jerremalm, A. & Johansson, J. (1981). Individual response patterns and the effects of different behavioral methods in the treatment of social phobia. *Behaviour Research and Therapy* **19**, 1–16.

Öst, L., Jerremalm, A. & Johansson, J. (1984). Individual response pattern and the effects of different behavioral methods in the treatment of agoraphobia. *Behaviour Research and Therapy* **22**, 697–707.

Patterson, G.R. & Forgath, M.S. (1985). Therapist behavior as a determinant for client noncompliance: a paradox for the behavior modifier. *Journal of Consulting and Clinical Psychology* **53**, 846–851.

Paul, G.S. (1967). Stray of outcome research in psychotherapy. *Journal of Consulting Psychology* **31**, 109–118.

Rachman, S.J. & Hodgson, R.S. (1980). *Obsessions and Compulsions*. Englewood Cliffs, NJ: Prentice-Hall.

Rachman, S.J. & Wilson, G.T. (1980). *The Effects of Psychological Therapy*. Oxford: Pergamon Press.

Rogers, L.E. & Bagarozzi, D.A. (1983). An overview of relational communication and implications for therapy. In *Marital and Family Therapy: New Perspectives in Theory, Research, and Practice* (D. A. Bagarozzi, A.P. Jurich & R.W. Jackson, eds). New York: Human Sciences.

Rogosa, D.R., Brandt, D. & Zimowski, M. (1982). A growth curve approach to the measurement of change. *Psychological Bulletin* **92**, 726–748.

Rogosa, D.R. & Willett, J.B. (1985). Understanding correlates of change by modeling individual differences in growth. *Psychometrika* **50**, 203–228.

Rohrbaugh, M. & Shean, G.D. (1988). Anxiety disorders: an interactional view of agoraphobia. In *Chronic Illness and the Family* (F. Walsh & C. Anderson, eds). New York: Brunner/Mazel, pp. 65–85.

Rosenthal, R. (1984). *Meta-analytic Procedures for Social Research*. Beverly Hills, CA: Sage.
Rosenthal, R. & Rubin, D.B. (1986). Meta-analytic procedures for combining studies with multiple effect sizes. *Psychological Bulletin* **99**, 400–406.
Shadish, W.R. (1992, August). *Mediators and Moderators in Psychotherapy Meta-analysis*. Paper presented at the 100th Annual Convention of the American Psychological Association, Washington DC.
Shoham-Salomon, V. (1991). Introduction to special series on client–therapy interaction research. *Journal of Consulting and Clinical Psychology*, **59**, 203–204.
Shoham, V. & Bootzin, R.R. (1991). *Outcome and Process Analyses of Therapeutic Paradoxes*. NIMH-funded grant proposal # 1RO1 MH47451 01A 1.
Shoham-Salomon, V., Avner, R. & Neeman, R. (1989). You're changed if you do and changed if you don't: mechanisms underlying paradoxical interventions. *Journal of Consulting and Clinical Psychology* **57**, 590–598.
Shoham-Salomon, V., Bice-Broussard, D., Tataryn, D. & Bootzin, R. (1990, June). *Individuals and Individual Change-curves in Search of Attention*. Paper presented at the Society for Psychotherapy Research Meeting, Wintergreen, Virginia.
Shoham-Salomon, V. & Hannah, M.T. (1991). Client–treatment interaction in the study of differential change processes. *Journal of Consulting and Clinical Psychology* **59**, 217–225.
Shoham-Salomon, V. & Rosenthal, R. (1987). Paradoxical interventions: a meta-analysis. *Journal of Consulting and Clinical Psychology* **55**, 22–27.
Shulman, L.S. (1981, August). *Educational Psychology Returns to School* (G. Stanley Hall Series). Paper presented at the 89th Annual Convention of the American Psychological Association, Los Angeles, CA.
Smith, B. & Sechrest, L. (1991). Treatment of aptitude × treatment interactions. *Journal of Consulting and Clinical Psychology* **59**, 233–244.
Smith, M.L., Glass, G.V. & Miller, T.I. (1980). *The Benefits of Psychotherapy*. Baltimore: Johns Hopkins University Press.
Snow, R.E. (1980). Aptitude processes. In *Aptitude, Learning, and Instruction: Vol. 1. Cognitive Processes Analyses of Aptitude* (R.E. Snow, P. Federico & W.E. Montague, eds). Hilldale, NJ: Erlbaum.
Snow, R.E. (1991). Aptitude-treatment interaction as a framework for research on individual differences in psychotherapy. *Journal of Consulting and Clinical Psychology* **59**, 205–216.
Steinglass, P., Bennett, L.A., Wolin, S.J. & Reiss, D. (1987). *The Alcoholic Family* New York: Basic Books.
Stiles, W.B., Shapiro, D.A. & Elliot, R K. (1986). Are all psychotherapies equivalent? *American Psychologist* **41**, 165–180.
Vallis, T.M. & Bucher, B. (1986). Individual difference factors in the efficacy of covert modeling and self-instructional training for fear reduction. *Canadian Journal of Behavioural Science* **18**, 146–158.
Watzlawick, P., Beavin, J. & Jackson, D.D. (1967). *Pragmatic of Human Communication*. New York: Norton
Wing, R.R., Marcus, M.D., Epstein, L.H. & Jawad, A. (1991). A 'family-based' approach to the treatment of obese type II diabetic patients. *Journal of Consulting and Clinical Psychology* **59**, 156–162.

4 Organizational Issues: Making Research Happen

GILLIAN E. HARDY
University of Sheffield, UK

INTRODUCTION

How can researchers ensure that their research proposals are taken up positively by their managers and organizations? Often good applied research projects that start out with great enthusiasm and commitment on the part of the investigators founder when they confront the barriers of a busy organization, challenge organizational practices or have insufficient authority to engage people in the research process (Beynon, 1988; Palazzoli *et al.*, 1986; Schein, 1987). Most psychotherapy practitioners do not work within organizations where research is a primary goal. Consequently in order to overcome organizational resistance and win co-operation, psycho-therapy researchers must find ways of aligning their research with the important goals of those within the organizations (recognizing that different groups may well have competing subgoals). The complexities of managing research within an organizational setting extend well beyond this however. Psychotherapy researchers need an explicit model or framework when planning research.

To describe the main issues surrounding the conduct of research within organizational settings, I have borrowed from Pettigrew's (1985) model of organizational change in *The Awakening Giant*. He argued for focusing on three principal elements of change: content, context and process. These elements together decide the strength of the impact of an intervention or introduction of new practices; if there is consistency between the elements then change is likely to be effective. Similarly, I propose that if there is consistency between one's research content, the context in which the research is to be conducted and the processes of implementation, then the research programme is more likely to be successful. In this chapter, the three elements of content, context and process are considered in detail.

Correspondence address: MRC/ESRC Social and Applied Psychology Unit, Department of Psychology, University of Sheffield, Sheffield S10 2TN, UK.

Research Foundations for Psychotherapy Practice. Edited by M. Aveline and D. A. Shapiro.
Copyright © 1995, Mental Health Foundation and Individual Contributors.
Published 1995 by John Wiley & Sons Ltd

How one can ensure consistency between the elements will then be discussed with illustrative examples that highlight the main issues. Finally, practical methods for working within organizations as a researcher using the three elements of change as a basic framework are discussed.

RESEARCH CONTENT

An example A group of researchers were investigating the viability of an extremely brief psychotherapy intervention. They decided to initially provide this for minimally depressed clients. Approaches to mental health, general practice and occupational health services although initially positive, never resulted in the referral of more than one or two clients.

The problem Part of the problem was that although the contacts were sympathetic to the aims of the research, neither they nor their colleagues felt that they were providing a service for clients suffering minimal or circumscribed distress. Most felt under pressure themselves to provide a basic service. As a consequence staff had little time to search for appropriate clients for the research as they thought that they neither came across such clients in their daily work, nor could see any benefits for themselves from the research.

The researchers then reconsidered their aims Whilst retaining their intervention, they redefined its use. They introduced it as an assessment and screening tool. This was taken up much more positively by other mental health workers. The researchers loosened their client intake requirements, and began to work alongside the clinicians. The referral problem reduced dramatically.

The content of the research will obviously depend upon the research aims and questions. For example, if the effectiveness of short-term psychotherapy is being evaluated in a clinical setting, the research will involve implementing new methods of treatment, scheduling and procedures for managing patient appointments. In addition, staff training will be an important element. Pre- and post-treatment assessments will also be necessary. However, in order to minimize or at least anticipate resistance, it is necessary to consider the extent to which the research aims are aligned with the aims of those affected by the research process. If clinicians are in favour of (say) achieving cost-effective minimal interventions, resistance to the research may be insignificant. On the other hand, if they are committed to high-quality psychotherapy without arbitrary limitations on treatment length, resistance is likely to be great. It is, therefore, important to know what organizational groups are likely to be affected by the research and to what extent their organizational aims are consistent with the aims of the research.

There are three main ways that the content of a research programme may impact upon the organization where the research is to be carried out:

1. aims of the research;

2. research procedures (what is going to happen);
3. research practicalities (what the research requires).

AIMS OF THE RESEARCH

When considering the aims of the research and the extent to which they are aligned with those who are affected by the research within the organization, it is important to effectively identify all the individuals, groups, teams and departments whose co-operation will be necessary for the success of the research. Careful consideration, therefore, needs to be given to those groups within the organization who may be affected by the research or some groups might react in an unexpected way to the content of the research proposal. Subgroups within organizations have differing and often competing roles (Beynon, 1988). Therefore framing the research proposal in a way that is acceptable to these different groups requires political sensitivity that can only be gained by careful consideration of the aims and different goals of the subgroups within the organization. For example, the management of a community health unit may wish to reduce the number of hours clinical psychologists spend in direct patient contact. Consequently they may be particularly attracted to research that evaluates the effectiveness of short-term psychotherapy. However, the psychologists themselves may be resistant to the research, seeing it as a threat to their clinical autonomy and integrity. Patients who also may be affected by the research, may have mixed views about the research. Some may be pleased at what they see as a streamlining of the service and a management interest in patient's views, others may feel they are going to be short-changed or that confidentiality issues are being violated by the repeated form filling. Without sufficient consideration, changing patient's expectations could have negative influences on their treatment and satisfaction with the organization. Although the importance of considering the alignment of the research aims and the aims of those affected by the research might seem self-evident, often researchers do not consider the organizational aims and concerns of those principal groupings affected by their activities. The kinds of problems that arise during a research programme can often be traced to the failure to consider sufficiently the form and degree of impact the research aims may have on others working in the organization.

The mistake that is often made by researchers is in underestimating or not considering the impact and consequences of the research they are proposing within the organization. Some of these have been mentioned above, but to contrast these, having said there needs to be compatibility between the research and organizational aims, if these are identical then the researchers run the risk of what Argyris (1979) described as the unintended consequences of research designs. Research participants and those assisting or affected by the research may begin to behave towards the research in the same way that they behave towards their work management. This may mean that parts of the research are effectively sabotaged.

Researchers, when aiming for compatibility between research aims and organizational goals, must not forget, therefore, to ensure the commitment and motivation of those participants. The best way of obtaining the commitment of those involved and of keeping them motivated is, in addition to the above considerations about research aims, involving people in the research process (cf. employee participation and job satisfaction, Trist et al., 1977; Wall & Lischeron, 1977). Giving information and listening to the concerns of relevant groups and individuals before starting the research is important (cf. Joos & Hickman, 1990 on patient–provider interactions and health care outcomes).

RESEARCH PROCEDURES

Organizations involved in the research will want to know not only the research aims, but what actually is involved in doing the research. They will be concerned with how the research procedures will affect the work of those involved. Often these questions will be related to clinical practice—how will the research affect client treatment? Some of these issues have been described by Firth et al. (1986), such as the effect of client or patient assessments, recording therapy sessions, strictness of treatment time-limits and use of manualized therapies. The costs and benefits of these constraints on therapy need to be discussed with clinicians and where possible resolved in advance. Such issues need ongoing work and support and should be tackled as clinical issues, rather than research problems, where they can be used positively to inform the therapy process.

RESEARCH PRACTICALITIES

The resource requirements of a research project again are going to be of particular interest to organizations. Will money or time be available for work attached to the research? Who is going to manage questionnaire distribution and collection? How is a clinician going to free up time to fill in post-session review forms? Where are clients going to fill in their questionnaires? Projects need adequate resources and commitment to resolve these issues so that they do not become the reason why incomplete date is gathered and research results become too messy to interpret.

RESEARCH CONTEXT

An example After conducting a multi-site research project into patients' satisfaction with the services they had received, one site withdrew its support and demanded that its questionnaire data be withdrawn from the study. The researchers tried unsuccessfully to change the organization's decision and had, therefore, to drop this data from their study.

The problem The researchers had failed to find out sufficient about this organization, which was undergoing a particularly difficult time, with threatened closure of its main site. In addition, the researchers had not sufficiently briefed staff, nor had sought support from the clinical director of the organization who was extremely influential and central to the running of the organization. When the clinical director found out about the research she was furious about not being informed and felt that the project was potentially harmful to the organization. She, therefore, successfully took action to prevent publication of the research findings.

Research context refers to the relevant features of the organization in which the research is to be conducted. These features constitute the institutional background to the research programme; the factors, processes and people that influence the research or that may be affected by the research. Below, seven aspects of organizational context are considered which are relevant to planning a research programme within an organizational setting.

ORGANIZATIONAL ENVIRONMENT

The organizational environment from a systemic point of view (Katz & Kahn, 1978) is very important since it plays a major role in the organizational functioning. Organizations, such as the NHS, who have experienced a great deal of change and uncertainty in the last 10 years, will often work to reduce the amount of future change and uncertainty (Burns & Stalker, 1961). Research that appears to decrease the likelihood of change will more likely to be seen positively by staff working within such organizations. For example, a study that looks at the organizational factors associated with staff mental health may well be seen as an opportunity by staff to show how constant change has had a detrimental effect on the workforce. An understanding of the organizational environment can help direct the researcher to salient organizational issues, which in turn will assist the researcher in improving the fit between the research plans and organizational issues, and therefore improve the chances of successful execution of the research.

ORGANIZATIONAL CULTURE

Organizational culture refers to the pattern of shared meanings which develop within organizations through staff selection, socialization and interaction processes (Pettigrew, 1979; Schein, 1985). At the deepest level, organizational culture refers to unspoken assumptions (not dissimilar to cognitive theorists' 'basic assumptions'; Beck, 1967), such as the assumption that patient care and wellbeing should come before career or political considerations. At the most superficial level organizational culture is expressed through rites and rituals, such as Christmas parties, leaving presents, etc. However, organizational culture also encompasses the concept of organizational defence mechanisms (Menzies Lyth, 1991; Hirschhorn, 1988). These are aspects of organizational functioning which are not discussable. Research which challenges such mechanisms is likely to be powerfully resisted. For example, research which aims to look at stress, coping and managerial supports within the caring professions may run into difficulties if it is seen to challenge an organizational assumption of the workplace being 'one happy, caring family'. In fact this is often more a myth than a reality, but an important defence against the enormous anxieties that surround working with ill and dying people.

ORGANIZATIONAL STRUCTURE

Organizational structure refers to the number of hierarchical levels within the organization as well as the extent to which work is conducted along functional or product lines (Robey, 1986). Researchers need to have an awareness of organizational structure and the level at which their research will be conducted. Failure to get approval and commitment for the research at the appropriate level could result in support for the research being withdrawn even though approval at another level was initially given. More junior staff may have concerns that completed questionnaires will be seen by management, which may affect their progress or employment within the organization. It is important, therefore, that the researcher remains 'neutral' with respect to the organizational hierarchy. Understanding the organizational structure is also important for considering the following aspects of an organizational context.

GROUPS WITHIN THE ORGANIZATON

The importance of recognizing and considering subgroups within the organization has already been mentioned. Groupings refer to both the formal task and professional groups, and to informal groups. Leaving out certain groups or ignoring the concerns of certain groups can lead to a good deal of resistance and perhaps to the ultimate failure of the research. For example, there may be informal groupings within an organization that are concerned with patients' rights. Research that contains questions to topics that are of importance to these informal groups, but which fails to consider patients' rights, may well be rejected by those concerned with these issues. Research that focuses on multi-disciplinary team functioning, but which fails to incorporate all staff groups within its programme, may well be undermined by these forgotten staff groups.

POWER AND POLITICS WITHIN ORGANIZATIONS

Organizations are essentially political; characterized by the exercise of power and the conflicting aims and interests of subgroups (Beynon, 1988; Mintzberg, 1983). Understanding of the power bases and the points of conflict is an important part of planning and conducting research within organizations. In addition, understanding the level at which it is necessary to gain approval for the research can be important in influencing the degree to which the research is implemented throughout the organization. Although in general the higher the level within the organization that gives approval, the greater the chance that the research will be successfully carried out, it should be remembered that suspicions of staff at the more junior levels about senior management's support for the research can also present a threat to the successful execution of the research.

Consideration of political issues is often absent in organizational research, yet can play a very powerful role in ensuring success and the implementation of recommendations suggested by the research findings. In negotiations with organizations, researchers need to have the authority and capacity to make strategic and tactical decisions.

STYLES OF DECISION-MAKING

A significant characteristic of organizations is the style of decision making employed. In some organizations, control is centralized and senior management make all the important decisions. In others, decentralization occurs, such that decision-making responsibility is devolved around the organization, so that individuals and groups have greater autonomy and control. Research which exposes or threatens a centralized style may be resisted by management, just as research that might provide management with opportunities for greater control might also be resisted by staff. For example, research designed to quantify clinical practice might well be resisted by clinicians, as this might provide management with ways of achieving greater control of the daily work of clinicians. Therefore, researchers need to take into account the extent to which their research results might threaten the position and interests of specific subgroups in relation to control over decision-making (Beynon, 1988).

INFORMATION AND CONTROL SYSTEMS

Information systems within organizations provide the data to aid and alter understanding and decision-making; control systems monitor staff functioning and provide greater control over staff activities. Researchers should consider the extent to which the research process and findings will release information which will reduce or increase managerial control. For example, research designed to monitor the management of a particular patient group as they pass through the health-care system, might expose patterns of staff–patient and staff–staff interactions that senior management consider inefficient, even though the clinicians involved see these as important procedures in the treatment process. Competing interests of subgroups could be differentially affected by the production of research data, thus producing conflict between staff and potentially be seen as a threat to one or more group.

RESEARCH PROCESSES

An example An evaluation of a psychotherapy service was being conducted. Questionnaires were given to clients when they had finished their therapy session by reception staff. The completed questionnaires were then handed to the receptionist who posted them to the researchers. One day, the researchers had a telephone call from the psychotherapy unit saying that they were suspending their involvement in the research until they had met with the researchers to discuss problems they had encountered.

The Problem The researchers had worked hard to design an evaluation study compatible with the psychotherapy unit's aims. All had gone well, until a receptionist had to deal with an angry and upset client who had found the questionnaires intrusive. The receptionist was given no support and refused to administer further questionnaires. The researchers had failed to consider sufficiently the process of implementation, especially support networks for staff and clients involved in the research process.

The researcher by considering content and context is likely to have a clear understanding of the research aims and the aims of the organization, and the meaning of the research from the organization's point of view, and the likely consequences of the research for the particular subgroups within it. How should she/he actually go about introducing the research plans? The research process is considered under the following four headings.

GAINING ACCESS

Initial meetings with an organization should involve discussing the content of the research. An initial question in terms of the process of the research in organizational settings, is the extent to which it is theory or problem-driven. Often conducting research within an organization requires compromise between the theoretical orientation of the researcher and the perceived problems of those within the organization. This may mean agreeing to conduct research on issues that are not part of the researchers' aims but are important to the organization. The balance between problem and theory-driven research will reflect to some extent the nature of the co-operation between researcher and organization. One research method that attempts to overcome this dichotomy is 'action research', where the research role is central to the relationship with participants and the organization, and emphasis is on the practical utility of the research for participants. One major difficulty of this research method is that the researchers' neutrality with respect to groups within the organization is often lost.

Another important issue is the amount of information the researcher discusses with the organization. This is not simply a matter of being honest. How the research is presented will depend on the understanding of those within the organization, and often downplaying complex theoretical issues and highlighting practical benefits will be the most successful strategy to employ. What, if any, consequences arise as a result of such 'marketing' should be considered (Bulmer, 1988).

Finally, there are a number of questions relating to the nature of the contract which need to be clarified for both the organization and research team: Who and what in the organization does the researcher have access to? What does the organization expect in return for co-operation, and how can this fit in with the research aims? Who owns the research? What is the extent of the commitment?

ETHICAL ISSUES

Ethical issues, such as those considered at local ethical committees, are usually concerned with the impact of the research on the participant. In the conduct of research most researchers will be aware of professional guidelines concerning confidentiality, protection of participants in research process, honesty vs. deception, debriefing, etc., (see, for example, British Psychological Society's, 1991, *Code of Conduct, Ethical Principles and Guidelines*). Very often the design of a research programme means that some clients will be excluded from a study. The reasons for this and what subsequently happens to such clients will be of concern to the organization. It is important that the delivery of a clinical service is seen to be sensible and sensitive to clients' requirements.

The ownership, storage, access to and use of research information also needs to be clarified with the organization. Does the organization give consent to publish? Does the organization and/or client own and have access to data and audio or video recordings? Is confidential information to be held securely 'on site' within a hospital or can such material be stored 'off-site'? All of these issues should be the concern of ethics committees from which approval will need to be obtained before beginning the research. When doing research in an organizational setting these issues, concerning the needs, rights and potential benefits to clients, need constant evaluation because of the complex relationship between researcher and participants.

Whilst the overall aim of clinical research is to benefit future patients, from an organizational perspective the management of the context in which the research is conducted also raises ethical issues. For example, how data regarding groupings within an organization are handled is a sensitive issue. Feedback to the organization can sometimes result in harm to particular groups, perhaps inadvertently. It is unwise to directly compare different clinical groups within an organization. There may be many reasons why differences may be found, some of which are not known to the researcher. It is preferable to compare each group with the overall results, so that no league table can be constructed between the groups.

If one treats the organization as a single unit in a research programme, then that organization should be given the same protection as one would give a research participant. Protecting the anonymity of an organization in research publications is only one aspect of researchers ethical responsibilities towards organizations. For a fuller discussion see Bryman (1988).

RELATIONSHIP WITH RESEARCH PARTICIPANTS

How the researcher conceptualizes the relationship with those involved in the research is an important part of the research process. Are those involved in the research subjects, participants or co-researchers? Will a steering group be appointed? Will participants have any influence over the research process? An important way of reducing resistance to the research process is to involve people. For example, when introducing evaluation of client outcome following counselling within a particular organization, key people should be identified. A workshop, briefing packs and the setting-up of telephone consultation and support links should be organized. These individuals can then repeat the process for their own staff, so that all staff affected by the research have a system for discussing issues and resolving problems.

CONDUCT OF THE RESEARCH

There are many issues connected with the conduct of the research in relation to the research process. Is the research primarily qualitative or quantitative? The decision to use a particular approach to the research is often an indication of the values of the researcher. Is the research longitudinal or cross-sectional? The former requires a great deal more energy for both the researcher and organization, and will entail a more substantial relationship between researcher and organization. For the researcher, this often requires greater commitment to the organization. Will data be collected from many sources and involve both subjective and objective measures? If so, how will conflicting views be managed, or used in feedback of findings? Will the researcher be a participant observer or attempt to remain objective? How will these approaches affect the relationship between researcher and organization? Will scientific rigour be maintained over practical expediency?

CONSISTENCY BETWEEN ELEMENTS

Pettigrew (1985), in his analysis of the concept of change, suggested that in order for change to be effective there needs to be consistency between content, context and process. Essentially this means that these elements are linked and therefore need to fit well together in fundamental aspects of their characteristics. For example, the content of a piece of research might be the implementation and evaluation of the introduction of patient treatment choice. The context is an organizational setting where management have devolved decision-making. The process of the research, in order to be consistent, should be a consultative, participative process in which those affected by the research should have some control over its process. A steering group of clinicians and patients might be set up along with some managers. Here the consistency between the three elements is apparent in their shared characteristic of participation in decision-making and autonomy. The aim of the research is to implement and evaluate participation in decision-making over treatment with patients. The context one where decision-making is devolved to local level and individual clinicians have a high level of autonomy; the research process is one which encourages local decision-making about the research process.

On the other hand an attempt to conduct the same research in an organization where there is centralized decision-making and a great deal of control, is likely to meet with resistance both at local and central levels. In such an organization, a research programme that would better fit, may be one where decisions are autocratically implemented once senior management have agreed to the programme and consultation is kept to a minimum. Such a research programme might be the decision to survey the type and needs of all patients referred to a department in order to improve service delivery and efficiency. This simple principle of attempting to achieve consistency between the fundamental characteristics of the research, i.e. content, context and process, is one that can help researchers when operating in complex organizational settings, or when faced with difficulties in the research planning or implementation phases. These ideas are not new to psychotherapists. Providing links and patterns to client's experiences are part of the therapeutic task. The treatment plan is developed and offered (process) in relation to the client's needs (content), the therapeutic allegiance of the therapist and service delivery considerations (context).

At times when the research programme seems to be stuck or problematic it is useful to consider how consistent are the main elements of the research as a way of generating hypotheses about the causes of the problem and alternatives to the way forward. For example, a psychodynamic psychotherapy outcome project was foundering because there were repeated failures in questionnaire completion by both the therapists and patients. The researchers tried, without success, to encourage better questionnaire completion. Finally, they decided to take time to discuss their difficulties with a sympathetic member of the organization. When the researchers stated the aims of the project, the member of staff replied that 'outcome', 'getting better' and 'helpful aspects of therapy' were little-used concepts in the organization. This then enabled the researchers to modify their research questions and the language of their questionnaires, which led to a better fit between the research content and context, and the questionnaire completion rate subsequently improved noticeably.

One of the most frustrating experiences for researchers working in organizational settings is the failure of research findings to be effectively utilized within organizations. The analysis presented here suggests that one of the reasons for this frustration is that researchers persist in the face of hostile organizational contexts to conduct research which according to the researcher' s values is important to understanding within their given field. But by insufficiently considering context and not attending to the compatibility of content, context and process, the process of research can often alienate those working with organizations and the results are consequently dismissed or unused. Pettigrew's model and the adaptation of the model to research within organizations suggested here, implies that if consistency can be achieved between content, context and process, the findings of the research are much more likely to be utilized within the organization than otherwise. Researchers expend a great deal of energy and effort ensuring their research is carried forward but the process often ends when the research ends. By achieving consistency between elements, it is argued, research in applied settings can have the impact the researchers hoped for at the start of their research.

PRACTICAL METHODS FOR DEALING WITH ORGANIZATIONS

There are essentially three stages in the research planning process where organizational issues should be considered. These match the constructs of content, context and process already discussed. The research team will find that they cycle through these stages and it is the case that one stage of the research will modify the previous stage and together these two stages will inform the next. A summary of these three stages highlighting the main action points for the researcher are shown in Table 4.1.

SITE SELECTION

In planning the research aims and methods, potential sites for the research will be considered. The design of research and evaluative studies and the introduction of these programmes are iterative. The circumstances of the investigator, their access to clients, therapists and other data sources will help determine the area of the research; the specific research aims and methodology will determine the impact of the programme on the organization. At this point, the aims, procedures and practicalities of the research need to be considered. The research team could usefully spend time considering what exactly they will require of the research site. Rather than be defeatist in considering how practical realities will modify theoretical and methodological aspirations, researchers should try to be opportunistic with what organizations present, and use these realities both to modify and further the research aims (Buchanan *et al.*, 1988).

PRE-PILOT WORK

At this stage the context of the research needs to be understood. Some form of organizational assessment should be conducted, and relevant features of the organization highlighted. This may take the form of informal discussions, or may involve more extensive interviewing and observing. When the research team think they have a clear view of the organization, they may wish to adapt and modify their research aims and content.

ACCESS AND IMPLEMENTATION

If the researchers have been careful in achieving consistency between the content and context of their research getting access should not be too difficult. When attempting to gain access, researchers should be sensitive to questions about their 'legitimacy' (Beynon, 1988); often researchers can be seen as coming from their 'ivory towers' with no real knowledge of the realities of organizational life. Careful preparation and a demonstration of inside knowledge are, therefore, vital. It is also important to be prepared to spend time negotiating access. Explaining the study clearly and without jargon, offering something to the organization and dealing positively with reservations from wherever they come are important parts of the process.

Table 4.1. Organizational steps in the research process

Step 1: Site selection (research content)

1.1 Aims:

    (a) Describe the research aims.
    (b) List all individuals and groups who will be affected by the research.
    (c) Consider each individual's/group's aims and goals.
    (d) Redefine and describe the research aims so that they are compatible with (c)

1.2 Procedures:

    (a) Consider clinical implications of the research.
    (b) Discuss these with staff involved with clients.
    (c) Provide ongoing support.

1.3 Practicalities:

    (a) List what the organization needs to provide in terms of money, space, staff-time, secretarial/administrative support.
    (b) List what the researchers will provide in terms of money, space, staff-time, secretarial/administrative support.

Step 2: Pre-piloting (research context)

2.1 Describe briefly the organizational environment. Is it stable/changing? clear/uncertain? competitive/co-operative? hostile/benign?

2.2 Describe the vision/aims of the organization. How does it view research(ers)? What does it emphasize about patient care? What do staff say is productive about the organization? What do staff say is bad about the organization?

2.3 (a) Draw the organizational managerial and professional structure.
    (b) Identify the level at which the research is to be conducted.
    (c) What other levels have important influences on the level identified in (b)?

2.4 Using 2.3 list the groups that will affect or be affected by the research. These should be both formal and informal.

2.5 (a) Where does the power lie within the organization?
    (b) Whose approval is needed for the research to be carried out?
    (c) Who gets things done within the organization and would be a useful ally?

2.6 Describe the decision-making style of the organization. Are decisions made centrally or are they devolved? How much autonomy do specific relevant groups have?

2.7 (a) What information systems exist in the organization?
    (b) How are staff and patients given information?
    (c) What monitoring systems are in place?
    (d) How can the researchers meet and exchange information with those involved in the research?

Step 3: Access and implementation (research process)

3.1 Using information gathered in Step 2, the research aims (Step 1.1) and practicalities (Step 1.3) should be modified so that consistency is achieved.

3.2 Draw up plan of implementation procedures. This should include:

    (a) Clear contract of actions and responsibilities between the organization and researchers.
    (b) Obtaining ethical approval from relevant ethical committees.
    (c) Setting up of a steering group.
    (d) Clarifying lines of communication; key people, etc. Setting up communications exercise.
    (e) Providing necessary support for those involved in the research process, including providing information and debriefing.
    (f) Listing those involved, their role and tasks with time-tables and deadlines.
    (g) Monitoring and evaluating the above mechanisms, especially with regard to consistency between these and the research and organizational aims. Modifying procedures where necessary.

Implementation also needs to be planned with the organizational context as the background. As the researchers become more familiar with the organization, so their understanding of context become richer and modifications to the implementation process made, and perhaps some of the research aims modified. Again the aim should be to achieve consistency with the research aims and organizational context.

METHODS

INFORMATION GATHERING

Individual interviews with 'experts' who are knowledgeable about the workings of the proposed research sites can be extremely useful to help set the wider context for the research, find out about potential sensitive issues and provide an understanding about how to approach the actual target site. Views on the proposed research, on the suitability of the proposed site(s), issues the research will raise for the organization and any other relevant issues could be discussed.

Interviews with representatives of those people within the organization who will be affected by the research will help inform both the research aims and feasibility, in addition to increasing commitment and motivation for the research. Negotiation for access is often best done initially through meeting individual managers and finding out their views and concerns.

A second method of information gathering, shadowing, involves the researcher being with an employee from the organization throughout their working day. The aim is to experience what the job is actually like, in order to inform the research aims and design. Whilst often it is unnecessary for researchers to be intimate with such level of detail, occasionally it can be invaluable in helping the researcher. For example, a study that aims to look at long-stay inpatient treatments, may find it valuable to shadow staff in the hospital in order to develop meaningful outcome measures for this group of patients.

Finally, it is usual when conducting research within organizations to set up a steering group or working party. The task for this group is to monitor the progress of the research for the organization. In addition, such a group can provide valuable information about the organization and its operations, and give the researchers increased credibility through being linked to key members within the organization.

INFORMATION DISSEMINATION

Wherever possible it is best to use the information structures already in place within an organization. So, if a hospital has a team-briefing system or a newsletter, and the researcher wishes to let staff either know about the proposed research or about the research findings, then this would be one method of dissemination. If staff are to assist in the research project then additional forms of information giving will be necessary. Willingness to participate will increase if those involved are given sufficient information about the project. It may be that the researcher decides that the best way to give the information is through a series of seminars or workshops, where staff can actually meet the research team, be given an overview of the project and concerns can be voiced.

As an alternative, or in addition to seminars, the research group could create an 'information pack' that includes a description of the research, samples of the research materials and a detailed description of the role and tasks of those within the organization who are to help with the research. Staff would need some form of briefing prior to distribution of the information pack in order to maximize co-operation.

The researchers should offer and may be required to give a presentation and provide a report of their research and subsequent findings. There are five areas that need to be covered when presenting a research project to an organization. These include:

1. the credentials of the researchers;
2. the aims of the research;
3. how information will be collected;
4. what will result from the research (including benefits to the organization);
5. what will be the relationship of the researcher to the organization and research participant.

The language used both in presentations and reports should be straight-forward and non-threatening. So, for example, instead of saying that participants will be interviewed, the researchers could say that the views of staff/patients will be sought.

Reports to organizations on research findings should not deal with theoretical issues. They generally are descriptive accounts of what the researcher has observed both from doing the research and from the research data, and should contain the minimum amount of statistical analysis. They should be written with the view of the reader(s) in mind. It is therefore good practice to ask the organization what sort of report they would like and for whom. Again, it is important that the report is 'neutral', it should not be evaluative, and should not attempt to make interpretations or recommendations that go beyond the research findings. The first report should be a draft report, to allow for feedback from the organization to be incorporated within the final report. Finally, it should be noted that the final report is often the point of exit from an organization for the research team, and should be managed as such. So, a clear deadline for the research report should be negotiated, and in the report positive feedback should be given to those who participated, and thanks given to both the organization and participants for their co-operation.

GAINING ACCESS

When working with an organization it is useful to have a named contact with whom access negotiations and future day-to-day issues of the research can be discussed. This person acts as a liaison between the research team and the organization, informing the research team of procedures they must adhere to, who they need to contact for specific issues and assisting with any negotiations. For the organization they act as a gatekeeper, keeping the organization informed of progress and any substantative issues, and presenting the organization's views to the research team. A good contact person is invaluable to the research team, and it is worth spending time clarifying with such a person the most effective way of working together.

Often researchers find themselves unprepared for their initial negotiations with management. For example, instead of a one-to-one meeting with a hospital's chief executive to discuss details of the research programme, the researchers find that they are expected to give a presentation at a Board meeting where they are required to promote themselves as well as their research programme. Preliminary enquiries and good groundwork should help prevent such occurrences.

Most managers have limited time and are not concerned with the fine details of a research programme. They usually want to know, as clearly and concisely as possible:

1. who the researchers are and what qualifies them to do the job well;
2. what are the research aims and methods;
3. why the research needs to be carried out in their organization, and what the researchers understand about their organization;
4. how these will impact on the organization;
5. how the organization will benefit from the research.

It is always a good idea to prepare brief summaries of the research proposal, and if possible agendas for the meeting and send these in advance to the relevant people. In addition, a summary of the main points of a meeting should be made and sent to those concerned.

CO-OPERATION AND SUPPORT

Workshops are perhaps the most popular way of increasing the motivation and improving co-operation of staff. The workshop format will be individually tailored to the requirements of the research and the organization.

There are however some ground rules to assist in the design of such an event.

The structure and the time-keeping of a workshop are the organizers responsibility. Participants should be clear of the aims of the workshop, be given a chance to express their hopes and fears of the workshop, and adequate space to raise concerns about the research, their role and potential research outcomes, and time for these issues to be dealt with. If the aim of the workshop is to obtain staff co-operation then information giving by the research team should be kept to a minimum. Time to listen to, and resolve where possible, the concerns of staff should be the focus of the workshop.

If the researchers wish to understand the impact their research may have on various groups within the organization, then a useful exercise is a 'stakeholder analysis'. Here the primary groups that could be affected by the research are identified. The research is then considered from each stakeholders' perspective. Ideas for improving the commitment of each stakeholder group are then discussed.

Potential resistances can be addressed using a procedure called forcefield analysis (Lewin, 1947). Workshop participants are asked to identify all the driving and restraining forces that accompany involvement in the research process. Driving forces are those forces that push for change and keep the process going; restraining forces act to decrease driving forces, such as apathy or hostility. Once listed, participants develop methods to strengthen the driving forces and, through understanding and negotiation, propose ways of reducing the restraining forces.

If staff are going to carry quite a substantial responsibility for data gathering, client selection and management of the project, it is useful to consider what help systems need to be in place, especially if the research team are not managing the day-to-day business of the research. There are two aspects to this:

1. Regular meetings should be established where staff are given feedback about their work and that of the research programme.
2. Mechanisms for contacting the research team if problems arise should be established and made known to staff within the organization.

Once the research is in progress, ways of checking and monitoring need to be established. Issues to do with resistance to the research programme can be dealt with through such mechanisms. Such monitoring systems include the provision of supervision time or quality checks. These meetings provide the opportunity for those involved in the research to check their performance in comparison to others, and for the research team to monitor that the research is being conducted correctly and to address any problems or failures to adhere to agreed ways of working.

CONCLUSION

The analysis presented here has ignored the other important context for research, that of theory and methodology. This latter context places constraints on the research that must not be lost or forgotten. The researcher has not only to consider the relationship between the content, context and process of the research from an organizational perspective, but also from that of the scientific community.

In addition, psychotherapy researchers have knowledge, skills and texts available to them from their clinical work which are invaluable in working with organizations. These include listening skills, interviewing skills, a framework for understanding interpersonal relationships, understanding group behaviours and hidden agendas. All of these are essential skills when negotiating and working with individuals and groups within organizations. These, combined with the framework and practical methods presented here, should provide a structure for approaching and working with organizations in the service of producing sound research.

REFERENCES

Argyris, C. (1979). Some unintended consequences of rigorous research. In *Research in Organizations: Issues and Controversies* (R.T. Mowday & R.M. Steers, eds) Santa Monica: Goodyear Publishing Inc, pp. 290–304.

Beck, A.T. (1967). *Depression: Clinical, Experimental and Theoretical Aspects.* London: Staples Press

Beynon, H. (1988). Regulating research: politics and decision making in industrial organizations. In *Doing Research in Organizations* (A. Bryman, ed.) London: Routledge, pp. 22–33.

British Psychological Society (1991). *Code of Conduct, Ethical Principles and Guidelines.* Leicester: The British Psychological Society.

Bryman, A. (ed.)(1988). *Doing Research in Organizations.* London: Routledge.

Buchanan, D., Boddy, D. & McCalman, J. (1988). Getting in, getting on, getting out, and getting back. In *Doing Research in Organizations.* (A. Bryman, ed.) London: Routledge, pp. 53–67.

Bulmer, M. (1988). Some reflections upon research in organizations. In *Doing Research in Organizations.* (A. Bryman, ed.) London: Routledge, pp. 151–161.

Burns, T. & Stalker, G.M. (1961). *The Management of Innovation.* London: Tavistock.

Firth, J., Shapiro, D.A. & Parry, G. (1986). The impact of research on the practice of psychotherapy. *British Journal of Psychotherapy* **2**, 169–179.

Hirschhorn, L. (1988) *The Workplace Within: Psychodynamics of Organizational Life.* Boston: MIT Press.

Joos, S.K. & Hickman, D.H. (1990). How health professionals influence health behavior: Patient-provider interaction and health care outcomes. In *Health Behavior and Health Education* (K. Glanz, F.M. Lewis & B.K. Rimer, eds) San Francisco: Jossey Bass, pp. 216–241.

Katz, D. & Kahn, R.L. (1978). *The Social Psychology of Organizations,* 2nd edn. New York: Wiley.

Lewin, K. (1947). Frontiers in group dynamics: concept, method and reality in social science; social equilibria and social change. *Human Relations* **1**, 5–41.

Menzies Lyth, I. (1991). Changing organizations and individuals: psychoanalytic insights for improving organizational health. In *Organizations on the Couch: Clinical Perspectives on Organizational Behavior and Change* (M.F.R. Kets de Vries, ed.) San Francisco: Jossey Bass, pp. 261–378.

Mintzberg, H. (1983). *Power In and Around Organizations*. Englewood Cliffs: Prentice-Hall.

Palazzoli, M.S., Anolli, I., Di Blasio, P., Giossi, L., Pisano, I., Ricci, C., Sacchi, M. & Ugazio, V. (1986). *The Hidden Games of Organizations*. New York: Pantheon Books.

Pettigrew, A.M. (1979). On studying organizational cultures. *Administrative Science Quarterly* **24**, 570–581.

Pettigrew, A.M. (1985). *The Awakening Giant*. Oxford: Blackwell.

Robey, D. (1986). *Designing Organizations*. Homewood, Illinois: Irwin.

Schein, E.H. (1985). *Organizational Culture and Leadership*. San Francisco: Jossey-Bass.

Schein, E.H. (1987). *The Clinical Perspective in Fieldwork*. Newbury Park, California: Sage Publications.

Trist, E.L., Susman, G. & Brown, G.W. (1977). An experiment in autonomous work groups in an American underground coal mine. *Human Relations* **1**, 3–38.

Wall, T.D. & Lischeron, J.A. (1977). *Worker Participation*. London: McGraw-Hill Book Company.

5 The Science of Health Care and Science for Health

MICHAEL. J. PECKHAM
Department of Health, London, UK

Science exerts a decisive and pervasive influence on the nature and disposition of the skills, facilities and interventions that collectively constitute health care. The transformation over the past four decades has been profound and the influence of science and technology gathers momentum as health services are presented with the output of a hugely expanded international capacity for technological innovation. Science relevant to health extends well beyond the conventional boundaries of medical research, encompassing biology, biotechnology, engineering, physical science and the social sciences. The conformation of health care has been reactive to the pressure of science and technology and while there is abundant evidence of progress, there are many anomalies with variations in practice, and the majority of diagnostic methods and treatments unevaluated. In short, the science of health care has been insufficiently distinguished from science for health. The emphasis on the latter and the neglect of the former is likely to be detrimental if there is not a rational basis for deploying scarce health resources to accommodate the fruits of research.

The output of science presents a threat and an opportunity. While some new developments are of worth, others are of marginal or transient value and some turn out to be inferior to existing methods. Hitherto, there has been no coherent approach to the scientific assessment of diagnostic procedures and treatment methods, no concerted attempt to make systematic use of research results, and little interest in the characterization of health issues as problems for research. Furthermore, the health sector has possessed little capacity for understanding the implications of scientific advances and for determining their likely impact on health care. Consequently an imbalance has grown up between investigator-led

Correspondence address: Department of Health, Richmond House, 79 Whitehall, London SW1A 2NS, UK.

Research Foundations for Psychotherapy Practice. Edited by M. Aveline and D. A. Shapiro.
Copyright © 1995, Mental Health Foundation and Individual Contributors.
Published 1995 by John Wiley & Sons Ltd

research and problem-focused research, which has resulted in insufficient attention being paid to issues which most pertain to health sector priorities. There is, for example, a dearth of information on the effectiveness and cost-effectiveness of health practice methods. Even simple unevaluated procedures that are widely and unnecessarily applied can result in the consumption of substantial resources. The diversity of approach in routine care will increasingly be difficult to defend unless different approaches to comparable clinical problems are supported by a sustainable and convincing rationale.

There is an assumption that the demand for health care has outstripped available resources, and that the gap between demand and supply is widening. It is increasingly the case that technological advances create new demands, sometimes very rapidly and unpredictably, for example, the explosive development in minimal access surgery since the late 1980s. Considerations of cost containment in health, in relation to the perceived gap between supply and demand, are fuelled by an uncertainty or even scepticism about the usefulness of some aspects of health care, which although consuming substantial resources are of uncertain benefit. There are a number of approaches to bridging the gap between supply and demand, including the provision of more resources for health, reduced demand on health services, greater efficiency of operation and enhanced effectiveness of the processes of health care. Rationalization, defined as 'the scientific organization of an industry', is an appropriate approach to securing the best outcomes for the investment in health services. It also offers the best prospect for sustaining clinical research and for taking advantage of scientific advances.

THE SCIENTIFIC ORGANIZATION OF HEALTH CARE

Rationalization requires a coherent strategy and the capacity to deliver the required information and to implement its use in practice. To provide a scientific basis for health care a Research and Development (R & D) infrastructure, strategy and programme has been introduced into the National Health Service (NHS). A prime objective is to base decision-making at all levels in the Health Service—clinical decisions, managerial decisions and the formulation of health policy—on reliable research-based information. A second objective is to provide the NHS with the capacity to identify problems that may be appropriate for research. A third objective is to improve the relations between the Health Service and the science base. Since the programme was launched in April 1991 there has been rapid progress. A national research and development infrastructure has been set in place throughout the health regions and a programme of work established with emphasis on the rapid provision of information relevant to health service problems and the setting in place of a medium and longer term strategy. It is important to emphasize that the R & D initiative is complementary to, and not in conflict with, basic science. Indeed, emphasis has been placed on close working relationships with the Medical Research Council, Economic and Social Research Council and two new research councils which will come into being in 1994, (biotechnology and biological sciences, engineering and physical sciences). Close working relationships between the health sector and the major charities have also been promoted through the creation of research liaison committees. Thus there is a commitment to encouraging scientists engaged in basic research to follow their instincts and judgements so long as their work is innovative and of high quality. The NHS R & D programme is not in conflict with that objective but seeks to create a more satisfactory balance between research driven by curiosity and research focused on solving specific problems. The attempt is to stimulate activity in applied health research, recognizing that we depend on speculative research for the occasional truly major advances in health care. The R & D initiative has required a substantial cultural shift on the part of the scientific community and on the part of health service staff, particularly management. The response of the scientific community has been extremely supportive. Outstanding individuals—predominantly from a biomedical background—have taken up the posts of Regional Directors of R & D. Although some have been in post for less than a year, rapid progress has been made. Networks with R & D contact personnel in health authorities and hospitals have been established and working links developed with universities, including not only those with medical schools, but those with strengths in health research and the social sciences. The Regional Directors have placed emphasis on *development* by encouraging the uptake and use of research in the contracts between providers and purchasers of health care and in clinical guidelines. Indeed some of the early achievements of the R & D programme are related to work on the analysis and practical use of existing research information.

There has also been a high level of support from NHS staff. In 1992 health service managers were asked for examples of decisions made during the preceding year in which they judged that research-based information had been lacking. The returns provided a substantial list of illustrative examples with one manager commenting that 'it wasn't so much that research information was lacking but that no-one in [his] health authority would have thought that such information could have contributed to decision making' . This observation served to highlight not only the lack of research data directly relevant to health service issues, but a lack of awareness that research could be directly pertinent to the solution of managerial problems. Since then there has been substantial progress and there is an understanding that R & D is not a luxury, but an essential component of a modern health care system. For example, the objective of separating the purchasers and providers of health care is to secure the largest volume of high quality care with available resources. This quest has made the need for research-based information explicit and there is a thirst for reliable data. There is also clear recognition that such information underpins effectiveness and cost-effectiveness in clinical practice. There is increasing understanding that R & D has potential applicability across the spectrum of health service activities contributing information, analytical methods and introducing a capacity for foresight and scientific intelligence. Previously research has been thought of predominantly in the context of clinical practice but there is a wider range of pertinent issues, for example, the design of hospitals and different models for delivering care.

FROM PROBLEM TO RESEARCH SOLUTION

A method for analysing and prioritizing health service problems has been developed, refined, and applied to a range of issues including mental health, cardiovascular disease, physical and complex disabilities and the interface between primary and secondary care. The task of identifying and prioritizing problems appropriate for research is undertaken by independent multidisciplinary groups who are provided with background data on prevalence, costs, available research knowledge and information on relevant developments in science. An essential feature is the conduct of an extensive consultation exercise with practitioners, researchers, managers and other—including lay—individuals and organizations. A key aspect is to secure input from those who are working with patients on a day-to-day basis. Input is derived from written consultation and from workshops. It is important that the groups responsible for setting the agenda for R & D are broadly based. For example, in a recently completed exercise on physical and complex disabilities, the membership of the group included rehabilitation engineers, medical physicists, rheumatologists and other medical specialists, representatives of the therapy professions, public health physicians, social scientists and representatives of the disabled including voluntary organizations. Often the members of such groups find that they are working for the first time in a truly multi-disciplinary context.

An important by-product of these exercises has been the identification of substantial gaps in routine information, for example, in health economics and in epidemiological data. In setting priorities the groups are asked to take into account the feasibility of research and the likely return from an investment in research.

Health service problems appropriate for R & D have been identified in relation to six overlapping perspectives: a disease perspective, management and organization of services, client groups, consumer issues, health technologies and research methodologies. Activities in these various areas are summarized in Table 5.1.

Table 5.1.

Perspective	Areas for review	Action
Disease related	Mental health and learning disability	Review complete —first projects commissioned
	Cardiovascular disease and stroke	Review complete —projects being commissioned
	Cancer	Starting October 1993
	Respiratory	Planned
	Dentistry	Regional review by Mersey Regional Health Authority
Management and organization	Interface between primary and secondary care	Report to the CRDC in October 1993
	Purchaser/provider contracting	Commissioned paper and workshop prior to the review
	Accident and emergency	Planned
Client groups	Physical and complex disabilities	Report to the CRDC in July 1993
	Mother and child health	Planned
	Elderly people	Planned, following current MRC field review
	Health of ethnic minorities	Regional review planned
Consumers	Focus on nature, role and input of users and potential users of the NHS to decision-making	Two papers commissioned, review being planned
Health technologies	Assessment of new and existing health technologies	Standing Group on Health Technologies established February 1993
Methodologies	Identification and development of appropriate methodologies to tackle the whole spectrum of NHS issues	Review planned, in discussion with MRC and ESRC

Reproduced from *Research for Health* (1993), Department of Health by permission.

The NHS R & D strategy includes a major programme on health technologies. The term 'health technology' in this context describes any method used by health professionals to promote health, to prevent and treat disease and to improve rehabilitation and long term care. The term 'health technology assessment' describes the systematic evaluation of these methods in terms of their costs, effectiveness and broader impact. A striking example of the challenge for technology assessment is minimal access surgery, where only nine of more than 100 procedures in current use are being systematically evaluated. The number of diagnostic procedures, drugs, biotechnology products, surgical devices and treatments is growing rapidly, and information on comparative costs and benefits is scanty. With so many interventions remaining unassessed, the effort must clearly be focused where the returns are likely to be the greatest. A new

NHS Standing Group on Health Technology convened in early 1993, will make its first report in December. The Standing Group is seen as the gateway into the NHS for health practice methods, with the eventual aim of registering technologies, and documenting whether they have or are being evaluated, and whether they have been shown to be effective and cost-effective. The Standing Group has been asked to advise the health service on new and existing technologies, which should be evaluated as a high priority, and to advise on those which should only be purchased if they have been or are being evaluated. The Group also has a foresight function, advising on new developments likely to arise from science and technology. The Group is advised by six panels dealing respectively with pharmaceuticals, acute sector technologies, chronic and primary care, diagnosis including imaging, population screening and evaluative method-ology. The work of the panels has, as for other areas of priority setting, been associated with wide consultation and regional workshops. The first consultation exercise resulted in the submission of more than 1000 priorities. From this list, the Standing Group has assembled information on 90 technologies from which a ranked list of priorities is being prepared. A 'vignette' has been prepared for each technology which includes back-ground information, data on benefits and cost implications along with the envisaged timescale of assessment, implementation issues, the potential returns on research funding and the urgency of the evaluation. The inten-tion is that those responsible for purchasing health care should be made aware of the assessments being undertaken in the R & D programme, and that those intending to buy unevaluated interventions that are currently subject to assessment will need to explain why they are taking such action. Examples of issues identified by the Standing Group are *repeat prescribing strategies* and the *effectiveness and cost-effectiveness of different hip prostheses.* Repeat prescribing relates to medication that on the authorization of a doctor can be repeatedly prescribed by non-medical staff on the request of patients at agreed intervals for an agreed period without requiring the patient to see the doctor. Current prescribing costs in the primary care sector account for approximately 10% of the health service budget and repeat prescribing is estimated to account for approximately two-thirds of prescribing costs. There has been an annual increase in the number of total hip replacements since 1967 with approximately 18% of the total financial burden accounted for by revision hip replacements. Whereas in 1970 there was only one type of prostheses, by 1991 this had risen to 34.

As mentioned above, the Standing Group on Health Technologies has a foresight function alerting the health service to likely new developments. In the rapidly evolving field of genetics, an NHS Genetics Group has been convened to advise the Standing Group on the health service implications of genome mapping, genetic interventions, genetic screening and genetic diagnosis.

Table 5.2. Lead directorates for NHS R&D management

Area	Lead directorates for NHS R&D
Mental health	Yorkshire
Cardiovascular disease and stroke	Northern
Cancer	South-Western
Health technology (HT) panel on methodologies	
Respiratory disease	South-East Thames
HT panel on acute sector	
Purchasing/contracting	Oxford
Accident and emergency	North-Western
Elderly	East Anglia
Interface primary/secondary care	North-East Thames
HT panel on chronic, community, primary care	
HT panel on pharmaceuticals	Mersey
HT panel on screening	North-West Thames
Mother and child health	South-West Thames
Consumer issues	Trent
HT panel on imaging	
Physical and complex disabilities	Wessex
Medical equipment	West Midlands

The identification of a problem should not automatically signal the commissioning of research. Two questions need to be asked:

1. Is there existing research information capable of answering the question?
2. Is there ongoing research relevant to the problem?

If there is either a lack of information or lack of relevant research, bids are invited from the research community to tackle clearly characterized problems. While much of this new work is supported by the NHS itself, some is supported by other bodies, particularly the Medical Research Council. The process of inviting competitive bids, conducting peer review and commissioning research is devolved to one of the Regional R & D Directors. As shown in Table 5.2, each directorate is responsible for managing one or more programmes on behalf of the health service. In this way the commissioning, management and conduct of the R & D programme is devolved from the centre while maintaining national coherence.

MAKING USE OF RESEARCH FINDINGS

There has been a lamentable lack of emphasis on the use of research information in routine clinical practice. There are many examples of where research findings have not been used promptly or uniformly. One of the most striking has been in thrombolytic therapy for acute myocardial infarction in which there was a 12-year delay between the publication of clinical trials demonstrating effectiveness and the recommendation by experts of thrombolytic therapy (Antman *et al.*, 1993). Another recent example has been a systematic overview of trials comparing the treatment of stroke patients in stroke units compared with their treatment on general medical wards (Langhorne *et al.*, 1993). As the *British Medical Journal* noted, the greatest leap forward in the treatment of stroke has not been a novel neuroprotective agent or a new way of imaging the ischaemic brain, but a publication showing that organized stroke care saves lives (Sandercock, 1993). Prior to this analysis, the consensus view had been that stroke units hasten recovery but do not reduce mortality or improve longer term outcomes. The review conducted on 10 trials carried out between 1962 and 1993 showed a reduction in mortality by almost 30%, reduced dependency at 6 months, reduced need for long-term institutional care and reduced length of hospital stay. This distillate of existing research information provides data highly relevant to the organization of services and also provides the basis for teasing out in further studies those elements of organized stroke care which most contribute to the beneficial effect.

In addition to the conduct of new research and the generation of new data in the medium and longer term, for example, from clinical trials, it is essential that a health R & D programme delivers practical information in the short term and that effective use is made of currently available research findings. Accordingly high priority has been given to the establishment of an information strategy to handle existing research data. The first component is the Cochrane Centre, opened in November 1992, which is concentrating on registers and systematic overviews of randomized clinical trials. The Centre has stimulated worldwide interest and activity with the creation of an international network known as the Cochrane Collaboration. This important initiative is designed to provide access to the output of high quality research from research groups based in many countries. In December 1993 a new unit at the University of York will open to extend the pioneering work of the Cochrane Centre by commissioning reviews of findings from across the whole range of research. The York Centre will also tackle the challenge of disseminating research information effectively, concentrating on the systematic transfer of research findings in appropriate formats to clinicians, managers, policy makers and other users. The objective is to ensure that research is used in purchasing contracts, in clinical guidelines and indeed in other vehicles that transfer new information into routine practice.

Regional Directors of R & D have also been provided with an information tool to allow research projects underway in their regions to be registered under a common format. This will lead to the establishment of a national register of research which will allow for the first time a comprehensive national picture of ongoing work to contribute to more effective decisions about the targeting of research resources.

Work on implementing the results of R & D in practice is receiving high priority. The goal is to make available R & D evidence to underpin decisions at all levels and in all sectors of the health service to inform policy, the development of new services, the commissioning and evaluation of existing services and to support routine clinical and management practice. The implementation strategy builds on a number of *development* initiatives. For example, one project involves collaboration with a number of health authorities. Here guidelines relating to cardiac services are being developed on subjects identified by health authority staff working with researchers. These subjects may apply to primary, secondary or tertiary care, or to the interfaces between services. The collaborating districts differ in terms of the number of tertiary providers, the availability of coronary care units and the number of physicians with an interest in cardiology. This project, in common with other development studies, is based on a formal protocol with prospective evaluation.

RESEARCH SKILLS

Invitations to the research community to bid for funds for various aspects of the programme have shown a high level of enthusiasm, but in some areas a shortage of research skills. For this reason attention is being given to the development of education and training initiatives. This includes training in health services research and analytical sciences for doctors and other health professionals, and the inclusion of R & D in modular training programmes for managers. A new centre for research and development in primary care is being established with training as an important part of its work. The intention here is to create a model dynamic centre for health services research in an aspect of care that previously has received relatively little attention.

CONCLUSION

A comprehensive R & D infrastructure, strategy and programme has been introduced into the National Health Service in order to generate the information and mechanisms necessary to develop knowledge-based health care. The goal is to place the health sector in the mainstream of research, to introduce a capacity for the identification of health problems appropriate for research, to focus on the systematic use of research findings in routine practice and to give emphasis to *development* in addition to research. The aim is to base decision making in the health service on reliable and relevant research-based information. The initiative seeks to establish a more appropriate balance between research driven by curiosity and research focused on specific problems. However, it is not in conflict with basic science but attempts to fill the lacuna between biomedical research and the health service bringing to bear the experience and commitment of professional leaders with strong track records in research. The R & D programme provides the basis for rationalizing health care in relation to the effectiveness of services. Since this is a new venture, the impact cannot be quantified, although anecdotal information gives ground for supposing that the potential for releasing and redeploying resources within the health service may be substantial.

REFERENCES

Antman, G.M., Lan, J. & Kupelnick, B. (1993). A comparison of results of meta-analysis of randomised control trials and recommendations of clinical experts: treatment of myocardial infarction. *Journal of the American Medical Association* **268**(2), 240–248.
Department of Health (1993). *Research for Health*. Department of Health, London.
Langhorne, P., Williams, B., Gilchrist, W. & Howie, K. (1993). Do stroke units save lives? *The Lancet* **342**(8868), 395–398.
Sandercock, P. (1993). Managing stroke: the way forward. *The British Medical Journal* **307**, 1297–1298.

Part II

INDIVIDUAL THERAPY
AND ITS EFFECTS

6 Assessing the Value of Brief Intervention at the Time of Assessment for Dynamic Psychotherapy

MARK AVELINE

Nottingham Psychotherapy Unit, UK

ABSTRACT   Robust relevant outcome research is possible in a specialist clinical psychotherapy service in the National Health Service. This chapter describes the rationale, feasibility and findings at the mid-point of a randomized controlled trial of the value of intervention at the time of assessment for dynamic psychotherapy. Brief intervention and follow-up (BRF), a *three plus one model*, is compared with our standard assessment (SA), a single session, in 136 consecutive referrals for dynamic psychotherapy, controlled for age, sex, age of completing education and severity of disorder. BRF patients receive extra therapist input at the time of referral when the problem is pressing and the patient's response may be less entrenched. Our results at the mid-point show that fewer BRF patients are being put on the waiting list than those in SA ($p = 0.12$); BRF patients show a significant reduction in symptomatology as measured by the General Severity Index of the Brief Symptom Inventory at 4-month independent follow-up ($p = 0.059$). If continued, the findings will assist the management of referrals and waiting lists in Psychotherapy Departments. When, as now, health policy is so dictated by short-termism, it should be borne in mind that results of this rigour of design take 5–7 years to accrue.

INTRODUCTION

The National Health Service in the UK is in the middle of a major reorganization whose final form is unclear. An uncontrolled experiment if ever there was one. The changes are being driven by two imperatives, both fiscal in inspiration. The Government hopes to contain costs through

Correspondence address: Nottingham Psychotherapy Unit, 114 Thorneywood Mount, Nottingham NG3 2PZ, UK.

Research Foundations for Psychotherapy Practice. Edited by M. Aveline and D. A. Shapiro.

competition between health providers and, having separated purchasers from providers, enjoins the former to buy cost-effective treatments. This reform substitutes a low-cost public service ethos for an administratively more expensive market economy. Even were this external change not the case, clinicians want to know empirically what relative benefit derives from different interventions at various stages of health delivery.

As a clinican providing a broad range of specialist psychotherapy service in the National Health Service, my *desiderata for psychotherapy research* are sevenfold. Psychotherapy research should be: *representative, relevant, rigorous, refined, realizable, resourced* and *revelatory*. These *Seven Rs* are discussed in Chapter 14.

This chapter explores the rationale, feasibility and potential findings from a randomized controlled trial with real patients in a specialist NHS Department of Psychotherapy. Preliminary findings from one such trial at the mid-point are reported. The trial was funded by the Mental Health Foundation. We hope that the findings will assist the management of referrals and waiting lists in Psychotherapy Departments.

CONTEXT

The Nottingham Psychotherapy Unit receives 450 referrals per year, roughly one-third of which are from primary care, i.e. general practice, and two-thirds from secondary care, i.e. consultant psychiatrists and other members of mental health teams. Referrals are screened and accepted by one of the three consultant psychotherapists. Based on the referrer's preference, scrutiny of the referral letter and case-note and questionnaire information, 300 of these referrals are allocated for assessment for dynamic psychotherapy in the first instance and the remainder for cognitive-behavioural psychotherapy. The Department offers a full range of therapies; all patients are considered for any of these. In the Dynamic Division, nearly all therapies are at a frequency of once a week. Individual therapy may be brief (less than 10 sessions), focal (16 to 25 sessions) and long-term (12 to 24 months, occasionally unlimited). Group therapy is in open and closed groups with an average duration of membership of 18 months. The consequence of not having enough therapist resource and our determination to provide most patients with the optimum therapy—given our position as a weekly outpatient Teaching Hospital treatment facility—is a waiting list.

In organizing our service, we decided to assess referrals relatively soon rather than wait till a vacancy for therapy occurs as some services do. This means two waits, the first briefer to assessment (3 to 5 months for dynamic psychotherapy)*, the second longer to therapy (6 to 18 months depending on type of therapy). Thus patients get to know sooner where they stand in terms of suitability for therapy, both generally and specifically, instead of waiting for a long period with the risk of being turned down with time wasted before making other arrangements for help.

In practice, we are concerned about the wait to therapy. Professionally, we feel it is too long. Certainly, many referees do not complete the course. Of 100 referees, only 75 return their pre-assessment psychotherapy questionnaire. In addition to beginning the reflective work of explorative psychotherapy, the questionnaire acts as a screening device for motivation. This attrition is an efficient use of our resource. Previous research showed that, overall, the same cumulative proportion of those referred, failed to attend their assessment appointment before and after the introduction of the questionnaire. The questionnaire appears to deter those who would not have attended anyway (Aveline & Smith, 1986). A further 10 do not show for their assessment, where 10 more are judged to be unsuitable for therapy. Patients then wait for therapy. For half, the wait is less than 6 months but some, the more damaged by the traumas of their life, wait 12 to 18 months, as they require longer therapy from more skilled therapists who are in short supply. Further attrition occurs on entry to therapy with 15 more lost. In total, for every 100 referred, 40 will enter therapy.

While as a Department we do not have the resource to cope with a 100% take-up of therapy, we would like to do more for more. This was the starting point for the research reported in this chapter.

BRIEF AND SUPER-BRIEF INTERVENTIONS

Over the last 30 years, there has been a considerable development of brief dynamic psychotherapy, grounded in empirical research (Malan, 1963; Strupp & Binder, 1984; Luborsky et al., 1988). Central to these approaches is the formulation of an operational understanding of the patient's difficulties based upon the narrative of the history and clinical observation of the relationship patterns shown at assessment and in the therapy sessions. Malan terms this the 'psychodynamic' or 'explanatory hypothesis', Strupp

---

* The waiting period to assessment in the Behavioural Division is shorter at 6 weeks. They achieve this through the device of offering a brief screening assessment of 20 minutes during which the cognitive-behavioural model is described to the patient. The patient is invited to decide to proceed or not with their referral for therapy, basing their decision on the information received and their perception of the model being potentially helpful for them. Detailed assessment of the problems is deferred to the early sessions of therapy. These two elements of assessment are combined in the 1 to 2 hours of the standard dynamic assessment but the longer duration of the assessment consumes more staff time at the assessment stage.

and Binder the 'dynamic focus' and Luborsky *et al*. the 'Core Conflictual Relationship Theme' (CCRT). In these brief psychotherapies, the therapist identifies at an early stage a dynamic focus which is of central importance in the genesis of the patient's difficulties and to which the closest attention is paid during the therapy. This selective focus, together with the urgency confirmed by the constraint of a time-limit, form the two principal ingredients in the demonstrated success of this approach. Successful outcome correlates with early positive therapeutic alliance (session 3), therapist activity, the prompt addressing of negative transference and focused work with intra-psychic and interpersonal conflicts of central importance to the patient. In Luborsky's model, Core Conflictual Relationship Themes (CCRT) are identified from analysis of narrative statements made by the patient about what has happened and does happen in his or her relationships. A tripartite formula of Wishes from the Other (WO), Reaction of the Other (RO) and consequent Reaction to Self (RS) is constructed. Research by Luborsky (Luborsky *et al*., 1988; Luborsky and Crits-Christoph, 1990) into psychoanalytic psychotherapy has shown:

1. it is possible to classify reliably Core Conflictual Relationship Themes;
2. addressing these themes in therapy, especially the element of expected Reaction of the Other, improves outcome.

It is these elements that are applied in the Nottingham research project in the form of a super-brief intervention.

Most work in brief dynamic psychotherapy has a duration of between 12 and 25 sessions. In the first Sheffield Psychotherapy Project (Shapiro and Firth, 1987), prescriptive and explorative psychotherapy were compared in a 16-week cross-over design. Both therapies were effective with a slight advantage in this work-related stress patient population for the directive therapy. Turning to super-brief intervention, Barkham (1989) and Barkham and Hobson (1989) have argued in favour of a *two plus one* model in which patients are seen for two sessions 1 week apart followed by a third session 3 months later. In their model, an immediate or key issue is addressed and an effort is made to enable the patient to experience themselves effecting change in their lives.

The Nottingham model follows this pattern but with the addition of a session at week 3. More than half the patients referred to the Nottingham Psychotherapy Unit for dynamic psychotherapy have histories of prolonged sexual or physical abuse; their capacity to form sustaining personal relationships is markedly impaired. Many have taken overdoses and have extensive psychiatric contact. Given this patient population, we considered a *two plus one* intervention unlikely to be a sufficient dose to effect useful change and so decided upon a *three plus one* design. As well as a 25% increase in dose, this design allows a traditional beginning, middle and end phase for therapy. The extra session reflects the greater severity of disturbance in our patients but is still sufficiently similar to the *two plus one* model to allow comparison of the results. The other extreme would have been to test the value of offering 10 sessions to all referrals, as is the practice in some NHS Psychotherapy Departments. Such an approach is expensive in terms of therapist time and may provide too much therapy for some patients and too little for others.

A RANDOMIZED CONTROLLED TRIAL

In our Standard Assessment (SA), patients are seen once for 1 to 2 hours by senior members of staff. This trial evaluates the effect of Brief Intervention and Follow-up (BRF), a *three plus one* intervention, and compares this with Standard Assessment (SA) for psychotherapy. All assessments are discussed in an assessment meeting of the research therapists, where core conflicts and central interpersonal themes are identified from the history using a special form developed in Nottingham and, in addition, the Global Assessment of Functioning (GAF) rating is made. The GAF is an external rating which measures symptomatic and inerpersonal functioning from gross morbidity to normality on a 100 point scale (APA, 1987); we confine our rating to functioning during the past week.

In Brief Intervention and Follow-up (BRF), patients are seen in week 1 for a standard assessment interview (BRF 1). Patients return in weeks 2 (BRF 2) and 3 (BRF 3) for a 45–50 minute therapy session. A 9-week break ensues with a 45–50 minute follow up at week 12 (BRF 4). At this final session, the benefits of the intervention are jointly evaluated by the patient and assessor, and the decision is made on need and suitability for psychotherapy in one or more of the variants offered by the Department. Through the BRF experience, the patient is in a better position to make an informed choice about therapy, especially dynamic psychotherapy. Our hope is that BRF will be sufficient for some and may usefully begin the work of explorative therapy for others. The style of BRF is active; a positive therapeutic alliance is promoted. Working within a psychodynamic frame, the assessor focuses the sessions on central interpersonal themes identified at the assessment meeting, especially negative expected responses from others expressed in the transference, and the issue of termination. Formally the research has two aims.

1. To evaluate the effect of brief intervention and follow-up and compare this with the standard assessment procedure for psychotherapy.
2. To identify the characteristics of those who do well and those who do badly with brief intervention and follow up (BRF).

Positively, we hypothesise that:

1. BRF will reduce the number of patients going on the waiting-list for psychotherapy.
2. BRF will deepen self-awareness and increase congruence with dynamic psychotherapy for patients placed on the waiting-list so that they have increased motivation and better outcome.

Negatively, we hypothesise that:

1. BRF will establish expectancies of therapist activity and focality that will not be met in subsequent psychotherapy.
2. The discrepancy will impair subsequent therapy.
3. Where childhood rejection was formative, the alliance formed in BRF and its termination will replicate the rejection and be marked by an increase in symptoms.
4. Engagement with any subsequent therapist will be difficult.

GETTING THE DESIGN RIGHT

A great deal of preparatory work was necessary to establish the research design and its feasibility. Developing the protocol which ran to 12 pages was an essential step; it forced us to be explicit, highlighted omissions and was a requirement of the funding body. We began with monthly meetings of the research group, working on 'To Do' items in between meetings. The monthly meetings continue and serve the functions of monitoring the progress of the research, refining the design, maintaining momentum and cohesion, and informing uninvolved members of the Department about the work and its potential value to our service. The last is an important function in sustaining a research project over a 5 or more year period in a Department whose primary *raison d'être* is clinical service and not research.

We wanted to conduct a randomized controlled trial with patients representative of a specialist NHS psychotherapy service and on a topic which was directly relevant to our role as service providers and trainers of specialist psychotherapists; the design had to be robust so that the findings would be credible, even revelatory. Our situation, the research literature, clinical judgement and statistical power calculations determined the shape of the trial.

Our first decision was not to have a no-treatment control group. Meta-analysis demonstrates significant effect size gains across the psycho-therapies in comparison with placebo (Smith *et al.*, 1980; Aveline, 1984). Of relevance to this study is the finding of similar gains in meta-analysis of trials of brief dynamic psychotherapy (Svartberg & Stiles, 1991;

Critts-Christoph, 1992). Psychotherapy researchers no longer have to demonstrate the value of psychotherapy in general; the task is to elucidate the particular benefits of specific interventions with defined patient groups.

Our second decision was not to standardize the BRF intervention through a manual to which adherence would be tested; in this, we went against the current tide where to enhance generalizability, therapy variability is minimized through manualization (Critts-Christoph, 1992; Krupnick & Pincus, 1992). Rather we adopted the naturalistic position of only using experienced senior members of staff whose normal role is to assess referrals; all work within a psychodynamic frame. Convergence in practice is fostered by presenting the assessments to an assessment meeting where core conflicts are identified, potential problematic and corrective interactions are anticipated, and peer-supervision of subsequent sessions is offered. The test in the research is of what experienced therapists, informed by a particular formulation but with the freedom to react creatively to the events of therapy and dealing with our profile of referrals, can achieve with a *three plus one* intervention. We appreciate that our decision may limit generalizability of our findings but argue that our non-manualized intervention is closer to representative practice than that with manualized procedures and could be a more powerful intervention through the inclusion of spontaneous elements.

Clearly we needed to control for clinical variables that might influence outcome. In our judgement, these were age, sex, age of completing education and severity of disorder: they also had the advantage of being relatively easy factors on which to stratify the sample. We surveyed a sample of 50 consecutive referrals for dynamic psychotherapy and found that the sample divided equally into patients aged less than 36 and 36 or over. The ratio of women to men was 5 : 3. Equal numbers of patients had completed their education before the age of 20 compared to those completing later. We now realize that this survey was not sufficiently detailed as we did not investigate the association between the stratifying variables and this is causing recruitment problems in a few cells (Figure 6.1). A pilot study of a three-point clinical rating of severity (mild, moderate and severe) based on the information in the referral letter and answers given by 46 patients to our Pre-assessment Psychotherapy Questionnaire gave two results (details of the questionnaire will be found in Aveline, in preparation). It showed that:

1. We were receiving equal numbers of patients in a combined category of mild and moderate compared with those in the severe category.
2. In a subset of 10 patients, the internal agreement on the divide was 80% between the two independent clinical raters who did the ratings in the first half of the main study and 70% between that group and a group of four of the researchers.

	Severity	Age	Education	Male	Female	
BRF	Mild/Moderate	<36	<20	4	6	68
			20+	3	4	
		36+	<20	4	6	
			20+	3	4	
	Severe	<36	<20	4	6	
			20+	3	4	
		36+	<20	4	6	
			20+	3	4	
SA	Mild/Moderate	<36	<20	4	6	68
			20+	3	4	
		36+	<20	4	6	
			20+	3	4	
	Severe	<36	<20	4	6	
			20+	3	4	
		36+	<20	4	6	
			20+	3	4	
				56	80	136

Figure 6.1. Stratification for brief intervention and follow-up (BRF) versus standard assessment (SA)

These frequencies for age, sex, age of finishing education and severity of problem determined the ratios in the stratifying cell sizes (see Figure 6.1).

To determine the sample size, power calculations were done. Our primary concern was with between-group comparisons in order to determine whether BRF was more beneficial than SA. We considered that, to be of practical importance, a between-group difference would have to equal or exceed 0.5 of a SD. To detect such a difference with a two-sided significance level of 5% with 80% power requires 64 patients per group or 128 in all (Cohen, 1977). To achieve the balanced design of Figure 6.1, a total of 136 patients is required.

To illustrate the sample sizes required to detect effects of different sizes, consider how the numbers alter depending on the assumptions made of what degree of change on the psychometric measures is to be accepted as clinically significant. On the assumption that a gain of 20 points on the Self-Concept Questionnaire (Robson, 1989), the first of our independent standardized measures, from time 0 to time 4 months is a significant and worthwhile gain, i.e. from an average score of 100 for psychotherapy patients in the normative literature to half way to the average score of 140 for normal controls (SD = 24), a sample size of 126 patients is required to demonstrate that a gain of 20 points is significant at the 5% level with a probability error of 1 in 20. To demonstrate that a gain of 10 points was similarly significant would require a cohort of 504 subjects, whereas a gain of 30 points would require only 58 subjects. Applying the same calculations to the second independent measure of outcome, the Brief Symptom Inventory (Derogatis and Melisaratos, 1983) and requiring a similar diminution in symptoms before accepting the results as significant, i.e. a shift of at least two-thirds of a SD on the three main subscales (0.5 on the raw score for GSI, 0.4 for PSDI and 10 for PST), requires a sample size of 108 for the GSI, 140 for PSDI and 88 for PST. We specified a sample of 136 subjects, divided equally between BRF and SA, and drawn as far as possible from consecutive referrals.

We established exclusion criteria (e.g. patients re-referred within the past year, those found to require immediate therapy at the first meeting and those, mainly staff referrals, whom it would be unethical to include in the research), devised the consent form and questionnaires, secured ethical approval and piloted the procedures, the identification of core conflicts and BRF intervention on a pre-trial sample of 50 patients.

We decided to have independent assessments of state on four occasions: time 0, 2 weeks before SA or BRF 1, time 4, 4 months later when there would be a clear month after BRF 4 and hence less distortion through the halo effect; time 15 at 15 months when most patients put on the waiting list would be or have been in therapy; and time 36 when most should be post-therapy. Patients gave written consent to participate in the research to improve our assessment procedure and to attend these four evaluations with the independent research psychologist. All this preparation took nearly 2 years.

SELECTION OF MEASURES

For the research to be credible in the eyes of clinicians and researchers, the measures needed to be refined, both in focus and sensitivity. There needs to be a mixture of self-report and standardized independent measures—

independently administered—to allow comparison with other samples, including normative groups.

The selected measures provide idiographic and nomothetic indices of subject state over time. They cover the three areas of interest to psychotherapists, that is self-esteem, psychiatric symptoms and quality of interpersonal relationships. These measures are our 'best buy' for a standard battery in psychotherapy research.

Target Problems (Battle *et al.*, 1966) are an idiographic way of delineating problems that the patient wishes to resolve in psychotherapy. At the first independent research assessment, three main problems that the patient wants help with in psychotherapy are identified in discussion with the patient and rank-ordered for importance, then rated for severity by the patient on five-point analogue scales for (1) distress, (2) interference with everyday activities and relationships and (3) how much the subject expects to change towards this goal in psychotherapy. In addition, the patient is asked to describe the way in which each problem would change if they were successful in therapy, i.e. to define a goal against which change can be assessed. At subsequent research assessments, patients are free to identify additional target problems; for administrative reasons, we set an upper limit of four problems.

Three nomothetic measures were used. The Self-Concept Questionnaire (Robson, 1989) derives from the work of Rosenberg on self-esteem but has been devised and validated in Britain. Examples of the 30 items, each to be answered on an eight-point analogue scale, are 'I can like myself even when others don't' and 'When I'm successful there's usually a lot of luck involved'. The questionnaire produces a total and five subscale scores. The Brief Symptom Inventory (Derogatis and Melisartos, 1983) is a widely used measure of psychiatric state. The full form, the SCL-90, was used in the Sheffield Psychotherapy Project but there is no effective loss of information in the cut-down form which has 53 items. Symptoms such as 'being suddenly scared for no reason', 'having to check and double check what you do' and 'feeling watched or talked about by others' are rated on five-point scales for the previous 7 days. Scoring produces three global indices of distress of which the General Severity Index (GSI) is the most widely used and nine sub-scales, e.g. Interpersonal Sensitivity, Depression, Hostility and Phobic Anxiety. Measures of interpersonal difficulties have been for the most part unsatisfactory. The Inventory of Interpersonal Problems (Horowitz *et al.*, 1988) is a promising measure with good test–retest reliability and high internal consistency. It has 127 items, 78 of which begin with the phrase, 'It is hard for me to...' and 49 with the words 'I am too...'. Scoring produces a total score and scores on 12 relatively independent subscales such as 'hard to be intimate" or 'supportive'and 'too aggressive' or 'eager to please'. The Inventory was used in the Second Sheffield Psychotherapy Project (Shapiro *et al.*, 1990).

While the focus of the research is on the impact of BRF and SA, the subject's state of mind at each independent assessment may be affected by other factors. We are investigating two: life events and help from other sources. Qualitative and quantitative estimation of life-events are derived from the administered Interview for Recent Life Events (Paykel & Mangen, 1980); this gives separate counts for addition, loss and conflict events and a total life events score. Each event is rated for objective and subjective negative impact. In addition, we ask about any events that have enhanced self-esteem. We ask about contact with other sources of help such as the church, alternative practitioners and counsellors, its timing and benefit. Finally we ask the subject to evaluate the utility of the assessment using 31 questions derived from the Therapy Impact Questionnaire (Llwellyn et al., 1988). Items include positive and negative elements such as 'I am looking at myself in a new way' and 'I felt the therapist did not understand me', each rated on a five-point scale. The assessor completes a complementary questionnaire after BRF 4 or SA. In total, we are collecting over 200000 items of information.

INTERIM RESULTS AND PROBLEMS

The main study commenced in April 1991. As at August 1993, 127 patients have been entered into the trial. Of these, 107 have completed the time 0 independent assessment in full and 61 the time 4 months assessment; 30 of the 61 received BRF and 31 SA. The results presented here refer to this subset where we have complete data for the two assessments. Figure 6.2 gives demographic information. The figures are as expected except that a greater proportion of subjects have been rated as having severe problems. This conforms to our clinical impression that the trend of recent years continues for ever more disturbed and disabled patients to be referred.

We are concerned about attrition from the study. A minority of subjects, variously, withdrew their consent, failed to attend the research assessment, did not complete all the questionnaires, missed the SA session or some of the BRF sessions, and did not carry through with the second research assessment. Attrition is a fact of life in clinical research but its degree means that we are having to continue to recruit subjects, even when stratifying cells had been filled. Consequently the time taken to complete the trial lengthens.

The attrition rate from 107 to 61 overstates the true picture as not all the 107 have completed their referral assessment. Logistic and personal factors contribute to the rate. The psychometric questionnaires are exhausting and disturbing to complete. They are sent to the patient for them to complete at their own pace a couple of weeks before they meet with the research psychologist for the administered measures. Patients report that it takes

	Total at T0 $N = 107$	BRF at T4 $N = 30$	SA at T4 $N = 31$
Gender			
males	42	12	11
females	64	18	19
Severity of problem			
mild/moderate	44	13	12
severe	63	17	19
Referral source			
consultant/MHT member	56	19	12
GP/other doctor	34	5	2
social work/DSH team	1	1	0
other	3	1	1
Mean age	35.8	37.4	35.7
Mean age left Education	20.19	19.4	21.27

Figure 6.2 Demographic information on sample with complete data at time 0 independent assessment (T0) and time 4 months assessment (T4)

1 to 3 hours, often distributed over several days, to complete the 227 questions of the BSI, SCS and IIP. Completing the questionnaires commonly forces that person to undertake a review of their life, facing them with much unhappiness and disappointed hope. They bring this emotional upheaval to the research assessment where the administered measures (target problems, life events, satisfaction with the assessment and receipt of other forms of help) take a further 1 to 1.5 hours to complete and which further highlights problems and adverse life events.

For the most part, the research psychologist is the first professional that the subject encounters after filling in the psychometrics and she has to contain the patient's distress while attempting not to be drawn into being a therapist. Inevitably the distinction between impartial researcher and concerned quasi-therapist is blurred. Unintentionally, the research assessment provides an opportunity for working through and, despite repeated briefing to the contrary, the contact is seen by many patients as being synonymous with therapy. Thus the dose of therapy in the study is increased by the interaction with the independent researcher. In our experience with this research psychologist, the research assessments constitute an important therapeutic experience in themselves, which fortunately for the statistical analysis is equally experienced by test and control groups. Other researchers may need to acknowledge the therapeutic impact of their independent assessments. In addition, proper attention needs to be given to supporting that person in their patient assigned therapist role.

Returning to factors contributing to subject attrition, the length of the psychometric questionnaires is daunting for the patient. *Inter alia*, if the research community can devise a reduced set of no more than 100 questions with no loss of sensitivity and reliability, they will be justly fêted. As an alternative, we are testing the acceptability to patients of computer administration of the psychometric questionnaires. This innovation will principally benefit the researchers as the second step in a manual procedure of entering the raw data into a computer will be obviated with twin benefits of improved accuracy of data-entry and speed of analysis, but subjects may find this method more stressful.

Patients give informed consent but the reality of what they are committing themselves to is not fully understood, especially the length of the follow-up. We suspect that some hold a fantasy of securing favoured status through consenting to participate in the research and that others comply with the initial phase but have no deep agreement. Some exit after they have secured a place on the waiting list, which is understandable as getting help from therapy is the hope of most. Even more understandable is the exit of those rejected for therapy, however professionally well-founded and well-explained is that decision. Finally, how a subject interacts with the assessment procedure expresses their character and, often, illustrates the problems for which they have been referred. We are trying to quantify these and other reasons by sending a brief questionnaire to those who do not attend the second research assessment (time 4).

Fortunately, from the point of view of interpreting the results, the reduced sample at time 4 months is representative of that at time 0 on the stratifying variables (age, sex, age of finishing education and clinical severity of problem) and the psychometric measures (Brief Symptom Inventory, Self-Concept Scale, Inventory of Interpersonal Problems) and number of negative life events. As Howard (Chapter 1) points out, this is not to say that the sample is equivalent to the larger cohort entered into the trial and, beyond that, of those refusing participation at the outset; further comparisons will need to be made to elucidate these aspects.

What we have known as clinicians is substantiated by the psychometric profile detailed in Figure 6.3. Referrals for dynamic psychotherapy are definitely the 'walking wounded', not the 'worried well' The mean score on the BSI, SCS and IIP is far removed from that of normals and documents greater disturbance than in normative British psychiatric outpatient populations. In itself, this is an important finding. While all levels of distress are of concern to the individual, in today's funding climate, it is important for specialist psychotherapy services to demonstrate that they are dealing—and dealing effectively—with the more disturbed.

	Total $N = 107$	BRF $N = 30$	SA $N = 31$
Mean total self-concept score	92.94	91.24	92.55
Psychiatric OPs Normals	100 140		
Mean IIP total score	1.859	1.816	1.909
Psychiatric OPs Normals	1.48 1.19		
Mean BSI total score	1.749	1.832	1.629
Psychiatric OPs Normals	1.32 0.44		
Life events mean objective score	9.991	10.022	10.032

Figure 6.3 Psychometric information at time 0 with norms

These findings raise two questions, one research directed and the other to do with service-contracts. How can one determine *clinically significant effects* and what relation might these bear to purchasing criteria? The answer, I suggest, is to be found in the recommendations of Jacobson and Truax (1991). Traditionally treatment effects have been inferred from statistical comparison of pre- and post-treatment mean scores. This method is deficient in two ways.

1. Mean score changes give no indication of the variability of response within the sample when this fact is crucially important to clinicians wanting to improve their practice.
2. Statistically significant change may have no direct relation to clinically significant change.

To compensate for these deficiencies, psychotherapy researchers recently have tended to use effect–sizes statistics, which express a difference between two means in terms of standard deviation units. This provides a common metric for appraising the extent of the difference between means in relation to variation between individuals at each point in time. However, this bears little direct relation to the practical or clinical significance of the extent of change shown by members of each group. Clinically, it would be important to know how many individuals improve enough to resemble members of the general population or those not requiring treatment. Accordingly, Jacobson and Truax have proposed quantitive criteria for establishing that an individual has shown clinically significant change with reference to normative data. The standard is the degree to which the patient, through the action of therapy, moves from dysfunction to normal functioning.

Figure 6.4 proposes three statistical models depending on the extent of knowledge of the distribution curves for the outcome measure in dysfunctional and normal populations. If only the norms for dysfunction are known, a clinically significant effect would be any change beyond two standard deviations from the mean in the direction of normality (point a). Should the norms for normal functioning be known and where they do not overlap with the dysfunctional group, the clinically significant effects cut-off point would be within two standard deviations of the mean for normal functioning (point b). When the curves overlap, the cut-off point would be at the point of overlap (point c). This method assumes normal distribution for the variable. Generally, point b is to be preferred followed by point c. The implication of this is that researchers should gather normative information in order to establish benchmarks for the success of interventions. Speculatively, health service contracts could be set to moving the patient a specified degree towards functional scores with therapy continuing while overall progress continues to be made. Such a system would have to take account of the starting-point of the patient—greater levels of disturbance going with greater difficulty in catalysing change—and the fact that in many disorders cure in the sense of attaining functional scores is unrealistic. Many of the disorders treated with psychotherapy are chronic and require long-term maintenance to minimize the mal-effects of permanently impaired function.

Jacobson and Truax propose two further statistical refinements, the first of which is shown in Figure 6.5. When dysfunctional and functional scores overlap, it is important to quantify the magnitude of change. This is done

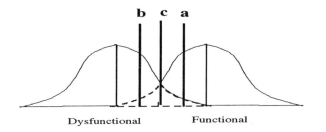

Dysfunctional   Functional

Assumes normal distribution

Figure 6.4 Three statistical models for clinically significant change
Reproduced from Jacobson & Truax (1991) by permission.

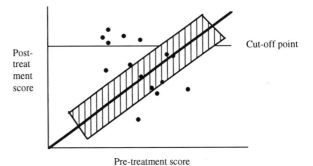

Figure 6.5 Reliable change index

through the *reliable change index* which takes account of the standard error of the difference between pre- and post-treatment scores. The index is used to define an area either side of the diagonal of change. Only scores outside the area are statistically reliable, i.e. the area excludes false positive and negative scores. Figure 6.5 shows the cut-off point determined by one of the above cut-off formulae. Only scores above the line in this example represent subjects who recovered during therapy. Displaying results in this way graphically illustrates the spread of findings as well as being a stringent test of effectiveness. Even greater stringency can be achieved by allowing for the imprecision in the outcome measure by plotting confidence limits based on the reliable change index around the cut-off line and only accepting scores outside and above the limit as true recovery.

Applying these principles to this study, what have we found at the mid-point of the trial? Dealing first with categorical outcome, fewer patients receiving BRF than those receiving SA have needed to be placed on the waiting-list for psychotherapy broadly defined (Figure 6.6). The result is significant at the 0.12 level. In line with our positive hypothesis, we have

	On waiting list	Discharged
BRF	18(21)	10(7)
SA	24(21)	4(7)

χ^2 square = 2.38, df = 1, $p = 0.12$
Waiting list: for any form of psychotherapy plus referrals elsewhere for psychotherapy.
Discharged: includes general psychiatry referrals.

Figure 6.6. Treatment decision outcome × intervention (expected values by chance in brackets)

been able to discharge more BRFs than SAs; these discharges include a minority of patients where the intervention has been sufficient and a majority where the person has been unsuitable for psychotherapy but the longer duration of the assessment has allowed a successful working through of the assessment decision which, we surmise, would not have been possible in a single meeting. Put another way, the SA group probably contains more patients who are inappropriately taken on for therapy. This assertion may be surprising at first sight but it reflects our position as a service of last resort for many of our referrals. We exist in a professional network where colleagues contend with intractable psychological problems and where professionally and personally it is hard for us to say 'no' to a referral after a single assessment session.

Normative data for dysfunctional and functional populations are available for the BSI (Derogatis & Melisartos, 1983; Francis *et al.*, 1990), SCS (Robson, 1989) and IIP (Horowitz *et al.*, 1988), so that we were able to set a cut-off point for clinically significant change at point c. Figure 6.7 shows for BRF patients a significant reduction in symptomatology as measured by the General Severity Index of the Brief Symptom Inventory at the 4-month independent follow-up (please note that the axes are reversed from that illustrated in Figure 6.5). It, also, shows the substantial variation in outcome within the sample; accurate extrapolation from group effects to individual patients will need further enquiry. Figure 6.8 expresses the

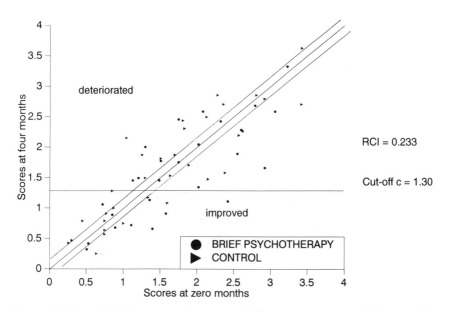

Figure 6.7 Mean Brief Symptom Inventory (BSI) scores at zero and four months

	BRF	SA
N	30	31
Mean GSI change score on BSI	0.2107	0.0505
SD	0.606	0.434

$T = 1.93$, $df = 59$, $p = 0.059$.

Figure 6.8. Change in mean Brief Symptom Inventory (BSI) score T0 to T4 × intervention

same result in more conventional form; the t-test shows a significant difference at the 0.059 level. These findings are tentative; they demonstrate a trend which may be revelatory once we reach the 'n' set in our power analysis.

Somewhat to our surprise, there was no significant change in total SCS score. We expected this to be more responsive to life-events, including the life-event of being assessed for therapy. No significant increase in self-concept score or reduction in severity of interpersonal problems with BRF or SA has been recorded, as we hoped would be the case. This may reflect a process wherein changes on these dimensions take longer to achieve; the results of the 15-month follow-up are awaited with great interest.

A research project of this magnitude has interesting effects on the service. Stemming from its different tempo, doing BRF assessments alters clinical practice. The assessor no longer has to gather all the relevant information in a single session and has much greater freedom to pace exploration and interpretation to individual capacity for constructive assimilation. Conversely, some assessors now find it harder to reach a conclusion when doing SA assessments. For the most part, assessors prefer the BRF format and, results permitting, look forward to introducing this or a modified version into routine practice once the study has been completed. For the Department as a whole, the research has given impetus to providing shorter therapies whenever possible.

On a more negative note, the fact that not all members of the Department are actively involved in the research and, of course, have varying degrees of commitment to it, offers opportunities for splitting and scapegoating. Shortfalls in secretarial and clinical resource have been attributed to the research draining off resources which should flow to the 'real work' of patient care (see Chapter 4). The greater goal of improving service to patients has at times been lost sight of in the day-to-day struggle to maintain standards in a service which year by year is trying to do more with less resource. Practically, we underestimated the administrative, data-entry and statistical requirements of the project. To illustrate the last point, preparation of the data reported in this chapter and entry onto the university mainframe took 146 hours. Writing the code for the statistics and analysis of the results required 138 hours of expert time.

CONCLUSIONS

Our experience demonstrates that it is possible to conduct robust relevant research in a clinical department. Maintaining the momentum of the work requires constant effort. We underestimated the amount of preparatory work that would be necessary and the way in which resource deficiencies in the Psychotherapy Unit would become conflated with the relatively modest resource requirements of the research itself. In retrospect, extra secretarial, data-entry and research administrator time would have facilitated the whole process. Good work takes longer and throws up more problems of detail than ever one hopes will be the case.

The profile of referrals identified in the pre-trial survey has proved to be valid but we made an error in not quantifying incidence in individual cells in the stratifying table (Figure 6.1). Randomization is producing equal entry to BRF and SA. Though most of the stratification cells are filling, there is a dearth of males with severe disorder completing education before the age of 20 and of women completing education after the age of 20. Clearly some stratifying variables are associated with each other in ways that we did not appreciate at the outset. We face a dilemma of selection. Either we extend the intake period of the trial in order to fill all the cells as planned but thereby distort the profile of referrals, or we loosen the design but gain in representativeness.

Recruitment to the trial is satisfactory. However keeping patients in the trial for follow-up at 4 months and beyond is difficult. Fortunately, so far representativeness has been maintained between those who stay in the trial and those who exit prematurely. While the reasons for the attrition are complex, one factor is that the measures are burdensome to complete, both in terms of the emotional disturbance they engender and the time they take to answer. The psychometric measures have good spread but the number of questions needs to be reduced from the present total of 227 to under 100 while retaining their discrimination and reliability.

Interesting results are beginning to emerge from this randomized controlled trial. At the mid-point stage, the findings are tentative and cannot be properly evaluated until we reach the 'n' determined by our power analysis. Also the great spread on the ratings recorded at each independent assessment makes it difficult to extrapolate from group results to individual response. The trend towards more discharges with BRF is in line with our positive hypothesis and is already benefiting our waiting list. The validity of this mid-point finding is buttressed by the demonstration that at the 4-month follow-up, BRF patients have significantly lower Brief Symptom Inventory scores than SAs. At that point, no significant increase in self-concept score or reduction in severity of interpersonal problems has been recorded, as we hoped would be the case. This may reflect a process wherein changes on these dimensions take longer to achieve; we await the results of the 15-month follow-up with great interest. An alternative view is that the dose of therapy is too little. Would *four plus two* be a better model for our patient population?

ACKNOWLEDGEMENTS

The realization of this research would not have been possible without the sustained commitment of the following:
Clinicians: Mark Aveline, Bernard Ratigan, Teresa Baker, Jane Taylor, Jill Staines, Sophia Hartland and Helen Lee.
Research psychologist: Sylvia Cooper.
Clinical raters of severity: Tricia Slack, Naomi Curry and Helen Lee.
Statisticians: John Cromby and Charlotte Sheard.
Secretaries and data coding and entry: Vicky Bartlett, Margaret Dolan-Shaw, Jill Robson, Barbie Gallimore and Mandy Dalgeish.
Funder: The Mental Health Foundation.

REFERENCES

APA (1987). *Diagnostic and Statistical Manual Of Mental Disorders*, 3rd edn, (revised version). American Psychiatric Association, Washington DC.
Aveline, M. & Smith, L. (1986). *The Impact of a Pre-assessment Questionnaire in Psychotherapy*. International Society for Psychotherapy Research Conference: Wellesley, Massachusetts.
Aveline, M.O. (1984). What price psychiatry without psychotherapy? *Lancet* (2), 856–859.
Aveline, M.O. (In preparation). The therapeutic impact of a pre-assessment psychotherapy questionnaire.
Barkham, M. (1989). Exploratory therapy in two-plus-one sessions 1—rationale for a brief psychotherapy model. *British Journal of Psychotherapy* 6, 81–88.
Barkham, M. & Hobson, R.F. (1989). Exploratory therapy in two-plus-one sessions 2—a single case study. *British Journal of Psychotherapy* 6, 89–100.
Battle, C.C., Imber, S.D., Hoehn-Saric, R., Stone, A.R., Nash, C. & Frank, J.D. (1966). Target complaints as criteria of improvement. *American Journal of Psychotherapy* 20, 184–192.
Cohen, J. (1977). *Statistical Power Analysis for the Behavioural Sciences*. Hillsdale, New Jersey: Erlbaum.
Critts-Christoph, P. (1992). The efficacy of brief dynamic psychotherapy: a meta-analysis. *American Journal of Psychiatry* 149(2); 151–158.
Derogatis, L.R. & Melisaratos, N. (1983). The Brief Symptom Inventory: an introductory report. *Psychological Medicine* 13, 595–605.

Francis, V.M., Rajan, P. & Turner, N. (1990). British community norms for the Brief Symptom Inventory. *British Journal of Clinical Psychology* **29**, 115–116.

Horowitz, L., Rosenberg, S.E., Baer, B.A., Ureno, G. & Violasenor, U.S. (1988). Inventory of interpersonal problems: psychometric properties and clinical applications. *Journal of Consulting and Clinical Psychology* **56**(6), 885–892.

Jacobson, N.S. & Truax, P. (1991). Clinical significance: a statistical approach to defining meaningful change in psychotherapy research. *Journal of Clinical and Consulting Psychology* **59**(1), 12–19.

Krupnick, J.L. & Pincus, H.A. (1992). The cost-effectiveness of psychotherapy: a plan for research. *American Journal of Psychiatry* **149**(10), 1295–1305.

Llwellyn, S.P., Elliott, R., Shapiro, P.A., Hardy, G.E. & Firth-Cozens, J. (1988). Clinical perception of significant events in prescriptive and explorative periods of individual therapy. *British Journal of Clinical Psychology* **27**, 105–114.

Luborsky, L. & Crits-Christoph, P. (1990). *Understanding Transference. The Core Conflictual Relationship Theme Method.* New York: Basic Books.

Luborsky, L., Crits-Christoph P., Mintz, J. & Auerbach, A. (1988). *Who will benefit from Psychotherapy? Predicting Therapeutic Outcomes.* New York: Basic Books.

Malan, D H. (1963). *A Study of Brief Psychotherapy.* London: Tavistock.

Paykel, E.S. & Mangen, S.P. (1980). *Interview for Life Events.* Department of Psychiatry, University of Cambridge.

Robson, P.J. (1989). Development of new self-report questionnaire to measure self-esteem. *Psychological Medicine* **19**, 513–518.

Shapiro, D.A., Barkham, M. *et al.* (1990). The Second Sheffield Psychotherapy Project: rationale, design and preliminary outcome data. *British Journal of Medical Psychology* **63**, 97–108

Shapiro, D.A. & Firth, J. (1987). Prescriptive vs exploratory psychotherapy: outcomes of the Sheffield Psychotherapy Project. *British Journal of Psychiatry* **151**, 790–799.

Smith, M.L., Glass, G.V. & Miller, T.I. (1980). *The Benefits of Psychotherapy.* Baltimore: Johns Hopkins University Press.

Strupp, H.H. & Binder, J.L. (1984). *Psychotherapy in a New Key.* New York: Basic Books.

Svartberg, M. & Stiles, T.C. (1991). Comparative effects of short-term psychodynamic psychotherapy: a meta-analysis. *Journal of Consulting and Clinical Psychology* **59**(5), 704–714.

7 Decisions, Decisions, Decisions: Determining the Effects of Treatment Method and Duration on the Outcome of Psychotherapy for Depression

DAVID A. SHAPIRO*, MICHAEL BARKHAM, ANNE REES,
GILLIAN E. HARDY, SHIRLEY REYNOLDS[a] AND
MIKE STARTUP[b]

*MRC/ESRC Social and Applied Psychology Unit, University of Sheffield,
[a]University of East Anglia, and [b]University College of North Wales, Bangor, UK*

ABSTRACT   Key decisions and choices made in designing the outcome phase of a large-scale comparative study of the psychotherapy of depression are reviewed. Many of these decisions took the form of trade-offs between conflicting desiderata, rather than unambiguous questions of good vs. bad design. A common conflict was between rigour and scientific control on the one hand, and generalizability to clinical practice on the other. Other decisions, however, were required to avoid pitfalls. Here, the conflict is typically between what is desirable and resource constraints. As an example, decisions required to ensure adequate statistical power are discussed. Key findings from the study are reviewed, with reference to their implications for clinical practice. More extensive treatment appeared warranted only for relatively severe depression, and the effects on outcome of the treatment method followed were, at most, limited.

INTRODUCTION

In this chapter, we review some of the decisions and choices made in designing the outcome phase of a large-scale comparative process-outcome study of the psychotherapy of depression (Shapiro, in press-a). The Second Sheffield Psychotherapy Project (SPP2) represents 'state of the art' research, requiring a level of resources only found in major research

---

* Correspondence address: MRC/ESRC Social and Applied Psychology Unit, Department of Psychology, University of Sheffield, Sheffield SI0 2TN, UK.

Research Foundations for Psychotherapy Practice. Edited by M. Aveline and D. A. Shapiro.
Copyright © 1995, Mental Health Foundation and Individual Contributors.
Published 1995 by John Wiley & Sons Ltd

institutions. We will present an overview of our study, interwoven with commentary selectively highlighting major design decisions we took along the way, and implications for research carried out in service settings. We hope that this will demystify the process of designing such studies, and encourage others to make their own contributions to comparative psychotherapy research.

Although any service delivery system has an interest in maximizing cost-effectiveness, a market-oriented health care system—such as the internal market developing in the British National Health Service—gives this added urgency. Purchasers and providers alike have a clear interest in evidence concerning the effectiveness of psychotherapy. In principle, it would be desirable for purchasing decisions to be guided by such evidence. Correspondingly, providers would do well to develop interventions backed by empirical data.

As noted by Parry (1992), important distinctions may be drawn between research, audit and evaluation. Research is constrained by the requirement to yield unequivocal answers to scientific questions. One of our concerns is with the extent to which comparative outcome research can inform practical decisions concerning the design of psychotherapy services.

RESEARCH QUESTIONS

The first step in any study is to formulate the research questions. These should be framed to maximize the informativeness of the work. Accordingly, they take account of what is known already, of any qualifications or ambiguities in such prior results requiring elucidation, and of the utility to interested parties of the range of alternative sets of results that is likely to be obtained (Horowitz, 1982).

One major issue addressed in our study concerned the contribution to effectiveness of the treatment method used. This is clearly important to the theory and practice of psychotherapy. Treatment approaches divide trainers and practitioners alike into groups with strong allegiances. Prior findings in the literature included the conclusion from several reviews that diverse psychotherapies appeared to yield broadly equivalent outcomes (Stiles et al., 1986), although some findings are suggestive of a possible advantage of cognitive-behavioural (CB) over psychodynamic-interpersonal (PI) psychotherapy in the treatment of major depression (Robinson et al., 1990; Shapiro & Firth, 1987). In relation to the latter findings, however, we were concerned by possible confounding variables, such as the allegiance of most investigators to CB (Robinson et al., 1990), inadequate measurement of changes in PI therapy, and insufficient treatment sessions to test PI therapy (Shapiro et al., 1990).

Other questions arose from the 'dose-effect' analysis by Howard *et al.*, (1986). These authors reviewed data from samples in which outcome was assessed more than once over treatment, and obtained a negatively accelerating dose–effect relationship, whereby outcome is improved with more treatment sessions, but with diminishing returns as the number of sessions increases. This suggested a direct test of the incremental effectiveness of assigning patients in advance to longer rather than to shorter treatment.

Interpretation of our own previous study (Shapiro & Firth, 1987) has been complicated by pretreatment differences in severity of disorder between the groups randomly assigned to receive two treatments in different orders. Accordingly, we decided to stratify the present sample into three severity groups. Whilst we were carrying out the present study, Elkin *et al.* (1989) reported differences between treatments that were confined to relatively severe depression, but their finding could have been an artefact of *post hoc* division of the sample into more and less severe cases. Serendipitously, our chosen design was ideal for the examination of the now-timely question of how treatment response varied with the severity of disorder. With an eye to efficient allocation of treatment resources according to need, we examined the relationship between severity and the extent of advantage conferred by longer treatment.

From the above considerations, four research questions can be derived:

1. Is CB more effective than PI in the hands of clinician-investigators holding no prior allegiance to CB?
2. Is CB more rapid in its effects than PI?
3. Do any differential effects of CB vs. PI vary according to the initial severity of depression?
4. Is 16-session therapy more effective than 8-session therapy, and does any difference vary with initial severity of depression?

Despite the logical primacy of research questions over research methods (Elliott, Chapter 2), in practice the planning of a study requires one to cycle between formulation of questions and designing methods to address them. Of course, not all questions that we might wish to ask are capable of being answered with existing methods and within the resources available to the investigator.

RESEARCH DESIGN AND METHODS

Even well-resourced researchers face complex and difficult choices in designing their studies. Research design is an art of creative compromise, since both practical constraints and the countervailing demands of different methodological criteria preclude the 'perfect' study (Shapiro,

1989). In reviewing some of the decisions we made in designing our study, we will comment on related choices faced by researchers in clinical service settings.

The design of the study is shown in Figure 7.1. Each of 120 clients would complete treatment in one of four conditions: equal numbers of clients were to receive either 8 or 16 sessions of either PI or CB therapy. Prior to randomization to one of the four conditions, clients were stratified into three ranges of severity on the Beck Depression Inventory (BDI) (Beck *et al.*, 1961). This ensured that 10 clients in each severity range were randomized to each of the four conditions.

Outcome was measured at comparable points in time for both 8- and 16-session treatments. A mid-treatment assessment of 16-session treatments corresponded to the post-treatment assessment of 8-session treatments. The post-treatment assessment of 16-session treatments corresponded to a 3-month follow-up of 8-session treatments. Only for the 3-month follow-up of 16- session treatments was there no assessment of 8-session cases at a corresponding point in time. A 1-year follow-up was also undertaken, but that is not reported here.

Some investigators have proposed that the 'equivalence paradox' of apparently similar outcomes of demonstrably different treatment methods can be resolved by improving the capacity of outcome research to detect true differences between methods (Stiles *et al.*, 1986). As detailed below, we strove to do this in several ways. However, in doing so we inevitably

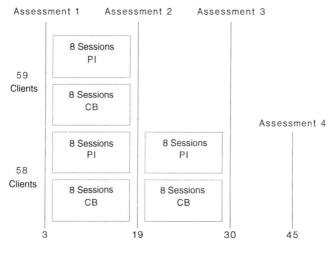

Figure 7.1 Design of the Second Sheffield Psychotherapy Project. Copyright © 1994, American Psychological Association, reproduced by permission

made our study less representative of treatment as routinely practised in service settings. This reflects a commonly found trade-off between external validity (generalizability) and other desirable characteristics of studies (Cook & Campbell, 1979; Shapiro, 1989).

STATISTICAL POWER

Increasing the sensitivity of outcome research to differences between methods involves consideration of statistical power (Cohen, 1977; Kazdin & Bass, 1989; Rossi, 1990), the probability of rejecting the null hypothesis of no difference between conditions when it is false. Power increases with sample size and with the size of the difference between groups that the study is seeking to detect, and decreases with increasing variation within the groups under comparison. Furthermore, the more rigorous the level of significance chosen, the lower the power of the test to detect a given effect. The convention of selecting the 0.05 level of significance (or 'alpha' level), whereby the probability of rejecting the null hypothesis when it is true (so-called 'Type I error') is just 1 in 20, is commonly followed and understood.

It is equally important, however, to set a limit to the probability of falsely accepting the null hypothesis (or 'Type II error'). There is a trade-off between Type I and Type II errors. The relative gravity of their respective consequences is a matter of judgment, according to circumstances. However, Cohen (1977), in a highly influential treatment of this topic, suggests that it is in general around four times more serious to reject a null hypothesis when true than it is to accept it when false. Accordingly, it has become conventional to aim, alongside the usual 0.05 significance or alpha level, for a 0.20 probability of erroneously accepting the null hypothesis (known as 'beta'). Power is defined as 1 minus beta. Hence, Cohen (1977) recommends a power level of 0.80, signifying an 80% chance of detecting a true difference between groups.

The larger the effect that a study is seeking to detect, the more readily will 0.80 power be achieved. Cohen (1977) describes effects around 0.20, 0.50, and 0.80 of a standard deviation (SD) as small, medium and large effects, respectively. To illustrate the practical importance of such effects, consider that these effect sizes account for approximately 1, 6 and 14%, respectively, of the variance among participants.

Sample size

The determinant of statistical power most fully discussed in the literature is the size of samples used. Kazdin and Bass (1989) found a median difference between alternative forms of psychotherapy of 0.47 of a standard deviation, a 'medium' effect in Cohen's (1977) terms. Only 34 of 75 studies had sufficiently large samples to yield adequate power (0.80) to detect this median difference. Fifty per cent of samples were sized between 10 and 19 patients per group. Adequate power requires samples four times this size.

The implications of this discussion for the design of clinical research are clear, and challenging (Rossi, 1990). The detection of differences between psychotherapies of the magnitude that is likely to be found requires at least 60 patients per group. If resources do not permit samples of this order, then an outcome study comparing active treatments should not be attempted. To obtain large enough samples, the service-based investigator may have to recruit the help of a broad range of clinical colleagues, possibly at several sites. But if this makes for greater variation within treatment groups (see below) then much of the power gained may be lost.

However, the larger expected difference between treatment and no-treatment (a 'large' effect of 0.80 of a SD) can be detected with much smaller samples. Given that a one-sided significance test may be used here (which assumes that any difference in favour of the untreated group must be a 'false positive' or Type I error), 0.80 power is attained with as few as 20 participants in each group. It is therefore much easier, although less informative, to complete a study comparing treatment with no treatment than one comparing active treatments.

To achieve maximum power within available resources, we designed our study to include 60 patients completing each of two treatments (PI and CB). Within each treatment condition, 30 of these received eight weekly sessions, and 30 received 16 sessions. Power for tests of both Treatment and Duration approached 0.80. Power to detect an interaction between these factors (to determine whether CB is more rapid in its effects than PI) was lower at 0.70. Thus, our sample was only just large enough for our purposes.

Reducing variation within groups

Statistical power depends on more than sample size. Efforts to reduce variation within treatment groups—sometimes described as increasing experimental control—are rewarded by greater power to detect between-group differences of a given magnitude in terms of scale points on an instrument. These efforts are costly, however. We screened patients carefully to ensure that they met rigorous criteria for entry into the study, provided thorough training and ongoing supervision of therapists, and demonstrated therapist adherence to treatment manuals (Startup & Shapiro, 1993).

Screening required extensive self-report (Beck *et al.*, 1961; Derogatis, 1983) and psychiatric interview assessments (Present State Examination; Wing *et al.*, 1974) of a total of 257 patients, involving a trained interviewer. The interview served both to reduce variation among patients and to ensure generalizability to a depressed population. Every included patient had to reach diagnostic criteria for a major depressive episode during the 3 months prior to assessment, and was currently rated as at least at the threshold of psychiatric caseness over the 1 month prior to interview. Patients were excluded if the PSE was scored for psychotic, manic or obsessional symptoms, or if depression was attributable to organic illness.

The five therapists treated at least two training cases in each of the four Treatment × Duration conditions. A total of 64 training cases were supported by weekly supervision. Throughout the study, every case was presented to the group of therapists for peer supervsion at least once in every eight treatment sessions. Therapist adherence was demonstrated by having trained raters listen to 220 sessions and rate these on 59 items describing the therapist's contributions to each type of treatment session (Startup & Shapiro, 1993).

Research in service settings confronts a stark trade-off between experimental control and homogeneity on the one hand, and both representativeness and feasibility on the other. Efforts to reduce within-group variation—to increase the power of a study to detect differences between groups—will result in clients, therapists and treatment procedures that are unrepresentative of everyday practice, limiting the extent to which its findings can be said to apply to that practice. Perhaps even more seriously, the feasibility of the study may be compromised. For example, if inclusion and exclusion criteria are too demanding, insufficient patients will be available. Accordingly, we decided not to require all patients to be currently in a major depressive episode at intake, which may compromise the applicability of our findings to such patients.

Severity

Severity of disorder is another powerful source of variation between patients that will reduce power, quite apart from its intrinsic interest as a focus of investigation. Traditional, unconstrained randomization was developed by early statisticians such as Fisher for agronomical research using larger samples than are typically available in psychotherapy studies. Smaller samples bring an increased probability of 'chance' bias resulting from arithmetically substantial differences between samples assigned to different conditions. These do not have to attain statistical significance to muddy the interpretation of data. Statistical control by partialling out pre-treatment scores (Mintz et al., 1979; Shapiro, 1989) is an incomplete remedy.

Based on the distribution of scores in pilot data, we stratified our sample into three ranges of severity on the pre-screening BDI mailed out to potential clients, before randomizing to Treatment and Duration. Ranges were 16–20 (Low), 21–26 (Medium), and 27 and above (High). Such a procedure is strongly recommended in any setting. However, it was not altogether successful, as, by chance, within the High Severity group (which had no upper limit) those assigned to receive 16 sessions had higher scores than those assigned to receive 8 sessions. With hindsight, a more rigorous procedure of narrower bands or matched sets of individuals assigned to the four conditions would have been preferable.

Therapist effects

There is mounting evidence that some therapists are more effective than others, even when adhering to a treatment manual (Lambert, 1989). This is another factor that can increase within-group variation. In addition, it opens up the possibility that when we think we are measuring the effects of different treatments, we are in fact measuring the effects of different therapists employed in each modality. Thus, therapist effects are potential sources of bias and unrepresentativeness, as well as reducing power.

Ideally, therefore, we would have a large pool of therapists in each treatment condition, and minimize the differences in their effectiveness via thorough training. However, this has implications for the costs of recruiting, training and monitoring the therapists. In practice, however, we had just five therapists available for our study.

There are two alternative ways of assigning therapists to treatments: most commonly, each therapist practises only one treatment (therapists nested within treatments); alternatively, each therapist conducts all treatments in the study (therapists crossed with treatments). The former design has the advantages of being more representative of usual practice, and allowing therapists to select themselves into methods that they find congenial. However, as already noted, it can compromise our ability to attribute differences to the treatment method itself rather than to stable personal characteristics of the therapist. We chose the latter design, for its strength, especially in studies of treatment process (Stiles et al., 1988a, 1988b), in equating personal characteristics of therapists across treatments. However, it remains possible for differential preference, suitability or skilfulness to influence the therapist's relative effectiveness in the different treatments (Shapiro et al., 1989; Stiles et al., 1989). Furthermore, therapists capable of offering different treatments may be unrepresentative of practitioners in general. No hard-and-fast recommendation can be made as to which of these design alternatives is preferable in clinical research.

Reliability of measures

Unreliable outcome measures add 'noise' to the within-group variation and hence reduce statistical power. Indices of reliability include the extent to which responses to the items making up a scale are intercorrelated, and the extent to which the ordering of individuals from high to low on the measure is stable over time. Up to a point, reliability improves with the number of items in a measure (which increases the time taken to collect the data, and may reduce client compliance with the demands of the research). There is therefore a trade-off between reliability and economy. However, most clinical instruments are longer than those used in (for example) occupational studies of stress. Clinical researchers could sensibly reduce the length of their instruments.

The pursuit of stability can result in a measure that is insensitive to the real changes associated with treatment. This argues in favour of selecting measures that have proved in previous treatment outcomes studies their ability to detect change. That is a strong argument for the development of a 'core outcome battery' of measures that have been tried and tested in change studies, in addition to the more obvious benefit of comparability across studies achieved by investigators using the same instrument.

REDUCING BIAS

In comparing treatments, researchers must be aware of, and seek to reduce, sources of bias that can invalidate the comparison. There are many different sources of bias. At a conceptual level, design decisions can unwittingly favour one treatment over another. For example, outcome measures may be more sensitive to the effects of one treatment, or the treatment protocol (e.g. number or scheduling of sessions) may be better suited to one method. Less obvious are biases that may creep in, not so much in the design of the study, as in its implementation. For example, if large numbers of participants are lost, those remaining may be a biased and unrepresentative sample of those entering the study. Thus, one treatment could appear more effective simply because many of those who did not find it helpful chose not to complete the treatment and evaluation phases of the study.

Choice of measures

One possible source of bias against PI therapies in previous research is the absence of psychometrically sound measurement of the interpersonal problems that are a major focus of this treatment. We therefore included the Inventory of Interpersonal Problems (IIP)(Horowitz et al., 1988), and the Social Adjustment Scale Self-report (Cooper et al., 1982). Other self-report outcome measures included the BDI, the Symptom Checklist-90R (SCL90R)(Derogatis, 1983), and a self-esteem scale (O'Malley & Backman, 1979). In view of the suggestion that the BDI may be biased in favour of CB treatment, we analysed the depression subscale of the SCL90R separately as a second index of depression. The psychiatric interview, the

Present State Examination (PSE)(Wing *et al.*, 1974) used in screening was also repeated once after treatment.

Number of sessions

It is commonly believed that CB's direct focus on symptoms is associated with more rapid relief than PI's focus on problem clarification and insight as preparatory to behavioural change. Our previous study's use of 8-session treatments (Shapiro & Firth, 1987), may therefore have biased it against PI. We tested this proposition directly by comparing both 8- and 16-session implementations of PI and CB therapy. Whilst some may argue that 16 sessions is still short for PI, we considered that this was a likely upper practical limit to routine NHS psychotherapy of depression.

Attrition

A further source of bias derives from attrition or loss of participants from the study (Howard *et al.*, 1986b). A progressive filtering occurs even before an individual is included in the study, as individuals through choice or circumstance do or do not find their way into it. This process continues through every stage from pre-screening to follow-up assessment, with a progressively increasing opportunity for bias as attrition cannot be dissociated from the treatment and participants' reactions to it. If attrition rates differ with the different treatments, the opportunity for bias is manifest. However, even similar attrition rates, if high, can yield bias if it is different people who leave the two treatments (e.g. more severely distressed people are lost to one treatment whilst less severely distressed people are lost to the other).

The potential impact of attrition is illustrated by the following account of our study. Five hundred and forty individuals returned postal pre-screening questionnaires, 291 were offered an intake interview, 257 completed it, and 169 met all inclusion and exclusion criteria. These criteria served to ensure major depression whilst excluding disorders (e.g. obsessive–compulsive disorder) not expected to respond to the treatments. The criteria also excluded those who had had prior psychological treatment within 5 years, or with a continuous history of disorder extending beyond 2 years, who were considered less likely to benefit from psychotherapy as brief as eight sessions. Medication *per se* did not exclude, but admission to the study of those whose medication had recently changed was delayed pending reassessment, to prevent symptom change consequent upon the changed drug regimen from distorting our findings. To ensure ethical and unbiased implementation of a protocol, it is essential to ensure that all clients admitted to the study are suitable for any of the conditions between which they are to be randomized.

Of the 169 clients suitable for randomization, 19 were diverted from the study because their severity cells were full. One hundred and fifty were randomized into treatment, 12 withdrew before the first treatment session (the numbers involved ranging from one for 8-session PI to six for 16-session CB), 15 did not complete treatment, four provided insufficient post-treatment data, and two were excluded because their depression had lifted between initial screening and commencing therapy. Comparing completers with non-completers on demographic characteristics, we found non-completers more likely to be single, and on average somewhat younger. Those excluded at intake or failing to complete treatment tended to be younger than completers, and to be somewhat less educated.

Investigator allegiance

Above and beyond specifically identified sources of bias such as those described above, the many decisions and actions taken by investigators in designing and implementing comparative studies present countless opportunities for the unwitting influence of the investigator's allegiance to one or other of the therapies under comparison. As noted by Elliott *et al.* (1993), most relevant comparative studies have been conducted by adherents of cognitive and behavioural methods, so that the relatively frequent findings favouring these methods remain inconclusive. One facet of the allegiance problem was noted by Hollon *et al.*, (1993): alternative psychosocial interventions with which CB has been favourably compared in prior research have been unconvincing quasi-control conditions, not fully representative of such treatments as practised. There is a dearth of comparative studies conducted by investigators with no prior allegiance (e.g. as inferred from introductory paragraphs of their reports by Robinson *et al.*, 1990). Our own prior position was that both PI and CB were likely to be effective depression treatments, and as therapists we believed ourselves equally committed to both treatments.

COMMENT

We have selectively reviewed some key design decisions in carrying out the Second Sheffield Psychotherapy Project. For a fuller and more formal account, the reader is referred to the original report (Shapiro *et al.*, 1994a). Two types of consideration emerge. First, there are design imperatives, such as sample size, which must be heeded at all costs. Second, there is a rather larger group of more complex decisions requiring trade-offs between incompatible desiderata. With respect to these, since we cannot have it both ways, we have to choose the most acceptable mix of gains and losses for a given study. That choice will depend upon the context and purpose of the research.

In addressing the implications of our design decisions for research in service settings, a recurrent theme has been that of the conflict between rigour and generalizability. Indeed, this conflict may render the comparative trial method unworkable as a means of answering a given question of interest in a given setting. This may tempt some to reject the comparative trial method, with its unique capacity to support causal inference.

Alternatives, such as naturalistic, descriptive studies exploiting covariation among variables within groups to answer substantive questions, usually appear more feasible in service settings. However, it would be mistaken to believe that such problems as statistical power do not apply equally to naturalistic studies. Here, the power issue is expressed, for example, as the ability of the study to detect the relatively small correlations among the variables of interest that it is reasonable to expect. And the loss of the capacity to infer cause and effect is a heavy price to pay, especially if the study seeks to persuade sceptical readers of the effectiveness of an intervention or of some component thereof. Furthermore, major conceptual difficulties arise in the use of correlations between process and outcome to identify effective process elements (Stiles, 1988; Stiles & Shapiro, 1989).

RESULTS

As in presenting the methodology, our account of the outcomes of the Second Sheffield Psychotherapy Project will highlight issues and choices in analysis and interpretation. Again, the data are presented more formally by Shapiro et al. (1994a).

RANDOMIZATION FAILURE

We were troubled by pre-treatment differences between groups assigned to different treatment conditions. For example, clients subsequently assigned to 16 sessions were significantly worse at the assessment interview than those subsequently assigned to 8 sessions. This difference was largely attributable to those in the High Severity group. For example, the adjusted mean BDI was 30.67 for High Severity clients assigned to 16 sessions, and 24.41 for High Severity clients assigned to 8 sessions.

How could we explain these differences? Could they reflect a systematically biased assessment or randomization procedure? The differences were obtained prior to randomization, which followed predetermined procedures preventing any account being taken of pre-screening data beyond identifying the severity range within which a case fell. Therefore, the differences could be attributed neither to investigator effects upon pre-treatment questionnaire completion (e.g. encouraging the client to believe that their depression must be severe because they were to receive 16 rather than 8 sessions), nor to covert decisions by the team to allocate clients according to need (e.g. assigning more severe cases to 16 sessions). We found these differences frankly inexplicable. We could only attribute them to 'randomization failure', the chance occurrence of between-group differences that bedevils research using samples measured in tens rather than hundreds. Interpretation of changes over treatment had to take account of the different starting-points of the different severity groups.

There were also significant pre-treatment interactions between Severity and Treatment. For example, on the BDI, Low Severity clients subsequently assigned to CB were a mean 3.5 points higher before treatment than Low Severity clients subsequently assigned to PI.

DECISIONS ABOUT DATA ANALYSIS

Planning the data analysis is part of the process of designing a study. Ideally, all strategic decisions about the form of the analysis should be taken before the data are collected. All investigators, but especially those doing their first outcome study, should discuss data analysis with an experienced outcome researcher and/or an expert statistician familiar with such studies, during the planning of the study. In practice, however, even the most seasoned researcher engages in 'play' with the data once they are available. At the level of detail, there will be many creative solutions, not all of which can be anticipated at the design stage. Only a small proportion of the data-analytic work carried out finds its way into final reports. There are always several defensible ways of analysing a given data set; the conscientious investigator reports analyses whose findings are representative of those of other analyses not reported.

To facilitate comparison with other studies using overlapping but not identical sets of instruments, we analysed each of seven outcome measures separately. For simplicity, the present account highlights findings with the BDI, the most commonly-used self-report depression measure. Recall from Figure 7.1 that Assessments 2 and 3 constituted end-of-treatment and follow-up, respectively, for those receiving 8 sessions. Meanwhile, Assessments 2 and 3 were administered midway through and at the end of treatment, respectively, to those receiving 16 sessions. The 3-month follow-up of 16-session cases was accomplished by an additional Assessment 4. Thus, our primary analyses considered end-of-treatment (Assessment 2 for 8-session cases and Assessment 3 for 16-session cases) and 3-month follow-up (Assessment 3 for 8-session cases and Assessment

4 for 16-session cases). Unless otherwise stated, means are averaged over end-of-treatment and 3-month follow-up Assessment Occasions.

However, these primary analyses entailed a confound between duration of treatment and elapsed time. In addition, they did not permit separation of changes occurring over the first and second halves of 16-session treatments. When required, we therefore performed secondary analyses of the data obtained, irrespective of treatment duration, at Assessments 2 and 3.

Analyses were conducted using SPSS MANOVA. Analyses were focussed on the effects of Duration (8 vs. 16 sessions) and Treatment (CB vs. PI) in relation to initial severity of depression (pre-screening BDI ranges 16–20 vs. 21–26 vs. 27 or higher). Assessment Occasion was a within-subjects factor. In the primary analyses, interactions with Assessment Occasion represented differential change between end-of-treatment and 3-month follow-up; in the secondary analyses, interactions with Assessment Occasion represented differential change from Assessment 2 to Assessment 3.

Correction for pre-treatment differences

Differences between individual clients that exist prior to treatment have the potential to distort findings. Accordingly, it is common to create 'adjusted scores' that are corrected statistically for the likely influence of these differences on the results obtained for each individual after treatment (analysis of covariance; Shapiro, 1989). On each measure, we did this to take account of Assessment 1 scores on the measure in question. In order to examine the effects of the severity factor, Assessment 1 scores were standardized within severity groups before computing adjusted scores. Scores were further adjusted for the impact of marital status, because analysis of the BDI showed that clients who were married or cohabiting were marginally less depressed at 3-month follow-up, $F(1, 112) = 2.62$, $p = 0.11$, than those living alone. To eliminate any overall differences in effectiveness among project therapists, we used residualized scores obtained by subtracting from each adjusted score the mean adjusted score obtained on that occasion by all patients seen by that therapist.

As discussed above, the statistical power of our study to detect effects of the magnitude it was reasonable to expect was only just sufficient. Since power depends on the alpha or significance level selected, there is a trade-off between the risks of Type I error (misinterpreting a 'chance' difference as showing an effect of the experimental conditions) and those of Type II errors (failing to detect a 'true' difference between conditions).

If statistical power is marginal, this creates a 'grey' area of differences that may or may not warrant interpretation as due to the experimental conditions. Accordingly, we adopted two significance criteria: effects with alphas below 0.05 were interpreted, in the conventional manner, as significant. Effects with alphas between 0.05 and 0.15 were presented as possibly worthy of interpretation. This increased the power of the study to detect, albeit cautiously, medium-sized interaction effects (Shapiro *et al.*, 1994b).

TREATMENT EFFECTS

Considering all seven outcome measures, there was little evidence that CB was generally more effective or acted more rapidly than PI, or vice versa, or that one treatment was more effective than the other with relatively more severe or less severe clients. However, there was a significant advantage on the BDI to CB, $M_{adj} = 7.99$, over PI, $M_{adj} = 11.51$. The remaining measures showed arithmetic differences in adjusted means favouring CB, but these differences were very small and none approached statistical significance.

Rapidity of treatment effects

The relative speed of change in CB and PI was assessed in two ways. First, we considered interactions between Treatment and Duration, and second, amongst 16-session cases alone, we examined interactions between Treatment and Assessment Occasion in the secondary analysis of Assessment 2 and Assessment 3 data.

Tests of the interaction between Treatment and Duration in end-of-treatment and 3-month follow-up data yielded F ratios below 1, indicating no evidence whatever for an effect, for six of the seven measures, including the BDI. Analysis of data from Assessments 2 and 3 for 16-session cases alone revealed no evidence of differential rates of change in CB and PI. Of the six measures administered on both occasions, five yielded F ratios below 1.

Differential treatment effects related to severity

Tests of the Severity × Treatment interaction yielded F ratios below 1 for four of the seven measures, and only for one measure was the p value below 0.15 for this interaction.

DURATION OF TREATMENT

Is 16-session treatment more effective than 8-session treatment, and does any difference vary with initial severity of depression? For each of our seven measures, adjusted means showed a numerical advantage to 16-session treatment. However, on five measures the F ratio was less than 1, and only for the BDI did the effect approach significance, $M_{adj} = 8.34$ and

11.16 for 16- and 8-session treatments, respectively, $F(1,97) = 2.43$, $p = 0.12$. Considering these main effects alone, there is scant evidence of added benefit from 16-treatment sessions over 8.

INTERACTIONS BETWEEN TREATMENT DURATION AND SEVERITY OF DEPRESSION

Does the effect of treatment duration vary with initial severity of depression? We examined Severity × Duration interactions for each measure. They were significant for nearly all measures. The High Severity group did better after 16 than after 8 sessions on every measure, and at least marginally significantly so on all except one. Amongst High Severity clients, those receiving 16 sessions, $M_{adj} = 9.29$, returned BDI scores a mean 10.5 points lower than those receiving 8 sessions, $M_{adj} = 19.83$. In contrast, Moderate Severity cases showed only a modest, statistically unreliable advantage to 16 sessions. On the BDI, for example, this was reduced to 4 points, $M_{adj} = 6.34$ and 10.38 for 16- and 8-session cases, respectively. In yet greater contrast, Low Severity clients showed arithmetic differences favouring 8-session treatment.

Figure 7.2 depicts the Severity × Duration interaction on the BDI. As shown in the top portion of Figure 7.2, the results are strengthened rather than weakened by the contrasting pre-treatment differences described above, whereby High Severity clients assigned to 16 sessions were *more* depressed at Assessment 1 than High Severity clients assigned to 8 sessions.

We examined raw data to ascertain the nature of the pre-treatment difference between High Severity clients assigned to 8 and 16 sessions. We found that only one 8-session client had an Assessment 1 BDI above 30, whilst 10 16-session clients fell in this range. This raised the possibility that the strong advantage of 16 sessions with High Severity clients was in some way related to the chance failure to randomize the very severest cases to 8 sessions. However, examination of clients in the highest range populated by both 8- and 16-session cases disproved this. There were 13 8-session cases and 10 16-session cases with Assessment 1 BDIs in the range 24–30. These subgroups yielded median Assessment 3 BDIs of 15 and 8 for 8 and 16 sessions, respectively.

A second finding was the contrast between 8- and 16-session treatments with respect to effects of severity. The ordering of High, Moderate, and Low Severity groups on all outcome measures persisted after completion of an 8-session treatment. In contrast, the effect of initial severity was abolished by 16-session treatments: the order of severity groups was on some measures no longer consistent with initial severity, and the differences among the groups were non-significant on most measures.

The abolition of initial severity differences by 16 sessions but not by 8 sessions could reflect the greater elapsed time associated with the former.

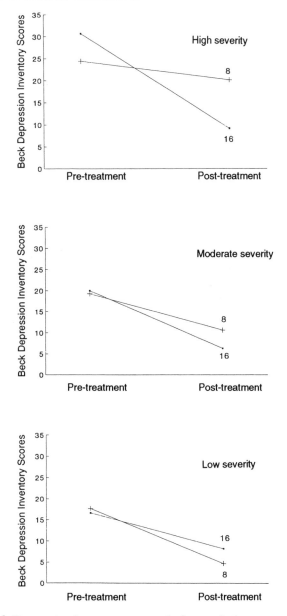

Figure 7.2 Beck Depression Inventory means before and after treatment for High, Moderate and Low Severity clients assigned to 16 and 8 sessions of treatment. *Note.* Pre-treatment scores were obtained at Assessment 1. Post-treatment scores were averaged over end-of-treatment and 3-month follow-up assessments. Pre-treatment means were adjusted for marital status. Post-treatment means were adjusted for marital status and Assessment 1 score. Copyright © 1994, American Psychological Association, reproduced by permission

To control this confound, we examined the Severity × Duration interaction in a secondary analysis of Assessment 3 data, focusing attention on the effects of Severity at each level of Duration. This yielded similar results to those described in the previous paragraph. Although the severity effect persisted uniformly at 3-month follow-up of 8-session treatment on all measures, it was substantially attenuated by 16-session treatment.

TIME COURSE OF CHANGE IN 8- VERSUS 16-SESSION TREATMENTS

The significant interactions between Treatment Duration and Assessment Occasion, typically reflected non-significant differences in opposite directions at Assessments 2 and 3. At Assessment 2, those who had completed an 8-session treatment were marginally ahead of those midway through a 16-session treatment, whereas at Assessment 3, those who had just completed a 16-session treatment were marginally ahead of those who earlier completed an 8-session treatment. The differential change between Assessments 2 and 3 presumably reflects continuing improvement of the 16-session group, who were still receiving treatment.

DISCUSSION

In this overview of the outcome phase of SPP2, our intention has been to convey greater insight into the research design process than can be obtained from scientific reports published in major journals. This may help demystify such 'state of the art' research and show practitioners and beginning researchers that many design decisions are choices reflecting trade-offs between conflicting desiderata, rather than clearcut issues of good vs. bad design. This is not to deny, however, that pitfalls await the unwary; sample size is perhaps the clearest case in point, with outcome trials comparing plausible psychotherapies with samples of 20 per group no longer justifiable.

This presentation highlights the scale of undertaking required to furnish a scientific basis for psychotherapy practice. The execution of this single study addressing a small number of specific research questions concerning the psychotherapy of just one disorder—albeit the 'common cold of psychiatry' depression (Seligman, 1975)—was by far the largest single task of our research group for several years. Specific, research-based guidance to purchasers and providers of psychotherapy services is likely to be restricted in scope, given the high costs of obtaining strong research evidence of the kind presented here.

Greater worldwide investment in high-quality research would assuredly increase the range and precision of 'treatment of choice' and other specific recommendations available to practitioners. Other research-based recommendations to purchasers and providers will perforce be cast in more general, strategic terms. For example, the present study illustrates a wider trend of evidence supporting the benefits and cost-efficiency of brief, structured treatment approaches to affective and anxiety disorders. It would be mistaken for purchasers and providers to await definitive findings from high-quality controlled research to inform many of their decisions. The cost, complexity and potential unrepresentativeness of such research argue strongly for audit and evaluation of clinical services as delivered.

That said, a strong practical recommendation did arise from this study: longer treatment—here represented specifically by 16 weekly sessions rather than 8—appears necessary only for those with relatively severe depression—here represented by a pre-screening BDI score of 27 or above. The effect of 16 sessions of treatment, in contrast to eight, was to abolish the pre-treatment differences amongst our three severity groups. Although such a finding must be replicated, especially in service settings, it shows that specific, practical implications can flow from scientific data.

The interaction between Severity and Duration could not be entirely explained with reference to the greater elapsed time associated with 16 sessions of treatment; it remained when elapsed time was equated. Nor was it plausibly seen as an artefact of pre-treatment differences in the opposite direction. Given data such as those at the top of Figure 7.2, where lines depicting changes by initially non-equivalent High Severity groups cross over to yield significant post-treatment differences in the opposite direction, rival explanations such as alternative scaling, selection-maturation and regression effects are implausible (Cook et al., 1990, pp 528–529).

The severer depressions, here identified via pre-screening BDI scores in the high twenties and above, may respond quite differently than milder disorders and participants in analogue studies. This reinforces the need identified by Hollon et al. (1993) for more work comparing cognitive therapy with other psychosocial treatments with clinically representative populations. More broadly, to the extent that secondary and tertiary psychotherapy services tend to be referred more severe cases than those successfully treated at less specialized levels of service, research findings on less distressed populations may not generalize to specialist service settings.

Turning to the differential effects of CB vs. PI treatment of depression, a question of enduring interest to providers espousing one or other of these methods, our data were rather more equivocal. Only on the BDI was there a significant treatment effect, amounting to 3.52 points or a medium effect size around 0.5 of a standard deviation. Given that the BDI is the most widely used self-report depression measure, this finding could be interpreted as revealing an advantage of CB over PI. However, the BDI may be seen by some as sufficiently grounded in a CB model of depression to predispose this instrument to favour CB therapy. Furthermore, other self-report instruments, including a second depression measure, did not even show trends towards such an effect.

These results are consistent with the suggestion that some at least of the advantage of CB over PI claimed in literature reviews may be due to investigators' predominant allegiance to CB. Our own allegiance could be described as to the 'equivalent outcomes' position (Stiles et al., 1986; Elliott et al., 1993). Although each may interpret our results in accordance with their own allegiance, we see the present findings as suggesting that an even-handed comparative evaluation yields, at best, a marginal advantage to CB.

We found broadly equivalent outcomes of CB and PI, despite adequate sample sizes, demonstrated therapist adherence, control of therapist effects and appropriate outcome measures. These findings suggest that equivalent outcomes cannot be attributed to methodological limitations alone. The treatment effect on the BDI was sufficiently modest to conceal wide variations. Even on this measure, substantial numbers of clients fared well in PI and others did relatively less well in CB. Given this variation, further analyses are warranted, to identify predictors of change in the two treatments and related process variables.

Considering the data as a whole, there was scant evidence of more rapid change in CB. Overall, 8-session CB treatment enjoyed no greater advantage over PI than did 16-session CB, nor was there any evidence of more improvement half-way through a 16 session treatment if that was CB rather than PI. For practical purposes, there appears to be no difference in the rate of change in depression in the two treatments.

There was no convincing evidence of differential response to the two treatments that varied with initial depression severity. Practically speaking, our findings offer no basis for differential recommendation of CB or PI according to the severity of depression. In this, they differ from the *post hoc* analyses of Elkin et al. (1989), in which the Hamilton Rating Scale for Depression showed somewhat better results of Interpersonal than of CB therapy in severe depression.

In practical terms, these findings would not warrant recommendation to depressed clients, purchasers or providers of CB as more effective than PI. Our data argue against the overriding importance attached to treatment method and orientation by many of those involved in training practitioners. They lend indirect support to theoreticians and practitioners favouring integration of CB and PI approaches, to the extent that this endows the practitioner with a broader repertoire of therapeutic strategies.

Confidence in the generalizability of these results (at least, to depressed patients involved in other research studies) is enhanced by the fact that the overall extent of improvement found in this study resembles the mean 14-point improvement over a mean 16 hours of psychotherapy found in a meta-analysis of depression studies by Nietzel et al., (1987).

METHODOLOGICAL ISSUES

The occurrence of 'randomization failure' has already been noted. The observed pre-treatment differences could not be readily explained by differential attrition, or by investigator effects. Our analysis was designed to minimize their effects on our conclusions; the most important differences were in the opposite direction to the reported post-treatment effects, and so do not appear to undermine our conclusions. However, such randomization failures appear an inherent hazard of studies with sample sizes typical of psychotherapy research, and matched designs may prove preferable.

As previously noted we used the same therapists for all treatment conditions, in an effort to minimize confounding between therapist characteristics and treatment methods. This is particularly valuable for the process–impact–outcome analyses that are integral to our research strategy (Shapiro et al., 1991). However, we acknowledge that this could in principle limit the generalizability of the present outcome findings to clinician-investigator therapists with the skills required to adhere at will to contrasting methods.

As already acknowledged, research questions addressed via interaction effects enjoyed marginal statistical power in this study; however, relevant findings were in general so clear-cut, showing either strong interactions for which power was adequate, or null effects characterized by F ratios below 1, that power limitations are unlikely to have compromised our conclusions.

Clients improved somewhat from pre-screening to Assessment 1. With hindsight, this could have been expected and it would have been preferable to set inclusion criteria and assign to severity group using pre-screening data. However, the range of BDI scores at Assessment 1 was well within the range of pre-treatment means reviewed by Nietzel et al. (1987). In addition, the detection of pre-treatment change via pre-screening is uncommon, so that our use of pre-treatment measures from Assessment 1 yields conservative estimates of initial severity and treatment effects relative to the literature.

FURTHER RESEARCH

Beutler (1991) and Shoham and Rohrbaugh (Chapter 3) recommend the study of interactions between client characteristics and treatment method, where these have firm conceptual foundations. This recommendation is supported by the present results in relation to severity of depression: conceptually, the obtained interaction with duration appears more cogent than the absent interaction with treatment.

Considering the present findings alongside those obtained by Elkin *et al.* (1989) and the literature reviewed by Robinson *et al.* (1990), it is clear that comparative studies of depression treatments need to include studies of the process of change in each treatment (Garfield, 1990; Greenberg, 1991; Marmar, 1990; Stiles *et al.*, 1988a, 1988b). In clinical practice, with its much wider range of patients, therapists and settings, the proportion of variance in outcomes due to the method of treatment alone is likely to be even lower than that found here. Future research should focus upon delineation of the mechanisms underlying change, to identify those which are common across different treatments, and those which are specific to each method. A continuum exists between such basic research on change processes, and more immediately service-related work on audit and evaluation. We hope that the present chapter will help strengthen the links between those working at different points along that continuum.

REFERENCES

Beck, A.T., Ward, C.H., Mendelson, M., Mock, J. & Erbaugh, J. (1961). An inventory for measuring depression. *Archives of General Psychiatry* **4**, 561–571.
Beutler, L.E. (1991). Have all won and must all have prizes? Revisiting Luborsky *et al.*'s verdict. *Journal of Consulting and Clinical Psychology* **59**, 226–232.
Cohen, J. (1977). *Statistical Power Analysis for the Behavioral Sciences*, 2nd edn. Hillsdale, NJ: Erlbaum.
Cook, T.D. & Campbell, D.T. (1979). *Quasi-experimentation: Design and Analysis Issues for Field Settings*. Chicago: Rand McNally.
Cook, T.D., Campbell, D.T. & Peracchio, L. (1990). Quasi Experimentation. In *Handbook of Industrial and Organizational Psychology*, 2nd edn., Vol. 1 (M.D. Dunnette & L M. Hough, eds). Palo Alto, CA: Consulting Psychologists Press, Inc., pp. 491–576.
Cooper, P., Osborn, M., Gath, D. & Feggetter, G. (1982). Evaluation of a modified self-report measure of social adjustment. *British Journal of Psychiatry* **141**, 68–75.
Derogatis, L.R. (1983). SCL-90R: Administration, scoring and procedures—Manual II. Towson, MD: Clinical Psychometric Research Inc.
Elkin, I., Shea, M.T, Watkins, J.T., Imber, S.D., Sotsky, S., Collins, J.F., Glass, D.R., Pilkoniz, P.A., Leber, W.R., Docherty, J.P., Fiester, S.J. & Parloff, M.B. (1989). National Institute of Mental Health Treatment of Depression Collaborative Research Program: general effectiveness of treatments. *Archives of General Psychiatry* **46**, 971–982.
Elliott, R., Stiles, W.B. & Shapiro, D.A. (1993). Are some psychotherapies more equivalent than others? In *Handbook of Effective Psychotherapy* (T.R. Giles, ed.). New York: Plenum press.
Garfield, S.L. (1990). Issues and methods in psychotherapy process research. *Journal of Consulting and Clinical Psychology* **58**, 273–280.
Giles, T.R. (ed.). (1993). *Handbook of Effective Psychotherapy*. New York: Plenum.

Greenberg, L.S. (1991). Research on the process of change. *Psychotherapy Research* **1**, 3–16.

Hollon, S.D., Shelton, R.C. & Davis, D.D. (1993). Cognitive therapy for depression: conceptual issues and clinical efficacy. *Journal of Consulting and Clinical Psychology* **61**, 270–275.

Horowitz, L.M., Rosenberg, S.E., Baer, B.A., Ureno, G. & Villasenor, V.S. (1988). Inventory of Interpersonal Problems: psychometric properties and clinical applications. *Journal of Consulting and Clinical Psychology* **56**, 885–892.

Horowitz, M.J. (1982). Strategic dilemmas and the socialization of psychotherapy researchers. *British Journal of Clinical Psychology* **21**, 119–127.

Howard, K.I., Kopta, S.M., Krause, M.S. & Orlinsky, D.E. (1986a). The dose–response relationship in psychotherapy. *American Psychologist* **41**, 159–164.

Howard, K.I., Krause, M.S. & Orlinsky, D.E. (1986b). The attrition dilemma: towards a new strategy for psychotherapy research. *Journal of Consulting and Clinical Psychology* **54**, 106–110.

Kazdin, A.E. & Bass, D. (1989). Power to detect differences between alternative treatments in comparative psychotherapy outcome research. *Journal of Consulting and Clinical Psychology* **57**, 138–147.

Lambert, M. (1989). The individual therapist's contribution to psychotherapy process and outcome. *Clinical Psychology Review* **9**. 469–486.

Marmar, C.R. (1990). Psychotherapy process research: progress, dilemmas, future directions. *Journal of Consulting and Clinical Psychology* **58**, 265–272.

Mintz, J., Luborsky, L. & Christoph, P. (1979). Measuring the outcomes of psychotherapy: findings of the Penn Psychotherapy Project. *Journal of Consulting and Clinical Psychology* **47**, 319–334.

Nietzel, M.T., Russell, R.L., Hemmings, K.A. & Gretter, M.L. (1987). The clinical significance of psychotherapy for unipolar depression: a meta-analytic approach to social comparison. *Journal of Consulting and Clinical Psychology* **55**, 156–161.

O'Malley, P.M. & Bachman, J.G. (1979). Self-esteem and education: sex and cohort comparisons among high school seniors. *Journal of Personality and Social Psychology* **37**, 1153–1159.

Parry, G. (1992). Improving psychotherapy services: applications of research, audit and evaluation. *British Journal of Clinical Psychology* **31**, 3–19.

Robinson, L.A., Berman, J.S. & Neimeyer, R.A. (1990). Psychotherapy for the treatment of depression: a comprehensive review of controlled outcome research. *Psychological Bulletin* **108**, 30–49.

Rossi, J.S. (1990). Statistical power of psychological research: what have we gained in 20 years? *Journal of Consulting and Clinical Psychology* **58**, 646–656.

Seligman, M.E.P. (1975). *Helplessness*. San Francisco: Freeman.

Shapiro, D.A. (1989). Outcome research. In *Behavioural and Mental Health Research: a Handbook of Skills and Methods* (G. Parry & F.N. Watts, eds). London: Lawrence Erlbaum Associates, pp. 163–189.

Shapiro, D.A., Barkham, M., Hardy, G.E. & Morrison, L.A. (1990). The Second Sheffield Psychotherapy Project: Rationale, design and preliminary outcome data. *British Journal of Medical Psychology* **63**, 97–108.

Shapiro, D.A., Barkham, M., Hardy, G.E., Morrison, L.A., Reynolds, S., Startup, M. & Harper, H. (1991). University of Sheffield Psychotherapy Research Program: Medical Research Council/Economic and Social Research Council Social and Applied Psychology Unit. In *Psychotherapy Research Programs* (L.E. Beutler & M. Crago, eds). Washington, DC: American Psychological Association, pp 234–242.

Shapiro, D.A., Barkham, M., Rees, A., Hardy, G.E., Reynolds, S. & Startup, M. (1994a). Effects of treatment duration and severity of depression on the effectiveness of cognitive/behavioral and psychodynamic/interpersonal psychotherapy. *Journal of Consulting and Clinical Psychology* **62**, 522–534.

Shapiro, D.A., Barkham, M., Rees, A., Hardy, G.E., Reynolds, S. & Startup, M. (1994b). Looking as strongly as we should in the right places in psychotherapy research. *Journal of Consulting and Clinical Psychology*, **62**, 539–542.

Shapiro, D.A. & Firth, J.A. (1987). Prescriptive vs. exploratory psychotherapy: outcomes of the Sheffield Psychotherapy Project. *British Journal of Psychiatry* **151**, 790–799.

Shapiro, D.A., Firth-Cozens, J. & Stiles, W.B. (1989). The question of therapists' differential effectiveness: a Sheffield Psychotherapy Project addendum. *British Journal of Psychiatry* **154**, 383–385.

Startup, M. & Shapiro, D.A. (1993). Therapist treatment fidelity in prescriptive vs. exploratory psychotherapy. *British Journal of Clinical Psychology* **32**, 443–456.

Stiles, W.B. (1988). Psychotherapy process-outcome correlations may be misleading. *Psychotherapy* **25**, 27–35.

Stiles, W.B. & Shapiro, D.A. (1989). Abuse of the drug metaphor in psychotherapy process-outcome research. *Clinical Psychology Review* **9**, 521–543.

Stiles, W.B., Shapiro, D.A. & Elliott, R. (1986). Are all psychotherapies equivalent? *American Psychologist* **41**, 165–180.

Stiles, W.B., Shapiro, D.A. & Firth-Cozens, J. (1988a). Do sessions of different treatments have different impacts? *Journal of Counseling Psychology* **35**, 391–396.

Stiles, W.B., Shapiro, D.A. & Firth-Cozens, J. (1988b). Verbal response mode use in contrasting psychotherapies: a within-subjects comparison. *Journal of Consulting and Clinical Psychology* **56**, 727–733.

Stiles, W.B., Shapiro, D.A. & Firth-Cozens, J.A. (1989). Therapist differences in the use of verbal response mode forms and intents. *Psychotherapy* **26**, 314–322.

Wing, J.K., Cooper, J.E. & Sartorius, N. (1974). *The Measurement and Classification of Psychiatric Symptoms*. Cambridge, UK: Cambridge University Press.

8 Questions to be Answered in the Evaluation of Long-term Therapy

CHESS DENMAN
The Cassel Hospital, Richmond, Surrey, UK

ABSTRACT   This paper sets out first to discuss the current pressing need for research into the central claims made for long-term therapy. Next it seeks to delineate the research questions which long-term therapy poses and to discuss the reasons for the paucity of research into long-term therapy. In the light of this discussion, existing research into long-term therapy is reviewed and its problems highlighted. Recommendations for future research are thereby generated.

INTRODUCTION

Currently there is a keen debate about the value, utility and efficacy of long-term therapy. Long-term therapy, i.e. therapy lasting upwards of 2 years at intensities varying from once to five times a week, is claimed by its practitioners to be at least an essential tool in the psychotherapeutic armoury. It is claimed to produce more reliable, more profound and more lasting change than brief therapies (Scott, 1993). The mechanism of this change is variously characterized as an alteration in underlying unconscious phantasy (Kleinian) the maturation and strengthening of the ego (Freudian), the dissolution of the need for a 'false self' (Winnicotian) and the resolution of developmental arrests in the ego–self axis (Jungian).

Common to all these descriptions of the change mechanism is the idea that a long-term relationship with an involved therapist fosters psychic development. Not surprisingly long-term therapy is claimed as particularly effective in dealing with patients whose psychopathologies are presumed to result from developmental delays—pre-eminently patients with personality disorders. Indeed, Bell (1992), a consultant at the Cassel Hospital, has argued that Freud's hysterical patients would now be classified as personality disordered patients of the same sort as those seen at the Cassel.

Correspondence address: The Cassel Hospital, 1 Ham Common, Richmond, Surrey, TW10 7JF, UK.

Research Foundations for Psychotherapy Practice. Edited by M. Aveline and D. A. Shapiro.

However, the value of long-term therapy in all conditions has been strongly questioned by critics of long-term psychodynamic psychotherapy (Andrews, 1993). Andrews particularly strongly refutes the suggestion that long-term therapy is effective in borderline personality disorders.

Clearly, therefore, research is needed if only to settle the debate about the claims of long-term therapy. The need for this research is especially pressing because the current political and funding situation in relation to health service provision, both in England and America, make it possible that long-term therapy (at least as a publicly funded option) may be eliminated by default. Public health authorities, seeing the costs of long-term therapy as unacceptably high, may act in the absence of research data as though negative findings existed.

WHAT ARE THE SPECIALLY IMPORTANT QUESTIONS WHICH NEED RESEARCH?

It is worth looking a little more closely at the sorts of questions which arise in relation to the practice of long-term therapy and for which clear, researched answers do not exist. These questions fall into three main groups.

The first group concern the application of long-term therapy as a treatment modality. One group of questions concerns selection for treatment and intensity. Who should receive long-term therapy? For whom is it contra-indicated? For how long should long-term therapy be continued and when should it be discontinued? How should the optimum intensity of therapy be set for a particular patient? These questions have important practical and resource implications for providers of psychotherapy. Without some answer to these questions, it will prove impossible to make informed allocations of resources.

A second group of questions concern technical issues in relation to ongoing therapy. When (if ever) is the use of parameters (i.e. alterations in the analytic frame such as extra sessions) justified? What is the value of different types of interpretations? How should regression be handled and what are the indicators of malignant regression? How should termination be managed? These questions have implications for the management of treatment. Answering them will, it is hoped, result in improvements to the efficacy of treatment again with resource implications albeit more distant.

A final group of research questions—intriguingly first proposed by Freud (1923)—arise from the way that long-term therapy may provide a unique setting for investigating particular psychological questions. Long-term therapy may offer a good setting in which to look at questions, such as the way the process of therapy develops, as well as issues about the stability or changeability of mental structure in a constant environment. Such questions amount to pure research (as opposed to applied research) with no obvious financial implications beyond the cost of doing the work. The traditional defence of pure research is that its findings do often feed back eventually into practical work as is likely to be the case in the case of long-term therapy.

WHAT ARE THE SPECIAL DIFFICULTIES OF DOING RESEARCH INTO LONG-TERM THERAPY?

Despite the strong claims for long-term therapy and the equally strong rebuttals of those claims, despite the pressing political need for research and despite the wealth of research questions which need answering, there is very little published formal scientific research on long-term therapy. This is especially true of psychoanalysis. Enquiry in the field is dominated by theoretical discussion, and illustrative description of single cases or of small groups of similar cases. There are a number of reasons for this lack of quantitative research, which can be divided into two main groups that interlock. The groups are:

1. Philosophical and cultural difficulties.
2. Special methodological problems combined with particular practical difficulties.

Taking philosophical difficulties first, there is a well established debate about the nature of legitimate evidence within psychoanalytic circles (see, for example, many of the papers in Wollheim & Hopkins, 1982) and some groups (following Freud) have been especially keen to argue that psycho-analytic methodology is in itself a form of scientific enquiry, which is therefore not in need of external experimental validation. Some have gone further and argued that psychoanalysis is not even capable of experimental validation, because being a field of study unrelated to physical sciences, it is logically distinct from them and so cannot be expected to benefit from the methods of validation appropriate to those sciences (Gardner, 1993).

However, these arguments can be rebutted to some extent because while it may be possible to see psychoanalysis as an interpretive or herme-neutical discipline rather than a science, this move then limits the types of conclusions which may be drawn from it. If generalizing claims as to cure or outcome are desired, or even if general statements about the later patho-logical sequelae of early events are made, then these cannot be insulated from calls for validation or refutation. That is, should psychoanalysis wish to claim special status then it may do so but only at the price of giving up most or all claims of a general therapeutic nature. Were psychoanalysis truly a purely interpretive art, then there would be no uncomplicated sense with which the claim could be made that it worked because interpretive disciplines, such as literary criticism, biblical hermeneutics and history, have no clear 'right' answers. In practice most psychoanalysts, especially those within health services, have been unwilling to give up such claims.

These sorts of philosophical considerations have informed a more general skeptical tendency to question the overall value and validity of the typical methodologies of psychological research. An illustration of this sort of skeptical critique of the the validity of some of these methodologies is a paper by Feldman and Taylor (1980). They criticize systems of symptom assessment which centre on behavioural measures at the expense of an account of the inner world of the patient's objects. However, their detailed analysis centres only on one psychometric method and, by highlighting its faults in a particular individual's case, Feldman and Taylor fail to take into account the possibility that, overall, despite some false results, it or other measures might turn out to correlate well with clinical judgments. There is also the widespread criticism that the investigations that research requires (tapes of sessions, reports from patients during therapy), contaminate therapy and may adversely alter its course.

In addition to this skeptical tendency there are, as well, what could be termed cultural difficulties, which involve a negative emotional response to the practice of research. It has often been characterized by long-term therapists as an oppositional activity and has often been seen, not so much as a method for discovery or justification, but as an activity of hostile competition. The subsequent splitting and polarization creates a situation which has many resemblances to that described in another setting by Norton (1992), when what he terms 'a culture of enquiry' is lost in a community. When this happens, Norton shows that on both sides of the divide positive and negative results are selected in isolation from a general understanding of the field and brandished or rubbished by the disputants. Findings are used only to bolster or undermine entrenched views rather than to analyse and investigate tentative positions

One important upshot of all of this is to create a situation where, when the results of an experiment or study appear to contradict established theory, it is normally the methodology which is questioned rather than the theory. Such a way of proceeding is, when routine, intensely hostile to the kind of two-way interaction of theory with experimental results which can generate active and fruitful research in a field. This is because central to the value of any empirical research endeavour is the way in which unexpected experimental results are allowed to challenge current theory and thus act as goads to the production of new theory.

Finally, for those who remain skeptical of the utility of more formal research methods in psychotherapy, it is worth noting that even very simple research into long term therapy may be quite revealing. Von Bendeck (1992) simply asked analysts about their attitudes to and predictions for their patients at the start of therapy and 1 year later. Importantly she found that analysts had often significantly changed their views over a year but tended to deny that they had done so, 'forgetting' their earlier judgments.

Now it is necessary to turn to the second group of problems, which are methodological and practical. Even when research into psychotherapy is believed to be both possible in principle and valuable in practice it must still face a number of more or less intractable methodological difficulties, which beset all psychotherapy research. These difficulties include: problems in setting clear criteria for outcome, unclarity about the likely natural history of the heterogeneous collection of 'diseases' treated, and difficulties in generalizing variables because of the interpreted or hermeneutic nature of the issues being judged. All these methodological difficulties, which are encountered when doing research into any form of therapy become more pressing in the case of long-term therapy.

In long-term therapy research, difficulties arising from the lack of predetermined focus and the extreme heterogeneity of conditions treated are at their most severe. In addition, the emphasis on spontaneity in the analytic endeavour makes manualization of therapy difficult. It is clear that, even within one school and consequently considering only the personal style or idiom of the psychoanalyst (Bolas, 1989), vastly different styles of therapy may be being done.

Added to the intensification of these general methodological difficulties are particular practical problems in researching long-term therapy. The long periods of time that any prospective study must run for in order to gather final outcome data creates yet another set of difficulties. Another group of practical difficulties centres round problems in handling the potentially huge quantities of data which it may seem relevant to gather.

PRACTICAL UPSHOTS OF THE SPECIAL DIFFICULTIES OF DOING RESEARCH INTO LONG-TERM THERAPY

The combination of the various factors listed above has led to a number of negative consequences for ongoing research endeavour in relation to long-term therapy. Attempts to simplify the difficulties that attend research into the kind of therapy which is done by most practicing long-term therapists has led to the use by researchers of models of therapy which are distant from the way that long-term therapists have of working. This means that long-term therapists find most of the research done to be largely irrelevant to their concerns. Morrow-Bradley and Elliott (1986) surveyed practicing psychotherapists concerning their use of research findings. They found that therapists wanted research that concentrated on the kind of therapy which they actually did and on the sort of patients they commonly treated. Therapists were also keen to read research which had good external validity (i.e. relevance and credibility) and which did not seek to over-simplify the nature of the therapy process. These findings were particularly marked for psychodynamic psychotherapists.

Another negative consequence of the above factors for research results from the way in which the cultural separation of research from practice, in the case of long-term therapy, has lead to the growth of research chiefly within the mainstream of psychiatric and psychological research. Consequently the research paradigms and standards used are often defined by a community mainly used to conducting research in fields other than psychotherapy and this may not always be appropriate. A good example of the effect of this state of affairs is the rise and fall of the comparative drug trial methodology as the chief paradigm for psycho-therapy outcome research. By analogy with drug trials, different therapies were compared either with placebo conditions or with each other. The results of these kinds of studies proved disappointingly inconclusive and it came to be realized that this might be a consequence of the inappro-priateness of the paradigm (Wilkins, 1986), chiefly because it had obscured some crucial disanalogies between drug treatment and psychotherapy. For example, it turned out to be very difficult to construct a suitably convincing placebo therapy which did not contain any of the active ingredients of ordinary therapy (Parloff, 1986). Even the tempting strategy of using comparative trials to get round the placebo problems proved difficult for a number of reasons but not least because of uncertainties about the true differences between, and the possible heterogeneity within, the different varieties of therapy being compared (Kazdin, 1986).

A final negative consequence of the factors which make research into long-term therapy difficult has been that, especially in England, the practical difficulties and the low esteem that, paradoxically, both the research community and the psychotherapeutic community hold for research into long-term therapy, has led to a failure to establish ongoing research programmes. This is a serious problem because such programmes may well hold the only possible solution to the long cycle times that research into long-term therapy requires.

Not surprisingly, given the difficulties outlined above, there is little research into long-term therapy. Most researchers into psychotherapy, pressed by budget constraints, informed by drug research paradigms and searching for a way out of seemingly intractable methodological problems, have adopted a combination of manoeuvres for dealing with their diffi-culties. These have almost always included shortening the length of therapy researched so that, perforce, most research is into brief treatments. Additional manoeuvres have included manualizing therapy, setting a predetermined focus and selecting patients who fall into some kind of homogeneous group. Such expediencies have seemed necessary to allow research to be done at all. However, they have led to research being condlcted on entities which are not clearly comparable with long-term therapy.

The consequences of the creation of brief dynamic therapies for research or other reasons are taken up by Altschuler (1989), who argues persuasively that the necessary modifications have generated forms of therapy which, despite their different brand names, are in many ways very similar, or at least share many common factors. He suggests that, in view of this, the finding that all therapies have roughly equal effectiveness is not surprising. He suggests that long-term therapy is likely to contain ingredients other than length which might give it a specific and different effect, particularly in its attention to resistance and unconscious conflict. He concludes gloomily that in future, because research into long-term therapy is unlikely, we are unlikely to be able to investigate these phenomena.

RECENT RESEARCH ON LONG-TERM THERAPY

It is worth reviewing recent work on long-term therapy to see which of the pressing questions outlined have begun to be investigated.

There are very few outcome studies for long-term therapy. Bachrach et al. (1991) review all of the studies of psychoanalysis conducted since 1917. They discuss the substantial methodological limitations in all these studies. In effect, all these studies—even, in some ways, the vast Meninger project—amount to audits of the practice of institutions involved in doing psychoanalysis. All do show that patients who are less severely ill benefit from psychoanalytic treatment but in the studies reviewed this judgment of benefit was often made by the treating analyst whose judgment might potentially be distorted by the hopes and fears of the treatment situation.

There are a considerably greater number of studies into long-term therapy which is short of psychoanalysis. Among them are two which are important partly for methodological innovations and partly as demonstrations that such research is possible despite the many difficulties. The study by Rosser et al. (1987) is important because it showed the value of using cost–benefit methods to evaluate the outcome of treatment at the Cassel hospital. These methods were able to demonstrate the cost-effectiveness of treatment at the hospital which, given the current economic climate, is a politically important finding. The research done on longer term outpatient psychotherapy with borderline patients by Stevenson and Meares (1992) demonstrates the efficacy of a relatively long-term intervention. Also, it shows that it is possible to do convincing research with a group of patients who are often difficult to persuade to do research evaluations and who can be very hard to follow-up.

Despite this promising work there are still glaring deficiencies in the field. Although there is some encouraging evidence in favour of a dose—response relationship (Howard et al., 1986), there are still no comparative studies which show that longer term, more intensive interventions are capable of producing differential benefits over shorter term, less intensive ones. Since many practitioners are convinced that long-term intensive treatment does produce such differential benefits, the lack of studies either positive or negative in this area is quite striking.

Because long-term therapy produces a stable setting over a long time period, it offers the maximum chance to measure change and process in a relationship. There is currently a great deal of interest in process work, although this has been mainly applied to brief therapies. A danger of pure process work is that beautifully measured and investigated processes are studied in some detail but they turn out to be unrelated to final outcome. However, the value of pure process work is that it may allow tests of the underlying psychological theory of therapy to be devised.

The work of the Mount Zion group in testing different psychoanalytic hypotheses about resistance (Weiss & Sampson, 1986), serves as a very good example. Independent judges were used to rate segments of a psychoanalysis for evidence both of anxiety and of the emergence of repassed ideas and they used the relation of anxiety to the emergence of such material to test two different hypotheses about the nature of the patient's unconscious engagement in the therapy process. Because the theoretical basis of the Mount Zion group's work is so clearly defined, they are able to make strong predictions about the progress of therapy in relation to the therapist's behaviour. Thus their work not only stands as a test of psychodynamic theory, it can also easily be linked to outcome.

In England, Moran and Fonagy (1987) used the session records of a psychoanalytic case to rate the presence of a set of analytic themes. Because the patient was a brittle diabetic who required intensive monitoring of urine glucose level, they would use an index of weekly glycosuria as a dependent intermediate outcome measure. They then used time series analysis to show that the emergence within the sessions of two particular themes related to conflict was correlated with short-term beneficial changes in diabetic control.

PROBLEMS WITH THE RESEARCH DONE SO FAR

The glaring problem which besets the research work done on questions of outcome is the lack of comparative studies of any sort. Until such studies are done there can be no answer to questions of differential efficacy, either those which arise within long-term therapy (such as issues of intensity) or where long-term therapy 'competes' with other therapeutic modalities.

The lack of work in this area is not accidental. It is worth considering the sort of difficulties that would be involved in conducting a comparative trial looking, for example, at differential efficacy between short- and long-term therapy. There would, for a start, be a considerable difficulty in deciding at what time point or points commensurable assessments of outcome between the two treatments could be made. Another problem would be drop-out during treatment and failure to attend follow-up appointments. These would almost certainly be different in each group (possibly in some systematic way), introducing inevitable and possibly fatal difficulties in statistical interpretation.

Another problem, which is more generally encountered in comparative studies of psychotherapies, is the preference most patients will have for one form of treatment over another. Patients cannot be blinded to their treatment group assignment as they would be in a drug trial and in a psychological treatment such psychological factors can be overwhelming. Parry (1992) goes so far as to argue that this is a conclusive argument against the use of random assignment based comparative trials in the evaluation of psychotherapies.

To this must be added a general difficulty for all comparative psycho-therapy research—the large group sizes needed. This must be particularly acute when one treatment is very long and labour intensive. In a com-parative study, the final size of the two treatment groups after drop-outs would probably need to be very large indeed (possibly 50–100 completers in each group), to give a reasonable likelihood of a statistically significant result.

A final difficulty which needs consideration is the delineation of outcomes which are relevant to long-term therapy. In long-term therapy, the lack of predetermined focus makes target problems often (but not always) difficult to specify and the length of time over which therapy takes place means that the life circumstances of the patient are unlikely to remain constant irrespective of the effects of treatment. These difficulties pose researchers the challenge of specifying outcomes which are more structural for the individual and less likely to be situation specific. Importantly, such outcomes must also be relevant to the considerations of practicing clinicians.

Turning to the work on the process of psychotherapy, one problem with this work, at least as it has been done so far, lies in the way that the questions asked and theoretical models used tend either to be linked very specifically to the work and theoretical position of a close knit possibly too idiosyncratic group or, conversely, they have been attempts at very general descriptions linked to psychological models of change not derived explicitly from psychodynamic origins. A good example of such a general model is the Assimulation model (Stiles *et al.*, 1992), which looks at the emergence of problematic themes and their resolution in therapy in ways which derive partly from Piaget's ideas. Although it is a valuable and general way of describing change in therapy which is independent of affiliation to a single theoretical school, its theoretical base is obscure to most psychodynamic psychotherapists. Consequently the ways in which the assimilation model will mesh with the kinds of process ideas which psychodynamic long-term therapists tend to use in thinking about their clinical work are not entirely clear.

Another broad process concept which has won general acceptance is that of the therapeutic alliance. However, even in this case some vital work of meshing and translation between research concepts and psychodynamic concepts needs doing. Safran et al.'s (1990) work on rupture and repair in the therapeutic alliance might be more easily recognized as relevant, if it were recognized that they were speaking in part of what many therapists would recognize as variations in the level of rapport. Malan (1979) had already pointed out the vital importance to success in therapy of tracking the level of rapport in the session.

The lack of meshing limits the apparent usefulness of many of the general models developed for practitioners in the psychodynamic field who, already skeptical of research, find that when they do come across the literature, the models and assumptions being tested seem irrelevant to them. This difficulty is augmented by the split between English and American psychoanalytic traditions. Because most of the research litera-ture is American and draws upon egopsychology, British therapists in the object relations tradition find it remote and jargonized.

Another problem with much current process work in long-term therapy is the frequent lack of links between process and outcome. Such links are vital if findings about process variables are to have practical upshots for therapeutic efficacy. The work of Luborsky and his team (Crits-Christoph et al., 1988; Crits Christoph et al., 1993), which shows that interventions which target the patient's central psychodynamic problem (the Core Conflictual Relationship Theme) are effective in building the thera-peutic alliance and in promoting beneficial outcome, is a shining exception.

Linking work on process with work on outcome must be an important feature of research, which has implications for changing practice, but particular care is needed when making these links in long-term therapy. It is not sufficient to demonstrate favourable intermediate results and to assume that these will necessarily accumulate to produce a favourable long-term outcome. Joseph (1989) has argued that if analysts concentrate on favourable or unfavourable results during the session-to-session work of analysis, they risk perturbing the analytic process to the detriment of long-term change.

Another difficulty related to the links between process and outcome lies in the way that process research may focus on variables which lie outside the control of the therapist. This arises from the way that process work often concentrates on transcript or video or audio tape data. The advantage of such access methods lies in the way that they bypass the difficulties of inaccurate, biased or highly interpreted reporting by the participants in therapy. A disadvantage is that, because it is possible to review the session in considerable detail, it can be tempting to measure elements of process between the participants which are at a level of subtlety that could never be achieved by a therapist in real time. From the point of view of pure research this is not problematic but it is not good, if research is going to link well with outcomes, because it will in consequence be difficult to construct an intervention in which one would train the therapist to alter their behaviour. This is a limitation of important and valuable process measures such as ratings of therapist verbal quality (Rice & Kerr, 1986), which looks at the therapist's tone of voice and, to a lesser extent, the verbal response mode system (Stiles, 1986), which looks at the type of verbal interventions therapists make—whether questions, interpretations, confirmations, disclosures, etc.

A final difficulty with process work is that while it may draw on psycho-dynamic theory developed in long-term therapy, it is often tested either on brief therapies or on only one long-term therapy. This is because of the potentially massive quantities of data which are generated by long-term therapies.

POSSIBLE FUTURE RESEARCH AVENUES

Armed wiltn an appreciation of current work in the field and its difficulties and limitations, it may now be possible to suggest some future avenues for research which might be capable of dealing with some of the difficulties outlined.

In relation to questions of differential efficacy and outcome, it would seem that two main areas need to be tackled.

1. Some way of overcoming the methodological difficulties which beset comparative trials needs to be devised.
2. Relevant outcome measures need to be specified for evaluating such comparisons as are undertaken.

It is possible that cost–benefit studies may offer an approach to some of the methodological problems outlined for comparative studies. Cost–benefit studies have the additional advantage of providing economic arguments to advocates of long-term therapy. In these studies the total costs of each treatment would be offset against the total savings or benefits which accrued from it. This procedure would, from the start, factor in the extra effort of an intensive treatment as an extra input cost. There is now more interest in cost-effectiveness as an outcome measure for therapy, and Krupnick and Pincus (1992) in their comprehensive review of this topic, call for this kind of study.

Cost–benefit studies are obviously relevant to those therapists who are also service providers within the state sector and are consequently faced with the distribution of scarce resources among a large pool of need. Additionally, the concept of cost–benefit could be expanded somewhat to include more general notions of enhanced capacity for living and working. If this was done it would be possible to reframe much of the current debate within the therapy world about the demarcations of different kinds of therapies on the basis of length intensity and therapeutic technique as a debate really founded on implicit questions of differential cost–benefit. The concept of the greater value of a deeper, longer and more intensive therapeutic experience, whether as a training requirement or a treatment recommendation, trades implicitly on the idea that the extra expended effort and investment will pay off in the end.

Relevant outcome measures will in the main need to draw on psychodynamic theory. A particularly interesting group of measures are those which can be locked onto the hypothesized psychodynamic difficulties of the patient as opposed to measuring only situation specific behavioural difficulties.

The work of Luborsky et al. (1986), which extracts a central problematic interpersonal relationship pattern (which they term the Core Conflictual Relationship Theme, CCRT) by a variety of methods—including examining transcripts of narrative episodes in psychotherapy sessions—seems potentially capable of yielding an individualized outcome measure.

The Target Problem Procedure concept used in Cognitive Analytic Therapy (CAT) (Ryle, 1990) is a similar although lesss restricted concept in which a patient's repertoire of problematic and maladaptive procedures for goal-directed action are systematically catalogued by therapist and patient. The move from overt problems, known as target problems, to a more psychodynamic (or in CAT terms procedural) understanding is made explicit in the change of term from target problem to target problem procedure. Already as part of this method's therapeutic procedure the patient and therapist seek to chart success in altering these patterns.

The advantage of each of these methods is that while they are individualized, they are also psychodynamically informed. A disadvantage that they all share is that they are to some extent inferred measures and rely on the accuracy of their judges. For this reason, the obvious way of making up in credibility what they lack in self-evident transparency is the use of multiple judges and the calculation of interrater reliabilities. This is a time-consuming exercise and requires the sort of established research programmes which do not yet exist in some cases.

There are also non-psychoanalytic sources of measures which seem promising. The concept of coping, which has developed out of the work of Brown and Harris (1978) on life events and difficulties, is an example. A measure of coping style has now been developed by this group (Bifulco & Brown, unpublished). It classifies coping into common sense categories, such as down-playing, self-blame, emotional expression and denial. Coping style has been shown to influence the risk of developing depression after a life event and consequently it would seem likely that beneficial changes in coping style might be a possible desired outcome of a successful treatment aimed at providing resilience to depression.

In relation to research into the therapy process the issues which need tackling are:

1. the quantities of data;
2. relevance to the concerns of clinicians;
3. measuring variables which clinicians could be trained to alter.

In relation to dealing with the large amount of process data which long-term therapy potentially generates, Katchele's (1992) work on creating a summary of a long-therapy, and also the similar work of Fonagy and Tallandini-Shallice (1993), represent starts in this direction which need following up.

Apart from the data overload problem, another important aspect of process research to concentrate on must centre round making it relevant to the current concerns of clinicians. Most interest has centred on the process detected in the patient and here there has been considerable success, particularly in the area of the therapeutic alliance (see, for example, Marmar et al., (1986). Another measure which seems to tap an important dimension of the therapy process is the experiencing scale (Klein et al., 1986) which tries to evaluate the level of the patient's involvement in describing their experience, assigning levels of increasing involvement ranging from account of external and seemingly unrelated events, through reactions to relevant events, towards emotionally involved accounts of current feelings and processes. However, these variables again cross the boundaries of different schools and so can seem difficult for some practitioners to assimilate. Therefore, in addition, measures which tap patient material in ways that are more specifically relevant to the areas of psychodynamic theorizing used by long-term therapists (e.g. regression, depression and mourning, projective identification and splitting), need to be devised.

Process research that focuses on variables which a therapist could be trained to alter, probably means concentrating on process variables that centre round the intentional actions of the therapist and relate to their likely or actual aim during the session. Such variables are most easily delineated within manualized therapies, where the theoretically desirable aims of the therapist are in any case prespecified. However, even non-manualized therapies set therapist aims and expectations (e.g. be non-directive, make transference interpretations) and it is probably these which should be measured.

CONCLUSION: MAKING A CASE FOR RESEARCH INTO LONG-TERM THERAPY

The reasons for a lack of research into long-term therapy have been outlined and the difficulties of doing research into long-term therapy have been discussed. The methodological difficulties are formidable but it also seems that the cultural divorce between long-term therapists and researchers may have combined synergistically with the methodological problems to unduly limit research thinking in the field. This is especially worrying because research in this area is a particularly pressing need, if a case for continuing public funding of long-term therapy is to be made. There is a risk that if it is not made, long-term therapy will vanish from public health provision. If this occurs, then patients with personality disorders (a group of patients who figure amongst the most vulnerable and easily neglectable in the mental health care system) will be the chief losers.

If there is sufficient reason to research long-term therapy, then the most pressing obstacles obstructing research will need to be removed. The ongoing distrust between the research community and the mainly psychodynamic psychotherapy community who are engaged in doing therapy, is the main obstacle. Another obstacle is the necessary funding. Doing research into long-term therapy will require considerable expenditure of resources over a long period of time. Such long-term programmes cannot depend only on enthusiastic individual or isolated students doing research projects as part of a course. Research programmes need to arise out of a collective decision that they should be engaged in.

REFERENCES

Altschuler, K.Z. (1989). Will the psychotherapies yield differential results?: a look at assumptions in therapy trials. *American Journal of Psychotherapy* **18**(3), 310–320.
Andrews, G. (1993). The essential psychotherapies. *British Journal of Psychiatry* **162**, 447–451.
Bachrach, H.M., Galatzer-Levy, R., Skolnikoff, A. & Waldron, S. (1991). On the efficacy of psychoanalysis. *Journal of the American Psychoanalytic Association* **39**, 871–916.
Bell, D. (1992). Hysteria—a contemporary Kleinian perspective. *British Journal of Psychotherapy* **9**(2), 169–180.

von Benedek, L. (1992). The mental activity of the psychoanalyst. *Psychotherapy Research* **2**(1), 63–72.

Bifulco, A. & Brown G.W. *Coping and the Onset of Clinical Depression: 1 The Coping Interview Schedule.* Unpublished manuscript.

Bolas, C. (1989). *The Forces of Destiny.* London: Free Associations.

Brown, G.W. & Harris, T.O. (1978). *Social Origins of Depression: a Study of Psychiatric Disorder in Women.* London: Tavistock.

Crits-Christoph, P., Barber, J.P. & Kurcias, J.S (1993). The accuracy of therapists' interpretations and the development of the therapeutic alliance. *Psychotherapy Research* **3**(1), 25–35.

Crits-Christoph, P., Cooper, A. & Luborsky, L. (1988). The accuracy of therapists interpretations and the outcome of dynamic psychotherapy. *The Journal of Consulting and Clinical Psychology* **56**(4), 490–495.

Feldman, M.M. & Taylor, D. (1980). *Some Problems in Psychotherapy Research.* Paper presented to the 1980 Cassel Conference.

Fonagy P. & Tallandini-Shallice, M. (1993). Problems of psychoanalytic research in practice. *The Bulletin of the Anna Freud Center* **16**(1), 5–22.

Freud, S. (1923). The ego and the id. In *Standard Edition of the Complete Psychological Works of Sigmund Freud,* vol. 19. Tr. Strachey, J.

Gardner S. (1993). Irrationality and the Philosophy of Psychoanalysis. Cambridge: Cambridge University Press.

Howard, K.I., Kopta, S.M., Krause M.S. & Orlinsky, D.E. (1986). The dose–effect relationship in psychotherapy. *American Psychologist* **41**(2), 159–164.

Joseph, B. (1989). Psychic change and the psychoanalytic process. In *Psychic Equilibrium and Psychic Change, Selected Papers of Betty Joseph* (M. Feldman & E.B. Spillius, eds). London: Routledge.

Katchele (1992). Narration and observation in psychotherapy research: reporting on a 20-year long journey from qualitative case reports to quantitative studies on the psychoanalytic process. *Psychotherapy Research* **2**(1), 1–15

Kazdin, A.E. (1986). Comparative outcome studies of psychotherapy: methodological issues and strategies. *Journal of Consulting and Clinical Psychology* **54**(1), 95–105.

Klein, M.H., Mathieu-Coughlan, P. & Kiesler, D.J. (1986). The experiencing scales. In *The Psychotherapeutic Process: a Research Handbook* (L.S. Greenberg & W.M. Pinsof, eds). New York: Guilford.

Krupnick, J.L. & Pincus, H.A. (1992). The cost–effectiveness of psychotherapy: a plan for research. *American Journal of Psychotherapy* **149**(10), 1295–1305.

Luborsky, L., Crits-Christoph, P. & Mellon, J. (1986). Advent of objective measures of the transference concept. *Journal of Consulting and Clinical Psychology* **54**, 39–47.

Malan, D.H. (1979). *Individual Psychotherapy and the Science of Psychodynamics.* London: Butterworth.

Marmar, C. R., Horowitz, M.J., Weiss, D.S. & Marziali, E. (1986). The development of the therapeutic alliance rating system. In *The Psychotherapeutic Process: a Research Handbook* (L.S. Greenberg & W.M. Pinsof, eds). New York: Guilford.

Moran, G.S. & Fonagy, P. (1987). Psychoanalysis and diabetes: an exploration of single case study methodology. *British Journal of Medical Psychology* **60**, 310–348.

Morrow-Bradley, C. & Elliott, R. (1986). Utilization of psychotherapy research by practicing psychotherapists. *American Psychologist* **41**(2), 188–197.

Norton, K. (1992). A culture of enquiry—its preservation or loss. *Therapeutic Communities* **13**(1), 3–25.

Parloff, M.B. (1986). Placebo controls in psychotherapy research: a sine qua non or a placebo for research problems. *Journal of Consulting and Clinical Psychology* **54**(1), 79–87.

Parry, G. (1992). Improving psychotherapy services: applications of research, audit and evaluation. *British Journal of Clinical Psychology* **31**, 3–19.

Rice, L.N. & Kerr G.P. (1986). Measures of client and therapist vocal quality. In *The Psychotherapeutic Process: a Research Handbook* (L.S. Greenberg & W.M. Pinsof, eds). New York: Guilford.

Rosser, R.M., Birch, S., Bond, H., Denford, J. & Schachter, J. (1987). Five-year follow-up of patients treated with in-patient psychotherapy at the Cassel hospital for nervous disorders. *Journal of the Royal Society of Medicine.* **80**, 549–555.

Ryle, A. (1990). *Cognitive Analytic Therapy: Active Participation in Change.* London: Wiley.

Safran, J.D., Mc Main, S., Crocker, P. & Murray, P. (1990), Therapeutic alliance rupture as a therapy event for empirical investigation. *Psychotherapy* **27**(2), 154–165.

Scott, A. (1993). Response to Anthony Ryle. *British Journal of Psychotherapy* **10**(1), 93–96.

Stevenson, J. & Meares, R. (1992). An outcome study of psychotherapy for patients with borderline personality disorder. *American Journal of Psychiatry* **149**(3), 358–362.

Stiles, W.B. (1986). Development of a taxonomy of verbal response modes. In *The Psychotherapeutic Process: a Research Handbook* (L.S. Greenberg & W.M. Pinsof, eds). New York: Guilford.

Stiles, W.B., Meshot, C.M., Anderson, T.M. & Sloan Jr, W.W. (1992). Assimilation of problematic experiences. The case of John Jones. *Psychotherapy Research* **2**(2), 81–101.

Weiss, J. & Sampson, H. (1986). *The Psychoanalytic Process Theory, Clinical Observations and Empirical Research.* New York: Guilford Press.

Wilkins, W. (1986). Placebo problems in psychotherapy research. *American Psychologist* **41**(5), 551–556.

Wollheim, R. & Hopkins, J. (1982). *Philosophical Essays on Freud.* Cambridge: Cambridge University Press.

9 Demonstrating Specific Effects in Cognitive and Behavioural Therapy

PAUL M. SALKOVSKIS
Department of Psychiatry, University of Oxford, UK

ABSTRACT Cognitive-behavioural teatment has made enormous advances in terms of improving both the scope of clinical treatment and the speed with which clinically significant results can be obtained. This process involves a reciprocal relationship between clinical treatment on the one hand and experimental studies and therapy research on the other. Such a relationship is important because cognitive-behavioural therapy is directly based on the cognitive theory of emotion; key implications of this theory include its normalizing (depathologizing) influence and the fact that different types of psychological problem are maintained by different and specific mechanisms, which effective treatment needs to address. According to this view, a good psychotherapeutic relationship is necessary but not sufficient for developing the most effective therapy. The parallel evolution of cognitive-behavioural theory and therapy in panic disorder is described as an example of rapid evolution of psychotherapy. The importance of research at different stages of development of such psychotherapy is emphasized; this not only applies to the validation of treatment efficacy, but also the actual development of treatment techniques. It is concluded that previous judgements of psychotherapy equivalence are no longer valid, and are likely to hamper the further development of more effective therapy.

INTRODUCTION

The most startling feature of the cognitive-behavioural approach to psychological problems is not the parallel development both of more effective and more rapidly implemented therapies. What is truly startling is the speed with which these developments have occurred (and continue to occur) and the rapid evolutionary progress of both theory and practice. Nevertheless, cognitive-behavioural theories (CBT) are recognized by their

Correspondence address: University of Oxford Dept. of Psychiatry, Warneford Hospital, Oxford OX3 7JX, UK.

Research Foundations for Psychotherapy Practice. Edited by M. Aveline and D. A. Shapiro.
Copyright © 1995, Mental Health Foundation and Individual Contributors.
Published 1995 by John Wiley & Sons Ltd

advocates as relatively poor approximations to complex clinical problems, which will certainly be overtaken by new developments and revised or replaced in the near future. Paradoxically, it is this recognition of the incomplete nature of the theoretical basis of CBT that represents the strength and vigour of the approach. Cognitive-behavioural research involves generalizing from experimentally verified general processes in psychopathology to the specific clinical manifestations, which are hypothesized to be involved in the maintenance of particular psychological problems. Modifications of the identified maintaining factors are therefore crucial to the efficient conduct of therapy research. Therapy research then may feed back into experimental research as part of a reciprocal relationship. It is the combination of theoretical approaches founded on empirical studies of process and outcome, together with a willingness to modify theory and therapy, which means that CBT represents a real hope for progress in psychotherapy. No other school of psychotherapy is showing such a rapid rate of advance. In this chapter, I will describe some specific examples of the way in which cognitive-behavioural research has resulted in not only a greater understanding, but also in treatment becoming more effective, quicker and more complete.

To begin with a now trivial example; in 1958, Joseph Wolpe described a new and apparently effective therapy for phobic problems, based on the theory of 'reciprocal inhibition' derived from animal experiments (Wolpe, 1958). Systematic desensitization, as the new treatment was called, required up to 80 treatment sessions, and significant but relatively modest improvements were described in the earliest outcome studies. This development met with considerable hostility from the psychotherapy establishment. Breger and McGaugh (1965) went as far as to make a specific prediction; that 'symptomatic treatment' (as they believed behaviour therapy to be) would result in 'symptom substitution'. Thirty-five years later, systematically conducted clinical and research work has resulted in the original theory and therapy being refined then replaced, the replacement refined, replaced and so on. The more recent theoretical view has been that *exposure to feared stimuli without avoidance or escape taking place* results in habituation (progressive and systematic decrease) of the anxiety response. The treatment itself has rapidly evolved to the point where therapy can now be conducted in a single session. Seventy per cent of people treated in this way show full remission of their problem (based on combined conservative clinical and psychometric criteria) (Ost, 1988; Ost *et al.*, 1991). Gains are maintained to at least 1-year follow-up. All this occurs without a scrap of 'symptom substitution'. Figure 9.1 shows the effects of one-session treatment on behavioural avoidance test and rating of phobic severity (degree of disability associated with the phobia) as compared to a significantly *greater* amount of self-exposure done using a self-help manual based on the one-session treatment.

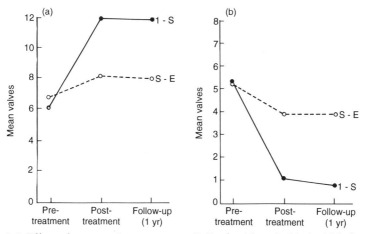

Figure 9.1 Effect of one-session treatment (1-S) of spider phobia in (a) behavioural avoidance test and (b) rating of phobic severity compared with self-exposure(S-E)

Quite apart from showing the extraordinary effectiveness of this briefest of treatments, this study has a further important implication. The comparison (self-exposure) group had significantly *more* exposure during the treatment period than the one-session group but improved significantly *less*. Together with the absence of any significant correlation between the amount of exposure and degree of clinical improvement, it seems likely that exposure-induced habituation *per se* probably does not account for the improvement observed. That is, the most recent *theoretical* descendant of Wolpe's reciprocal inhibition hypothesis is now being seriously challenged by research data. These data and other research and development work are most consistent with alternative cognitive theories, which assert that exposure may be effective because of belief changes such as disconfirmation of the expectation of threat (Salkovskis, 1991). Doubts about the validity of exposure/habituation theory open the way to yet further developments in the understanding of specific phobias, and therefore to yet more effective therapy. The psychotherapeutic approach which is currently labelled 'cognitive-behavioural' can thus be seen as the next stage in a process of iteration, i.e. progress.

CONDUCTING PSYCHOTHERAPY RESEARCH: PROBLEMS AND AN 'HOURGLASS' MODEL

There are two main types of problems in psychotherapy research. First, and most obvious, there are technical problems. For example, difficulty in obtaining subjects for studies, equipment not working, missing data and so on. In essence, these types of problems are soluble, often by putting greater resources (time) into the study. The second type of problem is conceptual, and concerns the stubborn refusal of data to conform to the expectations and hypotheses of the researcher. This type of problem can arise for at least two reasons.

1. The methodology used is inadequate, resulting in 'false negative' results (the so-called Type I error).
2. The least often considered reason is that the *hypothesis being investigated is incorrect*.

Clearly, it can be difficult to disentangle these two possible reasons; it is also crucially important.

The overall structure and sequencing of clinically relevant research can be set up in ways which minimize these types of issues. The metaphor of an 'hourglass model' can be applied. The foundations of research are usually a clinical problem which is of significant concern to large number of practitioners (how to make treatment better, faster, more efficient or even effective where it is found to be ineffective). Potential answers to such problems are often some combination of an informed clinical hunch and a theoretical framework for harnessing such hunches in experimental studies of processes involved and then working with the clinical problem through new variants of therapy. Identification of a problem and its potential theoretical solution leads to exploratory studies in which technical standards of design and implementation are relatively relaxed. At this stage, resources are often very limited (often small numbers or solitary clinicians engaged in routine service provision), sample sizes are small and measures unsubtle. Experimental designs are flexible, often involving quasi-experimental and single case designs with *ad hoc* measures. By the same token, effect sizes often need to be large to emerge from this 'messy' research; usually, this carries the advantage that effects are more likely to be of sufficient size to be clinically relevant. This kind of exploratory analysis then allows a narrowed focus on key effects, progressively moving towards the more rarefied research which is required by the 'academically respectable' journals. This more narrowly focused research requires a fuller range of control groups, more stringent measurement and statistical techniques and careful specification of samples to ensure replicability. Theoretical and practical problems are raised and dealt with. Often, there are important interactions between theoretical, experimental and clinical developments. This type of research is almost invariably better resourced (usually from grant funding), in that the design features required of such research can seldom be justified in routine clinical practice. Paradoxically, the factors which disambiguate the experimental design (such as careful subject selection) may have the effect of limiting the clinical applicability of the research. However, the foundations of the research in the earlier 'messy' phase can overcome such reservations. Nevertheless, the final stage of this type of research programme is again to broaden out the scope of the research, applying the findings to a less selected sample (or to samples selected on a different basis), using different therapists, reducing the amount of time required to conduct therapy and so on. This is the stage when the cycle begins again, as the limitations of the new approach become apparent in clinical practice. Note that this overall 'hourglass' strategy is one which tends to ensure strong effects in the first instance, so that if at the more refined stage effects are not noted, then one may be reasonably confident that negative findings are truly negative if the appropriate measures have been taken to ensure treatment and measurement integrity. At this stage, measures (particularly results of experimental studies on specific psychopathology) relevant to psychological processes believed to be important in the maintenance of a disorder become particularly important in refining the focus of treatment.

It is sometimes argued by some schools of psychotherapy that the effects of treatment are not measurable, or that the process of measurement would irreparably damage the therapeutic effort. Developments in the field have now overtaken such excuses; without research there is no chance of progress, and no treatment is sufficiently well developed to be immune to the need for progress. This is true on both scientific and ethical grounds. The potential for harm in psychotherapy is now too clear to ignore; a treatment of unknowable effectiveness but known potential to cause harm cannot be supported. Add to this that many treatment approaches are now of *known* effectiveness. An important requirement for clinicians and researchers alike is that of being able to accept contrary findings and abandon cherished positions, i.e. to be truly responsive to research findings.

MAKING PROGRESS THROUGH RESEARCH INTO 'FAILURES'

In the cognitive-behavioural approach, an important set of examples of the way in which the 'hourglass' sequence has been triggered and real progress has occurred comes from attempts to help people whose problems do not respond (or respond to only a limited degree) to the best available existing treatments. This is, of course, the origin of current work on problems such as obsessions, drug-resistant schizophrenia, with people who repeatedly attempt suicide and so on. The cognitive-behavioural response to areas where the effectiveness of treatment is limited is to assume that the specific factors involved in the maintenance of these problems have not been correctly identified. Without identification of such factors, individualized formulations will be inadequate and treatment cannot be targeted within the broader context of the therapeutic relationship.

Thus, intrinsic to the development process in cognitive-behavioural therapy has been the identification, evaluation and empirical demonstration of specific factors hypothesized to be involved in the maintenance of particular types of psychological problem. If such factors are identified, further research into their relationship to both specific effects and effectiveness of therapy can then follow. Results of this research can then be applied to the refinement of theory and therapy. *Failure* to find effects should inevitably lead to the search for new hypotheses and therefore new lines of therapy research. The results of research therefore influence treatment and vice versa as part of the evolving reciprocal relationship described above.

In the rest of this chapter, the general theoretical foundations of the cognitive-behavioural approach will be outlined, then it will be shown how this theory has been applied to the development of cognitive therapy for panic attacks. This work began as an attempt to deal with 'failures' in the otherwise effective treatment of agoraphobia by exposure therapies. Once the probable effectiveness of a clinically applicable treatment was established in 'messy' clinical research, a more sophisticated programme of focused experimental and treatment research was used to extend these findings. This has then been broadened out to increase the clinical applicability of the treatment studies. In panic, the evolution from no treatment to rapid, highly effective and clinically generalizable treatment has taken rather less than 10 years; the research strategy which has contributed to this quickening of progress will be described below in some detail. Finally, the apparently even more 'hopeless' example of people who repeatedly attempt suicide is also described as an illustration of the way in which cognitive specificity transcends the severity of impairment.

THE GENERAL COGNITIVE THEORY OF EMOTIONS AND EMOTIONAL DISORDERS

It is interesting to note that the response to a contrary research finding prompted the development of a key strand of present-day cognitive therapy. Aaron T. Beck was a committed psychoanalyst who decided that he wished to conduct research into motivational aspects of Freudian theory. In a study designed to investigate the content of dreams, the results he obtained were completely contrary to the psychoanalytic hypothesis which he was testing. He found that the content of dreams tended to reflect daytime concerns. This finding appears to have caused Beck some initial confusion and even distress, but the experience led him to consider more carefully the way in which *conscious meaning* may be directly involved in emotional responding. Beck's (1976) cognitive theory thus proposes that emotions, such as anxiety and depression, are a result of the immediate appraisal of the situation in which the emotion arises. Thus, it is not the situation *per se* which produces emotions but rather what a person thinks and believes about the situation. The specific emotion experienced depends on the particular idiosyncratic *meaning* attached to the situation, which in turn arises from learned attitudes and assumptions learned earlier in the person's life interacting with present circumstances. In general, *anxiety* is associated with thoughts of *danger or threat*; this perceived danger may be physical (e.g. thinking that one is about to be physically attacked) or social (e.g. thinking that one is about to be publicly ridiculed). Feelings of depression or despondency relate to thoughts of *loss*, including 'real' loss (e.g. loss of a loved one) and abstract loss (e.g. loss of self-respect or ability). Anger is the result of perceived *unfairness or the breaking of one's personal rules* by another, and so on. The key factor is what the person *perceives* rather than any notion of 'truth'. Thus, someone who (erroneously) is convinced that he is about to die of a heart attack will experience the same terror as the person who is convinced that a masked raider is about to shoot him (or the person who is *actually* dying of a heart attack). Note that anxiety would be expected whether or not the threat is 'real'; the gun might be plastic, and the incident a practical joke, but the emotions experienced relate directly to one's perceptions and beliefs about the situation.

The central notion of the cognitive hypothesis is thus highly normalizing and, if used correctly, empowering for people suffering from emotional problems. Therapy seeks to convey that anyone who found themselves in this situation with similar experience and beliefs would experience similar emotions. The cognitive view of emotional 'disorders' is that the experience of emotion is not the problem; rather, it is the *persistence* of extreme and unhelpful beliefs in ways which interfere excessively and inappropriately with the person's ability to live their life in the way in which they would like to which distinguishes people seeking help from those who do not. The cognitive theory proposes that extreme emotional reactions are largely maintained by coping attempts and mechanisms initiated by the particular appraisal made. Sometimes, coping mechanisms can create a series of vicious circles which bolster unhelpful beliefs and distorted patterns of thinking (see Teasdale, 1983 and Teasdale & Barnard, 1993, for examples of specifically cognitive and information processing examples). For example, the depressed person who believes that he or she is a failure tries to avoid failure by restricting his or her activities. The person then notes that they have had no recent successes or enjoyment. The person who believes that he has a heart condition closely attends to cardiac sensations; on doing so, he notes to his alarm that he frequently experiences fluctuations in his heart rate, and occasionally his heart skips a beat, thus fuelling his belief that there is indeed something wrong with his heart. The pattern of response which maintains the distorted thinking characteristic of emotional problems is idiosyncratic and relates closely to the focus of the person's concern (e.g. loss, threat, unfairness). Therapy aims to identify and modify distorted thinking and the factors which are involved in maintaining beliefs giving rise to distress. Research is therefore aimed at:

1. the identification of beliefs and patterns of thinking characteristic of particular emotional problems;
2. identifying specific factors which are involved in maintaining distorted thinking;
3. devising and evaluating the most effective ways of modifying both distorted thinking and factors which maintain it.

The way in which the key maintaining role of anxiety-related *behaviour* (e.g. avoidance and escape) has now been incorporated into the central notions of the cognitive theory (rather than being an uneasy hybrid of cognitive and behavioural theories) will be described below on pp. 207–8.

PANIC ATTACKS

In the late 1980s it was well-established that exposure-based therapy was highly effective in the treatment of agoraphobia (e.g. Mathews *et al.*, 1981). It was also clear that the co-occurrence of panic attacks was associated with a considerably poorer outcome. Previous work on panic had suggested a biological basis of panic (Klein & Fink, 1962). The suggestion that panic was 'biologically distinct' from other types of anxiety was highly influential in the way the anxiety disorders were defined in the third edition of the American Psychiatric Association's 1987 manual (DSM-III). Biological psychiatrists took the view that panic attacks *per se* represented a type of endogenous anxiety characterized by strong physical symptoms and imperviousness to psychological influences. It was suggested that associated phenomena (such as anticipatory anxiety and avoidance) *were* subject to psychological intervention, but panic attacks required biological intervention, particularly drugs with noradrenergic action, such as the tricyclic antidepressant, imipramine. It was argued that panic attacks are symptomatic of a hypersensitivity of the biological 'alarm' system, resulting in spontaneous discharges ('spontaneous panic attacks') and hypersensitivity to incoming stimuli, so that relatively low-intensity stimulation could provoke 'cued attacks' (e.g. in an agoraphobic situation). Table 9.1 shows the more recent diagnostic criteria of DSM-III (Revised) (American Psychiatric Association, 1987).

Table 9.1. Characteristics of panic attacks (DSM-III-R)

Symptoms (criteria specify an increase in at least four)	
Physical	Shortness of breath
	Choking
	Palpitations or tachycardia
	Chest pain or pressure
	Sweating
	Nausea or abdominal distress
	Numbness or tingling sensations
	Depersonalization or derealization
	Flushes or chills
	Trembling or shaking
	Dizziness, unsteady feelings, faintness
Cognitive	Fear of dying
	Fear of going crazy or doing something uncontrolled
Other characteristics	At least four symptoms occurred within 10 minutes of the beginning of the first symptom noticed
	Can be associated with avoidance behaviour (especially agoraphobia) and anticipatory anxiety
	Onset can be associated with physical disorder or factors; diagnosis depends on the persistence beyond the cessation of precipitating organic factors (e.g. hyperthyroidism, caffeine intoxication and so on)

From a cognitive perspective, the combination of somatic sensations in Table 9.1 and cognitive correlates of panic suggested an alternative hypothesis. As described above, the appraisal of threat is central to the cognitive theory of anxiety (Beck *et al.*, 1985); an obvious possibility was that panic could be occurring because panic patients were appraising the normal somatic symptoms of anxiety as excessively threatening, indicating a terrible disaster. In people experiencing repeated panic attacks, the negative appraisal of symptoms would therefore be more constantly present, resulting in a pattern of vigilance for potentially panic-provoking situations and avoidance of such situations, punctuated by occasional full-blown panic attacks. Panic could be seen as anxiety which has as its focus those somatic sensations which are increased by the experience of anxiety. The cognitive hypothesis of panic (Clark, 1986, 1988; Salkovskis & Clark, 1986; Salkovskis, 1988a; Beck, 1988) has been developed, based on the cognitive theory of anxiety described by Beck *et al.*, (1985). This cognitive hypothesis is based on the simple premise that people who experience recurring panic attacks do so because they have *an enduring tendency* to misinterpret certain bodily sensations as a sign of imminent disaster. The way in which this accounts for acute panic attacks is summarized in Figure 9.2.

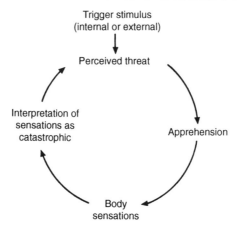

Figure 9.2 The suggested sequence of events in a panic attack. Reprinted with permisson from Clark (1986) p 463

Acute panic results from the misinterpretation of bodily or mental sensations as signs of imminent personal disaster. The anxiety engendered by
the misinterpretation of sensations produces an increase in sensations,
which can in turn feed back to the misinterpretation, rapidly culminating
in a full-blown panic attack. Thus, according to the cognitive hypothesis,
panic occurs if, and only if, sensations are misinterpreted; patients who
experience repeated panic attacks are particularly vulnerable to making
such misinterpretations by virtue of a pre-existing set of beliefs. A variety
of things may trigger off the panic vicious circle in the first instance; it may
be triggered by anxiety from general stress, by the anxiety of entering a
situation where panic previously occurred, by a sensation from an
unrelated source such as a hangover or too much coffee, by a frightening
thought and so on. It is important to note that the whole process of panic
generation can occur within a few seconds, as the person responds physio-
logically and psychologically to what seems, at that time, to be a profound
threat to his or her safety.

For example, patients often interpret palpitations and racing heart as a
sign that they are having a heart attack. This misinterpretation in turn
results in intense anxiety and thus the sensations associated with physiological arousal, including palpitations, tachycardia, dizziness and pre-
cordial pain. This increase in intensity and range of symptoms appears to
confirm the original misinterpretation, further increasing anxiety and
symptoms and so on, rapidly culminating in an acute panic attack.

A wide range of symptoms can give rise to misinterpretations, but the
cognitive hypothesis proposes that those involved in panic attacks are particularly likely to be sensations which can result from autonomic or central
nervous system responses to anxiety itself, given the immediate nature of
the feedback loop. A further factor in the readiness to misinterpret bodily

sensations should be the extent to which the person has available any innocuous alternative explanations of the intense and wide-ranging sensations associated with panic attacks. For example, if someone were to notice their heart racing, being short of breath, feeling dizzy and found they were sweating a great deal immediately after running to catch a train, misinterpretation is unlikely. The same sensations experienced without any obvious cause would, at best, be bewildering and at worst be taken as a sign of serious illness.

The catastrophic beliefs specified by the cognitive hypothesis account for the intensity of anxiety and the associated physical symptoms, insofar as the anxiety reaction experienced would not be considered abnormal if occurring in the context of a truly life threatening situation. Typically, patients express very strong beliefs in the possibility of feared disaster *during the attacks themselves*, although in the clinician's office (when the symptoms are not present and a trusted individual is) patients will report that they do not, at that time, feel convinced that the disasters would happen. The apparent unexpectedness or 'spontaneousness' of some panic attacks can also be accounted for; although the panic may start with a mild (and often normal) physical sensation, if such a sensation triggers off the thoughts of disaster and hence a panic, the originating sensation may then be lost in the surge of much more intense, anxiety generated symptoms which follow. Often, there is no specific explanation for the triggering symptom; it was simply one of the many ordinary physical fluctuations which most people experience every day.

This cognitive hypothesis can thus account for a number of otherwise puzzling features of the phenomenology of panic and agoraphobia with panic, including night-time panic, spontaneous attacks and panic provocation by biochemical agents such as sodium lactate, carbon dioxide, yohimbine and so on (Clark, 1986). However, the ability of the theory to *explain* observations is less important than predictions which can be *derived* from the hypothesis and its correlates. A key prediction upon which the utility of the theory rests is that therapy which modifies the misinterpretations of bodily sensations should be effective in stopping panic attacks. The treatment research which followed this idea is described below on p. 218. Before considering this, additional theoretical issues and how these were tackled will be dealt with; in particular, the way in which experimental studies of theoretical problems contributed to the further development treatment will be highlighted.

TESTS OF PREDICTIONS FROM THE COGNITIVE THEORY OF PANIC

First, if the cognitive theory is correct, then it should be possible to show that panic patients are more likely to misinterpret bodily sensations than people not having panic attacks, including anxious patients not suffering from panic. Of course, it is possible that panic patients might misinterpret *everything* in a more negative way, so the tendency to misinterpret stimuli other than anxiety-related bodily sensations also had to be assessed. This was done (Clark *et al.*, 1988) using a questionnaire measure in which subjects were asked to respond to questions such as, 'Your heart is beating quickly and pounding. Why?'. Following an open-ended response, subjects rank-ordered three options; in this instance it would be:

1. Because you are excited.
2. Because you have been exercising.
3. Because you are about to have a heart attack.

The results of this study are shown in Figure 9.3. They indicate that:

1. Panic patients are particularly likely to misinterpret bodily sensations compared to other anxious patients.
2. The tendency to misinterpret relative to controls is specific to the experience of bodily sensations which can be increased by anxiety as opposed to other potentially threatening situations.

It is also important to note that, after effective treatment, the specific misinterpretation is no longer detectable, consistent with the idea that misinterpretation may play a key role in the maintenance of panic attacks. However, the results of questionnaire studies are often criticized because

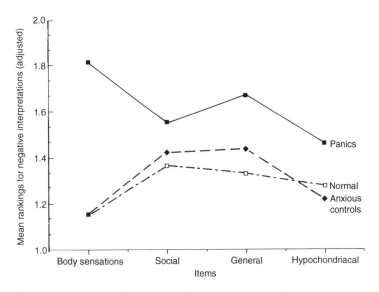

Figure 9.3 Interpretation of body sensations by panic patients, anxious patients and normal patients

it may be that patients are aware of how they are expected to respond. Furthermore, the theory requires that the misinterpretations made are almost instantaneous. In a study designed to evaluate the interpretation question without possible subjective biases, a study using a technique called Contextual Priming was used (Clark *et al.*, 1988). This technique, often used in psycholinguistics, is designed to determine pre-existing patterns of meaning implied by particular contexts. An incomplete sentence is displayed on a computer screen; for example,

The cat drank the

The sentence fades and a word flashes up where the end of the sentence had been; the person watching simply has to say the word as quickly as possible, and the speed is measured using an electronic device triggered by the subjects voice. In the above example, subjects would be appreciably quicker with the word 'milk' than the word 'vodka', revealing something about the persons belief about cats' drinking habits. In the experiment with panic patients, the incomplete sentences used were panic misinterpretation relevant; for example,

My heart is beating quickly and pounding because I am

Two possible completions were used: 'dying' or 'excited'. Consistent with the cognitive theory, panic patients were significantly faster to say catastrophic completions (such as 'dying' in this example), whilst people who do not suffer from panic did not show this difference. This shows that panic patients have specific negative beliefs, and that misinterpretations are rapid and automatic. After effective treatment by cognitive therapy, patients no longer show the contextual priming effect, again consistent with the view that therapy may be effective by modifying misinterpretations.

Although the demonstration that panic patients specifically misinterpret bodily sensations, it could still be argued that these patients show such misinterpretations because they are experiencing panic attacks rather than vice versa. More convincing and less ambiguous evidence for the theory comes from the prediction that *activating misinterpretations should make panic patients (but not anxious controls) experience panic-like states*. The task which was devised to do this was a simple word-reading task, where subjects were asked to read words such as,

breathlessness–suffocate
palpitations–heart attack
unreality–insane

slowly, as pairs, pausing to dwell on each pair. Of the panic patients asked to do this, 75% experienced a panic attack (as defined by DSM-III, viz. experiencing an increase in at least four physical symptoms and a sudden increase in anxiety), as opposed to 17% of anxious controls and no non-clinical controls. Another stressful task (backwards serial sevens) did not differentiate panic patients from anxious controls; although anxiety increased substantially in both groups, none of the subjects experienced a panic attack. Interestingly, no panic attacks occurred when the word-pair task was carried out with a group of panic patients who had been success-fully treated by cognitive therapy.

A range of other experimental investigations has now been completed by the Oxford group and others. Interestingly, the cognitive model has been able to explain findings from the biological field (e.g. how biological challenges such as sodium lactate provoke panic) and experimentally test these findings (Clark, 1993). On the other hand, biological theorists have been unable to account for findings in the cognitive field. Many specific predictions arising from biological research have been falsified, e.g. Gorman *et al.* (1989) predicted that psychological treatments will not reduce panic anxiety, but will only have effects on avoidance.

THEORETICAL PROBLEMS FOR THE COGNITIVE HYPOTHESIS OF PANIC

HYPERVENTILATION: HOW DO PHYSIOLOGY AND COGNITION RELATE OR INTERACT?

Early in the development of the cognitive theory, hyperventilation (breathing in excess of metabolic requirements so as to induce a symptomatic respiratory alkalosis) was employed in treatment. This led some theorists (e.g. Ley, 1985a, 1985b) to suggest that panic attacks were, in fact, hyper-ventilatory episodes. It is certainly true that the symptoms of hyper-ventilation bear some resemblance to those experienced during panic (Lum, 1976). However, these symptoms are also indistinguishable from those induced by other stressful tasks (Hornsveld & Garssen, 1992). In a review of the data on the link between hyperventilation and anxiety, Salkovskis (1988b) suggested that the link was best understood from a cognitive perspective, in which hyperventilation is a *potentially* important source of symptoms; however, panic will occur if, and only if, the symptoms are misinterpreted as a sign of imminent disaster. Figure 9.4 shows the way in which this works.

Thus the cognitive hypothesis proposes that hyperventilation is only one possible source of symptoms. It is unlikely ever to be *solely* responsible for symptoms during panic, in that the person who is catastrophically mis-interpreting symptoms is almost certain to be experiencing the effects of

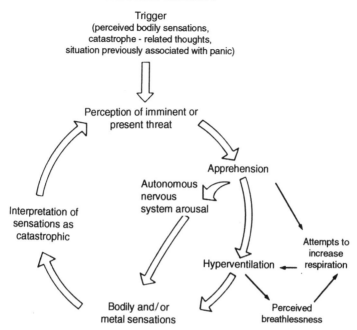

Figure 9.4 The principal components of the cognitive model applying to panic attacks in which hyperventilation is involved in the production of symptoms

autonomic arousal in parallel with the physiological effects of hyperventilation. There are two identifiable subgroups of patients particularly likely to experience hyperventilation as part of panic.

1. Patients who interpret breathlessness as a sign that they are not getting enough air during panic attacks (and who therefore deliberately try to breathe more to prevent suffocation).
2. Those patients who inadvertently hyperventilate during attempts to calm down by 'breathing deeply'.

These instances have important treatment implications (see below, p. 216).

To verify the proposed cognitive link between anxiety and hyperventilation, an experiment on normal subjects was carried out (Salkovskis & Clark, 1990). In this experiment, subjects were allocated to one of two instruction groups. All were given an account of the effects of overbreathing designed to lead to *expectations* of (1) mild positive affect, (2) experience of particular bodily sensations, chosen to sound plausible but actually occurring very infrequently. In order to influence subjects' *interpretation* of the effects of a brief period of hyperventilation, the instructions also contained an apparently incidental account of 'rare' sensations (actually the two most commonly reported bodily effects of overbreathing). These

were either described as being indications of risk of fainting (negative interpretation group), or as signs of good adjustment and a higher state of
consciousness (positive interpretation group). Following a brief period of
hyperventilation, the experience of bodily sensations and affect was rated.

Results showed that the interpretation given at the beginning of the
experiment did *not* influence the degree of hyperventilation, the *expectancy*
of affect, nor the somatic sensations reported. The affective response to
hyperventilation *was* strongly and significantly influenced by the interpretation provided (see Figure 9.5)

The cognitive hypothesis of panic strongly predicts that the *interpretation*
of bodily sensations is an important determinant of the affective response
to hyperventilation. The model would therefore predict that, within each
instructional group, individuals who experience the strongest and widest
range of bodily sensations will also experience the strongest induced effect.
Specifically, the higher the bodily sensations score, the higher the effect
score *appropriate to the particular interpretation condition* is likely to be. To
evaluate this, the correlations between the scores obtained for bodily
sensations and positive and negative effect were calculated separately for
the two interpretation conditions. Within the negative interpretation
group, body sensations score was significantly correlated with negative
effect, $(r = 0.72, p < 0.005)$ and not with positive effect score, $(r = 0.245,
p > 0.05)$. On the other hand, in the positive interpretation condition, body

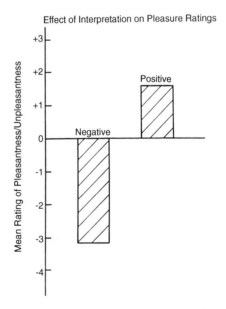

Figure 9.5 Effect of positive and negative interpretations on rating of the
pleasantness/unpleasantness of the experience of hyperventilation

sensations score was not significantly correlated with negative effect ($r = 0.106$, $p > 0.05$) but did correlate significantly with positive effect, ($r = 0.554$, $p < 0.05$). Thus, we were able to demonstrate that high body sensations scores are associated with high scores only on the effect appropriate to the interpretation provided.

The data on the link between panic and hyperventilation therefore is consistent with the cognitive account proposed by Clark and Salkovskis (Clark, 1986; Salkovskis & Clark 1986; Salkovskis, 1988a, 1988b). Thus, although hyperventilation is not totally irrelevant to panic, it is neither necessary nor sufficient, but is a potential source of symptoms which may subsequently be misinterpreted in the same way as some vestibular symptoms may be.

THE PROBLEM OF 'IRRATIONAL' PERSISTENCE OF PANIC: HOW DO BEHAVIOUR AND COGNITION RELATE AND INTERACT?

In a detailed discussion of psychological perspectives on panic, Seligman (1988) posed an important question, which he described as 'a central weakness in both the Cognitive and Pavlovian theories of the anxiety disorders: neither theory clearly distinguishes the rational from the irrational, the conscious from the unconscious'[*]. Seligman argued the need for 'two distinguishable processes, obeying different laws'. Particularly relevant to anxiety disorders, he proposed, is 'prepared' learning (which is biologically relevant, irrational and not readily modified by cognitive means). By implication, prepared learning is also relatively unconscious. He proposed, as an example, the case of the patient who has experienced regular panic attacks for a decade or more.

> [This person] may have had about 1000 panic attacks. In each one, on the cognitive account, he misinterpreted his racing heart as meaning that he was about to have a heart attack, and this was disconfirmed. Under the laws of disconfirmation that I know, he received ample evidence that his belief was false, and he should have given it up. On the Pavlovian account, he has had 1000 extinction trials in which the CS was not followed by the US-UR...His panics should have extinguished long ago...neither theory explains why the belief did not extinguish in the face of disconfirmation long ago. What is it

[*] At least some of these issues concerning the relative inaccessibility of cognitive processes, discussed in detail by Williams et al., (1988) can be resolved by making the important distinction between the measurement of cognitive *processes* and the cognitive *events* which can be regarded as the outcome of those processes. It should be possible to reliably measure both aspects of cognitive functioning.

about cognitive therapeutic procedures which makes them effective disconfirmations, and about the Pavlovian exposure procedures that make them
effective extinction procedures? (Seligman, 1988, p326).

Seligman thus highlighted the apparent failure of people experiencing
frequent panic attacks to take advantage of naturally-occurring disconfirmations (extinction experiences). He suggested that his theory of
preparedness accounts for the persistence of panic; he argued that
prepared learning follows different rules to unprepared associations. Panic
and most phobias, he suggested, involve highly prepared associations
which are particularly resistant to extinction.

To account for the failure of anxious patients to take advantage of
naturally occurring disconfirmations, the cognitive hypothesis postulates a
functional and internally logical link between cognition and behaviour
(Clark, 1988; Salkovskis, 1988a, 1989b, 1991). Assuming for the moment
that the same rules of logic apply to panic as to other areas of human
behaviour, then the logical response to threat is to take action designed
to prevent perceived imminent negative outcomes. That is, a person pan-
icking because he believes that a catastrophe is imminent will do anything
he believes he can to prevent the *catastrophe*. The person afraid of fainting
sits, the person afraid of having a heart attack refrains from exercising, and
so on. By doing so, the patient not only experiences immediate relief, but
'protects' his or her belief in the potential for disaster associated with particular sensations. Each panic attack, rather than being a disconfirmation,
becomes another example of being *nearly* overtaken by a disaster; 'I have
been close to fainting so many times: I have to be careful, or one of these
times I won't be able to catch it'. Thus, the apparent failure of panic
patients to take advantage of naturally occurring disconfirmations may be
because the non-occurrence of feared catastrophes is involved in maintaining and 'confirming' the person's fears.

Some of the beliefs held by anxiety patients are specific and idiosyncratic, e.g. patients may overestimate their personal anxiety sensitivity,
such that they believe that even a small amount of anxiety could result in
heart failure for themselves. Although it may be difficult for an observer
to understand the fear of an agoraphobic patient about entering a supermarket, this simply indicates that the observer does not share the idiosyncratic beliefs of the patient. It is easier to understand the avoidance of
agoraphobic patients if one simply reflects upon why he or she will not
enter a situation where he or she believes that death, loss of control or
insanity may occur as a result.

THEORY AND RESEARCH CONTRIBUTE TO TREATMENT

Both testable theoretical refinements (such as the importance of safety seeking behaviour as described above) and experimental studies (such as the use of the pairs of words described on p. 203) can make contributions to the refinement of treatment techniques and strategies. The way that these examples have had an impact on treatment will be described next.

SAFETY SEEKING BEHAVIOUR

Some of the most important implications of the cognition-behaviour links described above concern treatment (Salkovskis, 1991). In general, cognitive behavioural treatment emphasizes the need to deal with idiosyncratic factors which are bolstering the continued misinterpretation of bodily sensations. For example, a patient who became confused during panic describes mentally 'holding on to my sanity'. In each successive panic he becomes *more* convinced that he would have gone mad were it not for this effort. Another example is the patient who, hundreds of very severe panic attacks later, still believes in each new attack that she is about to go crazy, pass out or die. Clinically, questioning of such patients reveals that no disconfirmation has occurred because the patients believe that they have, in every instance, been able to take successful preventative action. Thus, each panic is a near miss and further confirmation of the risk. Once such behavioural responses are identified, the patient can be helped to begin the process of re-appraisal by withholding such protective responses and learning the true extent of risk. Typically, this may involve suggesting to the patient that they challenge their worries by *actively trying* to cause the feared disaster, for example, going into the supermarket and trying to faint or try to go mad. This helps the patient to discover that their efforts to prevent these disasters have been misdirected; this can also help the patient to then re-interpret their usually considerable past experiences of such anxiety-provoking situations as true disconfirmations instead of 'near misses'. This particular type of association might be a fruitful one for investigation, as controlling one's mind is a common response in panic patients as well as other groups, including non-clinical subjects. Given the readiness with which patients can reproduce this type of behaviour as either safety seeking or coping when requested in the laboratory, experiments concerning the use of the same degree of effort directed at different targets (i.e. reduction of anxiety vs. preservation of sanity) could be used to assess the validity of the cognitive basis of such behaviour and its putative anxiety-preserving effects when used as a safety-seeking behaviour. A further implication is that for the most complete and enduring treatment effects, *control* is not emphasized. Helping people to change their beliefs about the meaning of their symptoms in a more enduring way is the target, and should reduce subsequent vigilance and pre-occupation.

The present analysis also suggests ways of combining cognitive procedures and brief exposure in a way which should be particularly effective in bringing about belief change withour *repeated* and *prolonged* exposure being necessary. Thus, exposure sessions are devised in the manner of behavioural experiments, intended as an information gathering exercise dirceted towards invalidation of threat-related interpretations. Most previous studies in which cognitive and behavioural treatments have been combined (such as those reviewed in Marks, 1987) have used cognitive procedures as a way of dealing with general and 'background' life stress, most of which tend not to be directly relevant to the specific experience of anxiety. The particular strategy of using exposure as an exercise in testing alternative non-threatening interpreations of experience would be predicted to succeed better than brief exposure with 'supplementary' (threat-irrelevant) cognitive change procedures. That is, such a study should show that 'general' cognitive therapy combined with exposure has an additive effect, whilst 'anxiety-focused' cognitive therapy would be expected to multiply the effect of exposure, resulting in maximal cognitive change through behavoural experiments. Thus, according to the cognitive hypothesis, the value of behavioural experiments transcends mere exposure; such experiments allow patient and therapist to collaborate in the gathering of new information assessing the validity of non-threatening explanations of anxiety and associated symptoms.

PAIRED ASSOCIATES

Since devising the paired associates task in order to test predictions made by the cognitive model, we have found that it is a particularly effective strategy as part of treatment. Cognitive therapy involves a process of re-attribution in which the panic sufferer is helped to understand that the symptoms which they experience do not arise from the catastrophes which he or she fears. The most effective way of doing this is to provide the person with an alternative, non-catastrophic explanation (the personalized vicious circle model). The paired associates can be a particularly powerful way of *convincing* the sufferer of the validity of the vicious circle model. A personalized version of the paired associates is used (derived from a careful assessment of the symptoms experienced and the misinterpretations made by that person). If the person panics when reading the pairs, they are asked what they make of the experience. If reading these particular pairs of words can induce panic, how does that fit with the idea that your symptoms mean that you are having a heart attack? How does this fit with the vicious circle explanation?

We have also discovered, in the context of therapy, that quite simple strategies will readily modify panic patients' response to the task. For example, a patient undergoing therapy was concerned that her initial panic when reading the paired associates meant that she was so out of control that even reading words on a card could make her have a panic. Then she was asked to read the same card, but where the hyphen appeared in the pair (e.g. breathless–suffocate) she was to say 'does not mean I will'. This procedure completely abolished her anxious response to the card and led her to conclude that panic could be controlled by changing the meaning of symptoms. This clinical illustration shows both the potential strengths of the technique and the dangers of its uncritical prescriptive use.

COGNITIVE THERAPY: GENERAL PRINCIPLES

A number of common elements are crucial to the cognitive-behavioural approach:

1. the adoption of a normalizing strategy;
2. a collaborative ('guided discovery') style in therapy sessions;
3. focusing on client-defined targets and goals in therapy;
4. the primary focus of interventions being on factors involved in the maintenance of problems;
5. measurement of progress as an integral part of treatment (both process and outcome);
6. the use of individualized and idiosyncratic conceptualization and intervention strategies.

Cognitive-behavioural therapies have been wrongly characterized as:

1. prescriptive;
2. exclusively concerned with the here and now;
3. ignoring the therapeutic relationship
4. purely verbal;
5. ignoring the social/family/cultural context;
6. focusing on rational thinking (see Hawton *et al.*, 1989).

Cognitive therapy has similar prerequisites to other forms of psychological treatment; primary amongst these is the formation of a good, collaborative therapeutic relationship. The style of therapy emphasizes guided discovery, so that the patient is helped to understand and try to modify their problems by using careful questioning. Therapy also makes extensive use of diaries and other aids; a particularly helpful adjunct to the most efficient use of treatment time is to tape record therapy sessions so the patient can listen to the tape at home. These elements are considered to be a minimum baseline against which any effective therapy needs to be judged. The general cognitive theory predicts that specific techniques will be of minimal effectiveness in the absence of a good therapeutic relationship as described above. Part of this relationship involves a proper assessment of the idiosyncratic problems experienced by the person in their own unique individual context. Therapy is based on the view that if a person holds a 'distorted' belief, this is because they have good present, contextual and historical reasons for doing so. This is equally true, not only in the obvious case of general feelings of low self-worth in a person whose main difficulty is low self-esteem, but also for specific misinterpretations of bodily sensations in someone suffering from frequent panic attacks. For example, it is often important not only to pinpoint just what the person means by the idea of losing control, but also *why* loss of control is a major issue for them. Most commonly, these types of concern reflect 'developmental' issues.

Evaluation of specific techniques or strategies must therefore take place against the *necessary* background of good and consistent general therapeutic skills. People (clients and therapists) seldom work well with those who they tend to distrust. Many of the negative findings in the field can be explained not only by the failure of the specific intended manipulation (which is, of course, a fatal error) but also by inappropriate therapy styles. One of the most common errors in this respect is the use of cognitive-behavioural treatments in a prescriptive fashion. Another example is to abstract a single element (e.g. the emphasis by some on 'correct breathing'; see p. 204 above). In order to ensure therapy integrity, therapy evaluations need to combine assessment of specific techniques with the evaluation of therapy style and quality of relationship. Other factors which are of key importance (and which need to be equated across any more refined comparisons of different therapies) are therapist time, the initial credibility and expectation of improvement engendered by the different treatments *prior to any differential treatment effects*, any assessments and self-monitoring, and so on.

COGNITIVE THERAPY AS SPECIFICALLY APPLIED TO PANIC: THEORY-GUIDED THERAPY

Once it has been established that panic is a major problem for the person and a more general assessment of the patient's situation has been carried out, a specific panic-oriented assessment is undertaken. This assessment focuses on eliciting an account of recent panic attacks and devising an individualized vicious circle explanation of the sequence of events during panic (loosely based on Figure 9.2, but always tailored to the idiosyncratic description of the patient). The links which comprise the vicious circle should be elicited by systematic questioning, thus ensuring both that the assessment is accurate and that the patient is clear about the way in which the formulation has been reached. The patient recalls the last major panic they experienced, and then describes, step by step, how it developed. It is helpful to use the sequence used in the vicious circle shown in Figure 9.2 to guide questioning; thus, if the first thing the patient noticed was a physical sensation, the next question should concern the way the patient interpreted it (e.g. When you noticed your heart pounding, what went through your mind at that time?). Another question which may be useful at this crucial stage in the assessment is, Right then, *at that time*, what did you think was the worst thing that could happen? Once a specific catastrophic thought (or image) is elicited, questioning shifts to the anxiety experienced (e.g. When you had the thought, 'I'm having a heart attack', how did that make you feel?). The next question concerns physical reactions to anxiety (e.g. When you became anxious about having a heart attack, how did that affect you physically?). Once the vicious circle has been elicited in this way, it is summarized verbally and outlined on paper for the patient. Thus, in the above example, the summary would go:

> Lets see if I have understood what you are telling me. During the attack you experienced on Wednesday night, you first noticed your heart pounding. You then had the thought 'I'm having a heart attack', and this, not surprisingly, frightened you. The anxiety made your heart pound more, which seemed to confirm the idea that you were having a heart attack, and so on round in a vicious circle as I've drawn here. Does that seem to you to be right?

In this way, a positive description of the factors involved in the production of panic is built up from information provided by the patient. Note that the emphasis is on providing a positive, non-catastrophic account of panic rather than on disproving feared catastrophes. The internal logic of panic is emphasized as a way of normalizing the intensity of the anxiety experienced (e.g. How do you think that anyone would feel if they noticed that they were feeling dizzy and believed that this meant they were about to faint? When someone becomes anxious because they have a very frightening thought, such as about having a heart attack, how do you think they would feel physically?).

Table 9.2. Examples of the association between symptoms and catastrophic misinterpretations occurring during panic attacks

Sensation	Typical misinterpretations
Heart racing, pounding, palpitations	I'm having a heart attack, my heart will stop, I'm dying
Breathlessness	I'm going to stop breathing, suffocate
Feeling unreal	I'm going to go crazy, lose my mind
Losing feeling in arms	I'm having a stroke
Feeling dizzy and faint	I'm going to faint, fall over, pass out
Feeling distant, tense, sweaty and confused	I'm about to lose control of my behaviour
Feeling dizzy, heart pounding, chest tight, palpitations, flushed and tingling	I'm dying

A helpful pointer for the assessment, and a further illustration of the logic of panic, is the way in which catastrophic interpretations tend to be meaningfully associated with particular symptoms. Examples of the way in which this association works are illustrated in Table 9.2.

TREATMENT

Assessment as described above is the first component of treatment, because it already involves a major element of decatastrophizing, helping patients to make sense of the panics which had previously been especially frightening because they appeared to be inexplicable. The procedures used for further decatastrophizing are divided into verbal techniques, in which discussion of the patients' past experience is used to change beliefs, and behavioural experiments, in which the patient and therapist devise experiments to test the way in which particular processes may be involved in panic attacks.

DISCUSSION

The vicious circle model is again the starting point for discussion aimed at helping the patient to attribute the symptoms they experience during panic away from catastrophic explanations and onto less threatening interpretations. The therapist starts from the assumption that, if the patient makes catastrophic interpretations, they do so because they have *evidence* which they find convincing. This evidence may be past experiences (e.g. a close relative or friend who experienced similar symptoms associated with serious illness), in generalized attitudes (e.g. If I can't control my thoughts I will go mad), in the intensity or nature of the physical sensations themselves (e.g. I feel so faint that I *must* be about to pass out) and so on.

Discussion involves helping the patient to evaluate such evidence and consider alternatives. For example, assessment of a patient revealed that, during panic, she interpreted dizziness, unsteadiness and feelings of unreality as a sign that she was about to faint or collapse. Attacks invariably started with an episode of dizziness, often triggered by postural change or feeling angry and frustrated. The evidence for her belief that she would faint was that the symptoms, especially faintness, became very intense when she was at the height of an attack, and that she had, on a previous occasion, fainted. During the first stage of this discussion, the patient and therapist went over the factors involved in a true faint, and agreed that a drop in blood pressure was required in order to faint. The patient, when asked what she thought might happen to her blood pressure during panic, said that she knew that it went up a little and this therefore must mean that she could *not* faint during panic. The discussion then moved on to her previous fainting episode, which had occurred after giving blood. The reasons for blood pressure drop in this specific situation were discussed. The patient was also asked to compare the experience of fainting with the experience of feeling faint in panic; she identified these as quite different. In particular, she noted that when she fainted she was not anxious, but instead became rather sleepy and distant. When asked whether that distant feeling resembled the unreality she experienced during panic, she said that it was quite different, and summarized by saying that she would easily recognize a true faint if it were to occur, and that fainting was particularly *unlikely* during panic attacks. This was an especially useful manoeuvre, because it meant that previous anxiety-provoking situations had now become safety signals. Following on from this, the discussion moved on to consideration of the nature of 'feeling faint', and whether such feelings were necessarily associated with the fact of fainting. The comparison with having a splitting headache was used; did a splitting headache mean that one's head could split? The patient agreed that this was unlikely and her ratings of belief in the idea that she might faint during a panic attack dropped substantially, but not completely. When questioned about the basis of her beliefs, she said that the symptoms were so intense at the time that she believed there had to be something wrong that could lead her to faint. The therapist then asked her if she had been excited recently. She said that she had. Three weeks previously, she had thought that she had won a very large sum of money (on the football pools). She described her bodily sensations at that time, which were almost identical in type and intensity to those she experienced during a panic, although she had not panicked and found the experience pleasurable and exciting. Careful questioning elicited the main difference to be the way in which she interpreted the sensations of excitement; she had not had any catastrophic thoughts. The conclusion she drew from this discussion was that the symptoms she experienced during panic attacks were the normal effects of adrenaline, and not a sign of something physically amiss. Finally, ways in which the patient could apply the reasoning outlined above during panic attacks were devised and discussed.

BEHAVIOURAL EXPERIMENTS

Discussion in which the evidence upon which patients base their catastrophic misinterpretations is reviewed continues throughout therapy. However, patients sometimes do not have ready access to information which would help them to challenge their misinterpretations, or they may understand the alternative explanations which have been generated in the course of discussion but require further evidence to support these alternatives. This is where behavioural experiments are particularly valuable as part of treatment. A wide range of behavioural experiments are useful.

One of the most commonly used involves the use of hyperventilation, which often occurs in association with panic when the associated symptoms are misinterpreted (as described on p. 204). The metabolic alkalosis which accompanies hyperventilation produces a wide range of physical symptoms, such as tachycardia, dizziness, paraesthesia, breathlessness and so on; these symptoms are a particularly potent basis of misinterpretation because they are often relatively unfamiliar to the patient in the context of his or her previous experience of anxiety (e.g. paraesthesia and dizziness).

Particularly common is the experience of breathlessness, which, if misinterpreted as a sign of impending suffocation, can have the effect of making the patient try to breathe more, worsening the breathlessness and so on. An especially helpful way of dealing with this is to ask the patient to show how he or she breathes (or tries to breathe) at the height of a panic attack. If the pattern the patient demonstrates appears to involve an element of hyperventilation, the therapist asks the patient to continue to breathe in the same way (or, if it is only a slight degree of hyperventilation, to exaggerate this) for a few minutes, then to describe the physical feelings they experience. This type of procedure reproduces the bodily sensations experienced during a panic attack in about 50% of cases; the fact that the patient is simply doing what they would normally be doing during a panic attack makes this demonstration of the role of hyperventilation particularly convincing to him or her. Discussion of the extent to which the experience of deliberate overbreathing was similar or different from the patient's naturally occurring panic will usually show that the *physical* sensations were similar, but the experience differed in the extent to which anxiety was present. This, of course, is because the patient in the therapist's office has an unambiguous and non-threatening explanation of their symptoms, and because the patient is in control of the onset (and therefore the offset) of the symptoms themselves. This difference can itself help pinpoint the crucial role of misinterpretations; thus, the patient is taught that hyperventilation does not cause panic, but is a likely source of symptoms which can be misinterpreted. Subsequently, patients may find it helpful to learn simple breathing control techniques which they can use during panic attacks as part of the general cognitive strategy. It is important to stress that breathing control is a helpful way of dealing with the nuisance value of symptoms, but only at times when the patient notices the physical symptoms. It is emphasized that the control of breathing is in no way essential; indeed, in the later stages of therapy the patient deliberately hyperventilates in situations which tend to make them anxious; this exercise provides further confirmation of the non-catastrophic nature of bodily sensations. Throughout the emphasis is on the idea that the key factor is misinterpretation; breathing control is most useful as evidence for a non-catastrophic interpretation of the symptoms. During an acute attack, the patient reminds him or herself of the following.

1. The physical symptoms of panic were reproduced by overbreathing in the therapist's office, suggesting that the symptoms are both a sign of anxiety induced by catastrophic misinterpretations and also a result of stress-induced hyperventilation;
2. Strategies which help them to control their thoughts and breathing, although unnecessary for their safety, have the effect of reducing the symptoms, thus providing further confirmation of the non-catastrophic interpretation.

Note that, despite encouraging early clinical reports, breathing control without the crucial cognitive element has been shown to be ineffective; the cognitive element appears to be necessary for the successful treatment using respiratory control.

SAFETY-SEEKING BEHAVIOURS

The therapist who wishes to deal effectively with the problems of patients suffering from panic attacks should also deal with those safety-seeking behaviours which the patient uses during panic attacks. Typically, three main types of behaviour are seen:

1. Avoidance of situations which the patient believes *might* provoke panic (e.g. avoiding supermarkets).
2. Escape from a situation when a panic attack occurs (e.g. leaving a shop once the symptoms of panic begin).
3. Safety seeking behaviours carried out during panic with the intention of actively preventing the feared catastrophe (e.g. when dizziness leads to the thought 'I'll faint', holding onto another person or shopping trolley or sitting down).

Each of these behaviours can have the effect of maintaining the panic-related beliefs; patients logically infer that they have prevented the occurrence of feared catastrophes by their behaviour (e.g. If I had gone to the supermarket yesterday, then I would have passed out; If I had not left immediately I would have fainted; If I had not sat down then I would have fainted). Thus, behaviour of the type described above prevents disconfirmation of the feared catastrophes, and transforms potential disconfirmations into 'near misses'. In cognitive therapy for panic, patients are helped to understand how their behaviour can maintain anxiety (in the same way as outlined above), and the therapist helps the patient to use this information to construct an explanation of the persistence of their beliefs. For example, a patient summarized this as

Although I believe that my behaviour has prevented disaster, maybe this is only because I have never tried to check this out, and have just assumed that my *not* fainting was because of the precautions I took. By continuing to behave *as if* I might have a heart attack, I have kept my anxiety going, because it was a constant reminder that I have to be careful, that I am vulnerable. I have never given myself the opportunity to test the truth of my fears.

The therapist uses the patient's understanding of how behaviours can prevent disconfirmation of feared disasters to devise direct tests; for example, a patient who felt weak during attacks and believed that she was going to fall would tense her muscles to prevent collapse. In the office therapy session, patient and therapist tried out just how weak muscles would have to be to result in collapse. The patient decided to test the possibility of falling at times when she felt panicky and weak by trying to *relax* her muscles in the same way she had in the therapy session. This exercise had the effect of completely changing her belief in the possibility of collapsing. This type of exercise, integrating cognitive challenges with short periods of exposure, is a particularly useful and rapid way of modifying both avoidance behaviour and catastrophic interpretations.

TREATMENT STUDIES

A variety of treatment studies have now been completed by the Oxford group and by other independent groups. Early studies were conducted in a relatively 'messy' way, with consecutive single case series conducted as part of routine clinical practice (see Salkovskis & Clark, 1986, for review of these). These studies used a quasi-experimental design which required the measurement of a baseline, followed by the demonstration of change associated with treatment that was greater in magnitude than any variation seen during the baseline period. Not only was it possible to show this, but it was also shown in these relatively unselected samples that the changes associated with treatment (most patients being panic-free at the end of treatment) persisted at the 2-year follow-up. These studies were criticized for not being sufficiently selective about the patients included and not adhering exactly to a pre-set treatment protocol. For example, several of the patients in the Clark *et al.* (1985) study required and were given additional treatment for non-panic related problems. As these people were treated as part of routine clinical practice, a certain (small) amount of time was devoted to dealing with problems such as assertiveness difficulties, marital problems and so on. This was fully documented in the published report, but was nevertheless a source of some criticism, and highlighted the need to conduct better controlled studies. A later single case series (Salkovskis *et al.*, 1991) was used in a quite different way. In this series, it was demonstrated that it was possible to make a substantial impact on

panic attack by using brief (2-hour) and purely cognitive (verbal) procedures. This study was consistent with the hypothesis that cognitive change
(focusing on the modification of the misinterpretation of bodily sensations)
could be an element in the effectiveness of cognitive therapy for panic in
reducing panic attack frequency. This study also gathered preliminary
evidence that cognitive procedures which do *not* target misinterpretations
may not be effective.

Following these clinically based studies, which demonstrated that the
effectiveness of cognitive therapy was unlikely to be due to spontaneous
remission or to non-specific factors (see also Beck, 1988), it was clearly
important to assess the effectiveness of cognitive therapy compared with
the best available alternative psychological and pharmacological treatments. In a recently completed study (Clark *et al.*, 1994), cognitive therapy
was compared to:

1. a waiting list control group;
2. imipramine delivered in the optimum fashion by an experienced
   biologically-oriented psychiatrist (gradually increased to high doses
   maintained to 6 months then gradually tapered);
3. applied relaxation, a specially modified form of relaxation for use with
   panic (Ost, 1988);
4. cognitive therapy.

All active treatments were delivered by experienced therapists in the
fashion currently considered optimum for each therapy; thus psychological treatments were 12 sessions over 12 weeks, pharmacotherapy was
continued for a total of 6 weeks.

Important features of this study were as follows.

1. The selection of patients who were particularly unlikely to show spontaneous remission (i.e. those patients experiencing frequent panic
   attacks with a minimum duration of 6 months).
2. Delivering each treatment in an optimum fashion, with considerable
   care being taken with treatment integrity (e.g. sessions audiotaped and
   checked against the competency scale for cognitive therapy adapted for
   panic treatment).
3. Ensuring that each treatment had an equally credible rationale and
   expectance of good outcome at the outset (i.e. before any differential
   effectiveness was apparent).
4. The inclusion of as many common elements as possible (e.g. daily
   recording/self-monitoring, printed therapy handouts early in treatment,
   self-exposure instructions in the later stages of treatment and so on).
5. The use of blind assessor ratings as well as self-ratings.
6. Inclusion of high-end-state functioning criteria in outcome assessment.
7. Maximum attention to follow-up completeness.

8. Inclusion of theoretically relevant variables to predict relapse.

The outcome of this study clearly indicated that

1. All three treatments were active (superior to the waiting list).
2. Both in the short term (12 weeks) and long-term (1-year post-treatment folllow-up), cognitive therapy was significantly superior to both applied relaxation and imipramine.
3. The degree to which catastrophic misinterpretations of bodily sensations remained at the end of treatment predicted relapse at 1-year follow-up, and that this remained the case when any residual symptomatology at the end of treatment was partialled out.
4. All treatments resulted in good levels of high end-state functioning at the end of treatment and at 1-year follow-up (cognitive therapy 80% and 70%; applied relaxation 25% and 32%; Imipramine 40% and 45%).

Particular problems had to be faced and dealt with in this study. For example, the issue of pre-trial medication, which potentially may complicate the interpretation of results. Should patients be 'washed out' prior to being included in trial treatments? The risk here is, of course, that a washout period may result in a rebound which would artificially elevate anxiety and panic levels during the crucial pre-treatment baseline assessments. On the other hand, concurrent medication may complicate the interpretation of results. The solution adopted in the trial was to exclude patients on effective doses of tricyclic antidepressants† (as this was one of the comparison conditions and is known to be effective in the treatment of panic); patients on other types of medication were asked to continue on pre-existing levels of medication for the first few weeks of treatment. A sub-analysis was also carried out in order to ensure that the outcome of patients taking previously prescribed medication did not differ from those not doing so.

The particular effectiveness of cognitive therapy has been replicated in a number of centres; a recent feature of this work has been the co-ordination of outcome measures allowing detailed comparisons across centres (Margraf et al., 1993).

The next stages in this line of research are as follows.

1. Improve cost-effectiveness, i.e. to evaluate whether it is possible to conduct treatment in a smaller number of sessions without sacrificing the outstanding results found by our own and other groups.
2. If a briefer treatment is found to be as effective, to return to the initial strategy of applying the therapy to a less selected clinical sample of people suffering from panic attacks.

---

† Previous clinical studies have showed that such patients responded equally well to cognitive treatment.

3. To evaluate the generalizability of cognitive methods and theory developed in panic by adapting these to the needs of other patient groups such as hypochondriasis (Salkovskis, 1989a; Salkovskis & Clark, 1993; Warwick & Salkovskis, 1990).

Work currently in progress suggests that it is possible, by the use of self-help manuals integrated with therapist contact, to reduce therapy time to around five sessions without loss efficacy, and that it is possible to extend some of the techniques used in panic to hypochondriasis, generalized anxiety and social phobia with suitable modifications. In these ways, the research effort is broadening out again to encompass more patients and therapists.

SUICIDE ATTEMPTERS, SOMATISERS AND COGNITIVE-BEHAVIOURAL PROBLEM SOLVING

It could be argued that both the specificity found and the effectiveness of cognitive therapy in specific phobias and panic relates to the homogeneity and severity of the problem studied rather than the adoption of the strategies integrating cognitive theory, research and treatment outlined here.

A major and very heterogeneous clinical group causing much concern at present are patients who repeatedly attempt suicide. Some patients (although surprisingly few) are clinically depressed or suffering from another identifiable psychiatric disorder; the rest appear to have a range of difficulties (Hawton & Catalan, 1987). Previous research suggests that there are specific cognitive problems in this group (MacLeod et al., 1992); overall, *a deficit in problem-solving ability* appears to be a common and prominent feature (Schotte & Clum, 1987). Previous intervention studies have been most disappointing (Hawton, 1989). However, careful examination of previous treatment studies suggests the following.

1. Interventions have not been targeted in terms of the known and hypothesized characteristics of repeated attempters.
2. Treatment has almost invariably focused on unselected patients who present with a parasuicidal attempt.

Salkovskis et al. (1990a) suggest that this second point has tended to result in a floor effect, so that studies involving unselected parasuicides have tended to find that psychopathology and repetition rate is low in the control group, making it virtually impossible to establish treatment effects relative to controls. However, extremely high probability of repetition can be predicted using Buglass and Horton's (1974) criteria; the important factor in this is that, in this especially disturbed and distressed group, it is known that the problems are particularly likely to persist under normal circumstances.

Other problems are presented in selecting this group, however; these patients are known to be relatively unlikely to attend the clinic for treatment appointments and even less so for research assessments. It was therefore decided to conduct a randomized controlled trial with this group, with brief problem-solving treatment conducted in the patients' own home, with the comparison group receiving treatment as usual together with the research assessments. Subsequent parasuicidal behaviour was also assessed.

Figure 9.6 shows the impact of cognitive-behavioural problem solving on these patients on two of the measures used.

1. problem-solving treatment was associated with a significant decrease in the main problem (i.e. the problem focused on during the five sessions of therapy);
2. this generalized to subsidiary problems;
3. treatment was associated with significant improvements in psychopathology as measured by depression (which was initially high) and hopelessness scores (which are known to be associated with suicidal behaviour).

Most encouraging of all, the proportion of patients repeating a suicide attempt in the 6 months after the index attempt was significantly less in the treatment group compared with controls (Fisher exact $p = 0.049$, two-tailed). In the treated group, mean time to repeat suicide attempt was 9.3 months, whilst in the controls it was three months.

A further treatment study, conducted as part of the same project, cast further interesting light on the issue of specificity. Accident and Emergency (A&E) departments are known to deal with a relatively high proportion of patients who repeatedly attend with relatively minor physical complaints. We were able to show that a very high proportion of such attenders could be identified as having psychiatric problems (Salkovskis et al., 1990b), and that a subgroup of the most disturbed who would be prepared to accept psychological intervention could be readily identified. On the hypothesis that these patients might be similar to those who attend A&E with repeated suicide attempts, an intervention study identical to that described above in the parasuicide group was carried out (Atha et al., 1992). The patients selected were those who fulfilled criteria which had previously been established as predicting a high probability of repeat A&E attendance. The patients showed substantial levels of psychological disturbance, although as a group they were slightly less severe than the parasuicide patients. Despite the lesser severity of psychopathology, the problem-solving intervention was found to have no effect at all relative to the treatment as usual control condition. The most likely explanation for this is that the treatment was simply not targeted specifically for this group; the attempted suicide patients' difficulties involved a problem-

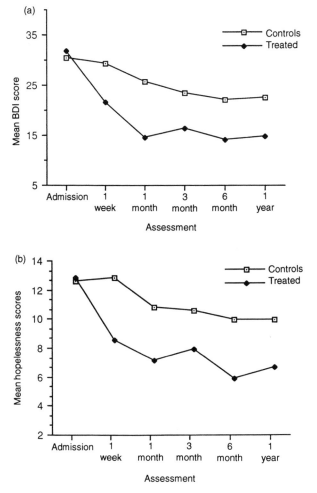

Figure 9.6 The impact of cognitive-behavioural problem solving on patients who repeatedly attempt suicide: (a) psychopathology (as measured by Beck Depression Inventory scores); (b) hopelessness scores (which are known to be associated with suicidal behaviour)

solving deficit, the A&E attenders' probably did not. The cognitive-behavioural research model suggests that research into the specific main-taining factors involved in this group is required in order to allow the appropriate focus of treatment.

SOME CONCLUSIONS

Research is a process of evolution, and is multi-faceted. In the cognitive-behavioural research described in this chapter, the psychotherapy research process can be regarded as taking an hourglass shape. At one end, clinicians with an interest in research conduct 'messy' exploratory investigations, often working with people who are not benefiting (or who are not benefiting enough) from currently available treatments. The research may be treatment development or focus on experimental psychopathology; most commonly it is an interacting combination of both. As a particular field evolves, it becomes increasingly important to refine and focus the research; a smaller number of more specialized research projects are carried out with more strictly selected patients, a 'purer culture' of the problem being tackled. The broad focus on specific effects and processes is provided by the earlier stage of research, but is further focused, refined and improved at this stage in an attempt to maximize the effectiveness of treatment. Beyond this more exacting stage, the research applications broaden out again, both in terms of the people conducting therapy, the people who can be helped and making the mode of delivery of the help itself more cost-effective without compromising the actual outcome. Each stage of this 'hourglass model' tends to require different research approaches and skills. All stages require consideration in terms of research funding; the process of therapeutic evolution is not a truly sequential process, but is a series of reciprocal interactions, with the key information coming, at each level, from a series of interactions between theory and clinically based research, as has been outlined above in this chapter.

The cognitive-behavioural approach assumes that helping people change involves:

1. helping them to identify why their problems persist;
2. working collaboratively to empower them to bring about desired changes by giving them the means to modify the factors involved in the maintenance of their problems.

The myth of psychotherapy equivalence is not helpful in the process of refinement. Like the idea that all swans are white, the range of exceptions is now too striking to be ignored. As more knowledge is acquired about maintaining factors involved in different types of psychological problems, the rate of progress increases. Of course, where research is not guided by attempts to understand the idiosyncratic nature of the factors involved in the maintenance of such problems, outcome studies may show that there is no difference between therapies. Where none of the therapies attempts to deal with the specific processes involved in psychological problems, poor outcomes compare favourably with other poor outcomes.

COMPARATIVE PSYCHOTHERAPY RESEARCH

A minimum requirement of comparative psychotherapy research *must* be the demonstration that the 'active' treatments are indeed effective themselves. This can be established by showing each treatment is, at worst, better than either a waiting list control or therapy controlling for non-specific factors; at best, better than a comparison treatment where the rationale for treatment is equally credible to the index treatments.

Further requirements are for finely tuned clinical theories integrated with experimental studies; for careful quality control over the independent variables (i.e. treatment integrity and distinctness as well as therapy fulfilling basic quality requirements in terms of so-called 'non-specific' factors, such as appropriateness of the therapeutic relationship and treatment credibility); for theoretically relevant manipulation checks, and for a combination of focused and generalized outcome measures. And, of course, the willingness to give up and replace cherished theories in the face of competing evidence. With these elements, progress and evolution may be possible in psychotherapy research.

According to the late Max Hamilton (personal communication), science proceeds in stages.

> In the first stage, we know nothing but believe that we know everything.
> In the second stage, we know nothing and believe that we know nothing.
> *There are no further stages.*

In order to progress, psychotherapy research must reach the second stage and begin to adopt strategies for navigating it in ways which could benefit those who depend on us for help. This should be a stimulus for never-ending attempts to push things forward rather than despairingly trying to find the lowest common denominator or integrate in forms of theory-free technical eclecticism.

REFERENCES

American Psychiatric Association. (1987). *Diagnostic and Statistical Manual of Mental Disorders*, 3rd edn revised. American Psychiatric Association, Washington DC: Author.

Barlow, D.H. & Craske, M.G. (1988). The phenomenology of panic. In *Panic: Psychological Perspectives* (S.J. Rachman & J. Maser, eds). Hillsdale, NJ: Erlbaum.

Beck, A.T. (1976). *Cognitive therapy and the emotional disorders*. New York: International University Press.

Beck, A.T. (1988), Cognitive approaches to panic disorder: theory and therapy. In *Panic: Psychological Perspectives* (S.J. Rachman & J. Maser, eds). Hillsdale, NJ: Erlbaum.

Beck, A.T., Emery, G. & Greenberg, R.L. (1985) *Anxiety Disorders and Phobias*. New York: Basic Books.

Beck, A.T., Laude, R. & Bohnert, M. (1974). Ideational components of anxiety neurosis. *Archives of General Psychiatry* **31**, 319–325.

Beck, A.T., Ward, C.H., Mendelson, M., Mock, J. & Erbaugh, J. (1961). An inventory for measuring depression. *Archives of General Psychiatry* 4, 561–571.

Breger, L. & McGaugh, J.L. (1965). Critique and reformulation of 'learning-theory' approaches to psychotherapy and neurosis. *Psychological Bulletins* 63, 338–358.

Buglass, D. & Horton, J. (1974). A scale for predicting subsequent suicidal behaviour. *British Journal of Psychiatry* 124, 573–578.

Clark, D.M. (1986). A cognitive approach to panic. *Behaviour Research and Therapy* 24, 461–470.

Clark, D.M. (1988). A cognitive model of panic. In *Panic: Psychological Perspectives* (S.J. Rachman & J. Maser, eds). Hillsdale, NJ: Erlbaum.

Clark, D.M. (1993). Cognitive mediation of panic attacks induced by biological challenge tests. *Advances in Behaviour Research and Therapy* 15, 75–84.

Clark, D.M. & Salkovskis, P.M. (In press). *Cognitive Therapy for Panic and Hypochondriasis.*

Clark, D.M., Salkovskis, P.M. & Chalkley, A.J. (1985). Respiratory control as a treatment for panic attacks. *Journal of Behaviour Therapy and Experimental Psychiatry* 16, 23–30.

Clark, D.M., Salkovskis, P.M., Gelder, M.G., Koehler, C., Martin, M., Anastasiades, P., Hackmann, A., Middleton, H. & Jeavons, A. (1988). Tests of a cognitive theory of panic. In *Panic and Phobias II* (I. Hand & H.U. Wittchen, eds). New York: Springer-Verlag.

Clark, D.M., Salkovskis, P.M., Hackmann, A., Middleton, H., Anastasiades, P. & Gelder, M. (1994). A comparison of cognitive therapy, applied relaxation and Imipramine in the treatment of panic disorder. *British Journal of Psychiatry*, 164, 759–769.

Garssen, B. (1986). Agoraphobia and the Hyperventilation syndrome—the role of interpretations of complaints. In *Panic and Phobias II.* (I. Hand & H.U. Wittchen, eds). Berlin: Springer.

Garssen, B., Veenendaal, W.V. & Bloemink, R. (1983). Agoraphobia and the hyperventilation syndrome. *Behaviour Research and Therapy.* 21, 643–649.

Gorman, J.M., Liebowitz, M.R., Fyer, A.J. & Stein, J. (1989). A neuroanatomical hypothesis for panic disorder. *American Journal of Psychiatry.* 146, 148–161.

Hawton, K. (1989). Controlled studies of psychosocial intervention following attempted suicide. In *Current Research in Suicide and Parasuicide* (N. Kreitman & S.D. Platt, eds). Edinburgh: Edinburgh University Press.

Hawton, K, & Catalan, J. (1987). *Attempted Suicide: a Practical Guide to its Nature and Management.* Oxford: Oxford University Press.

Hawton, K., Salkovskis, P., Kirk, J. & Clark, D.M. (1989). *Cognitive, Behaviour Therapy for Psychiatric Problems: a Practical Guide.* Oxford University Press.

Hornsveld, H., Garssen, B., Dop, M.F. & van Spiegel, P. (1992). Symptom reporting during voluntary hyperventilation and mental load: implications for diagnosing hyperventilation syndrome. *Journal of Psychosomatic Research.*

Klein, D.F. & Fink, M. (1962). Psychiatric reaction pattern to imipramine. *American Journal of Psychiatry* 119, 438.

Kopp, M., Milhaly, K. & Vadasz, P (1986). Agoraphobics es panikneurotikus betegek legzesi kontroll keyelese. *Ideggyogyaszati Szemle.* 39, 185–196.

Lader, M. & Mathews, A.M. (1968). A physiological model of phobia anxiety and desensitisation. *Behaviour Research and Therapy* 6, 411–418.

Ley, R. (1985a). Agoraphobia, the panic attack and the hyperventilation syndrome. *Behaviour Research and Therapy* 23, 79–82.

Ley, R. (1985b). Blood, breath and fears: a hyperventilation theory of panic attacks and agoraphobia. *Clinical Psychology Review* 5, 271–285.

Lum, L.C. (1976). Hyperventilation syndrome. In *Modern trends in psychosomatic medicine* (O.W. Hill, ed.). London: Butterworths.

MacLeod, A.K., Williams, J.M.G. & Linehan, M.M. (1992). New developments in the understanding and treatment of suicidal behaviour. *Behavioural Psychotherapy* **20**, 193–219.

Margraf, J., Barlow, D.H., Clark, D.M. & Telch, M.J. (1993). Psychological treatment of panic: work in progress on outcome, active ingredients and follow up. *Behaviour Research and Therapy* **31**, 1–8.

Marks, I.M. (1987). *Fears, Phobias and Rituals*. Oxford University Press: New York.

Mathews, A.M., Gelder, M.G. & Johnston, D.W. (1981). *Agoraphobia: Nature and Treatment*. Guildford Press, New York.

Ost, L.G. (1988). Applied relaxation vs progressive relaxation in the treatment of panic disorder. *Behaviour Research and Therapy*. **26**, 13–22.

Ost, L.G. (1989). One session treatment for specific Phobias. *Behaviour Research and Therapy* **27**, 1–7.

Ost, L.G., Salkovskis, P.M. & Hellstrom, K. (1991). One-session therapist-directed exposure vs self-exposure in the treatment of spider phobia. *Behavior Therapy* **22**, 407–422.

Paul, G.L. (1966). *Insight versus desensitisation in psychotherapy*. Stanford CA: Stanford University Press.

Salkovskis, P.M. (1986). The cognitive revolution: new way forward, backward somersault or full circle? *Behavioural Psychotherapy* **14**, 278–282.

Salkovskis, P.M. (1988a). Phenomenology, assessment and the cognitive model of panic. In *Panic: Psychological Perspectives* (S.J. Rachman & J. Maser, eds). Hillsdale, NJ: Erlbaum.

Salkovskis, P.M. (1988b). Hyperventilation and anxiety. *Current Opinion in Psychiatry* **1**, 76–82.

Salkovskis, P.M. (1989a). Somatic Disorders. In *Cognitive-behavioural Approaches to Adult Psychological Disorder: a Practical Guide* (K. Hawton, P.M. Salkovskis, J.W. Kirk & D.M. Clark, eds). Oxford University Press: Oxford.

Salkovskis, P.M. (1989b). Cognitive models and interventions in anxiety. *Current Opinion in Psychiatry* **2**, 795–800.

Salkovskis, P.M. (1991). The importance of behaviour in the maintenance of anxiety and panic: a cognitive account. *Behavioural Psychotherapy* **19**, 6–19.

Salkovskis, P.M., Atha, C. & Storer, D. (1990a). Cognitive-behavioural problem solving in the treatment of patients who repeatedly attempt suicide: a controlled trial. *British Journal of Psvchiatry* **157**, 871–876.

Salkovskis, P.M. & Clark, D.M. (1986). Cognitive and physiological processes in the maintenance and treatment of panic attacks. In *Panic and Phobias* (I. Hand & U.H. Wittchen, eds). Berlin: Springer-Verlag.

Salkovskis, P.M. & Clark, D.M. (1990). Affective response to hyperventilation: a test of the cognitive model of panic. *Behaviour Research and Therapy* **28**, 51–61.

Salkovskis, P.M. & Clark, D.M. (1993). Panic disorder and hypochondriasis. *Advances in Behaviour Research and Therapy* **15**, 23–48.

Salkovskis, P.M., Clark, D.M. & Hackmann, A. (1991). Treatment of panic attacks using cognitive therapy without exposure to feared situations or bodily sensations. *Behaviour Research and Therapy* **29**, 161–166.

Salkovskis, P.M., Clark, D.M. & Jones, D.R.O. (1986a). A psychosomatic mechanism in anxiety attacks: the role of hyperventilation in social anxiety and cardiac neurosis. In *Proceedings of the 15th European Conference on Psychosomatic Medicine* (H. Lacey & J. Sturgeon, eds). London: Libby.

Salkovskis, P.M., Jones, D.R.O. & Clark, D.M. (1986b). Respiratory control in the treatment of panic attacks: replication and extension with concurrent measurement of behaviour and pCO_2. *British Journal of Psychiatry* **148**, 526–532.

Salkovskis, P.M., Storer, D., Atha, C. & Warwick, H.M.C. (1990b). Psychiatric morbidity in an accident and emergency department: characteristics of patients at presentation and one month follow-up. *British Journal of Psychiatry* **156**, 483–487.

Salkovskis, P.M. & Warwick, H.M.C. (1986). Morbid Preoccupations, health anxiety and reassurance: a cognitive behavioural approach to hypochondriasis. *Behaviour Research and Therapy* **24**, 597–602.

Salkovskis, P.M., Warwick, H.M.C. & Clark, D.M. (1990c). *Hypochondriasis, Illness, Phobia and other Anxiety Disorders*. Paper submitted to the American Psychiatric Association working group on DSM-IV (Anxiety, Disorders and Simple Phobias).

Salkovskis, P.M., Warwick, H.M.C., Clark, D.M. & Wessels, D.J. (1986c). A demonstration of acute hyperventilation during naturally occurring panic attacks. *Behaviour Research and Therapy* **24**, 91–94.

Schotte, D.E., & Clum, G.A. (1987). Problem solving skills in suicidal psychiatric patients. *Journal of Consulting and Clinical Psychology* **55**, 49–54.

Seligman, M.E.P. (1988). Competing theories of panic. In *Panic: Psychological Perspectives* (S.J. Rachman & J.D. Maser, eds). Hillsdale, NJ: Erlbaum.

de Silva, P. (1984). Does escape behaviour strengthen agoraphobic avoidance? A preliminary study. *Behaviour Research and Therapy* **22**, 87–91.

Sokol, L., Beck, A.T., Greenberg, R.L., Wright, F. & Berchick, R.J. (1989). Cognitive therapy of panic disorder: a non-pharmacological alternative. *Journal of Nervous and Mental Disease* **177**, 711–716.

Teasdale, J.D. (1983). Negative thinking in depression: cause, effect or reciprocal relationship? *Advances in Behaviour Research and Therapy* **5**, 3–25.

Teasdale, J.D. & Barnard, P.J. (1993). *Affect, Cognition and Change*. Hillsdale, NJ: LEA.

Warwick, H.M.C. & Salkovskis, P.M. (1990). Hypochondriasis. *Behaviour Research and Therapy* **28**, 105–118.

Williams, J.M.G., Watts, F.N., MacLeod, C. & Mathews, A. (1988). *Cognitive Psychology and the Emotional Disorders*. Wiley: Chichester, UK.

Wolpe, J. (1958). *Psychotherapy by reciprocal inhibition*. Stanford, CA: Stanford University Press.

Part III

CURRENT ISSUES: COMBINED THERAPY, SEXUAL ABUSE VICTIMS AND COUNSELLING IN GENERAL PRACTICE

10 The Relationship Between Drug and Psychotherapy Effects

IVY-MARIE BLACKBURN
Newcastle Cognitive Therapy Centre, UK

ABSTRACT This chapter examines the relationship between pharmacotherapy and one type of psychotherapy, cognitive therapy, with emphasis on the treatment of depression. It is expected that the comments can be generalized to other disorders and other psychotherapies. Although antidepressant medication and cognitive therapy have generally been shown to be equally efficacious in the treatment of depressed outpatients satisfying set criteria for major unipolar depression, little is known about the specific effects of these two treatment modalities. Outcome research using overinclusive diagnostic subgroups does not give much indication about which treatment would best suit the individual patient seen in the clinic. Similarly, little is known about the process of action of the two treatments and when they are best delivered singly or in combination. The methodological problems inherent in comparing psychotherapy and pharmacotherapy are discussed and possible ways forward for future research signposted.

INTRODUCTION

Pharmacotherapy and psychotherapy for psychiatric disorders have often been contrasted and compared for their efficacy in a competitive and antagonistic spirit. The competitiveness is probably a healthy reflection of the keen interest of researchers and clinicians working from different theoretical approaches, but the antagonism reflects the conceptual difficulties which derive from the cartesian dualism which is still deeply anchored in our thinking. There are evident core differences between these two treatment modalities and the natural inference is that their mode of action must be different and/or that they cannot be equally effective for

Correspondence address: Newcastle Cognitive Therapy Centre, Newcastle Mental Health NHS Trust, Collingwood Clinic, St Nicholas Hospital, Gosforth, Newcastle Upon Tyne NE3 3XT, UK.

Research Foundations for Psychotherapy Practice. Edited by M. Aveline and D. A. Shapiro.
Copyright © 1995, Mental Health Foundation and Individual Contributors.
Published 1995 by John Wiley & Sons Ltd

the same type of patient. The questions which arise are, therefore, primarily:

1. Are drugs and psychotherapy equally effective for the same group of patients?
2. When faced with individual patients, can we decide with some degree of certainty who would best benefit from medication and who from psychotherapy?
3. Is the process of action different in the two treatments?

In this paper, I shall address these three questions and the methodological problems related to research in these areas. I shall also discuss possible ways forward in future research. The primary emphasis will be on the effects of cognitive therapy and of pharmacotherapy in the treatment of depression, as this has been the most fruitful area of research to date. However, I think that my comments could be generalized to other disorders and other psychotherapies.

ARE DRUGS AND PSYCHOTHERAPY EQUALLY EFFECTIVE FOR THE SAME GROUP OF PATIENTS?

Over the last 30 years, hundreds of studies have examined the relative efficacy of these two modes of treatment. With regard to depression, the first comparative outcome study of Rush *et al.* (1977), comparing cognitive therapy and imipramine, had an immense impact in that it was the first study which had ever indicated that a psychotherapeutic approach could be at least as effective as antidepressant medication in major depression. Since then, controlled and uncontrolled outcome studies have been published at regular intervals, the latest being in October 1992 (Hollon *et al.*). Some of these studies have dealt with in-patients (e.g. Miller *et al.*, 1985), but most have been concerned with outpatients satisfying set criteria for major, unipolar depression. Examining controlled studies in this particular group of outpatients and only those which have used a specified type of cognitive therapy (namely, that of Beck *et al.*, 1979), I recently counted 15 studies. These are summarized in Table 10.1.

It is not necessary to examine these studies in detail here, as the aims of this paper are not to determine who wins the 'horse race', but rather to discuss methodological issues. (For detailed reviews, see Blackburn, 1988; Stravynski & Greenberg, 1992).

The studies in Table 10.1 were selected because they are all relatively methodologically sound, although all have weaknesses of their own. The general findings are that:

1. Cognitive therapy alone is never inferior to medication alone.

Table 10.1. Comparisons of cognitive therapy (CT) with pharmacotherapy in unipolar major depression—outpatients

Study	Outcome
Rush *et al.* (1977) (N = 41)	CT > imipramine
Beck *et al.* (1979) (N = 26)	CT = CT + amitriptyline
Dunn (1979) (N = 20	CT + imipramine > supportive therapy and imipramine
McLean & Hakstian (1979) (N = 154)	CT > amitriptyline = relaxation > insight therapy
Blackburn *et al.* (1981) (N = 64)	Hospital clinic: COM > CT = antidepressants General practice: CT = COM > antidepressants
Rush & Watkins (1981) (N = 38)	Individual CT = individual CT + antidepressants > group
Murphy *et al.* (1984) (N = 70)	CT = nortriptyline = CT + nortriptyline = CT + placebo
Teasdale *et al.* (1984) (N = 34) (GP)	CT + treatment as usual > treatment as usual
Beck *et al.* (1985) (N = 33)	CT = CT + amitriptyline
Ross & Scott (1985) (N = 51) (GP)	CT (individual) = CT (group) > treatment as usual
Beutler *et al.* (1987) (N = 56 (elderly)	Group CT = CT + alprazolam = group CT + placebo = placebo
Covi & Lipman (1987) (N = 90)	Group CT = group CT + imipramine > psychodynamic
Elkin *et al.* (1989) (N = 239)	Imipramine + clinical management = CT = IPT. Imipramine + clinical management > placebo + clinical management. For HRSD > 20, GAS < 50 Imipramine + clinical management > IPT > CT > placebo + clinical management
Scott & Freeman (1992) (N = 121) (GP)	CT = counselling = amitriptyline = treatment as usual
Hollon *et al.* (1992) (N = 154)	CT = imipramine = CT + imipramine

COM = combination of antidepressant medication and cognitive therapy.
GP = general practice.
HRSD = Hamilton rating Scale for Depression (Hamilton, 1960).
GAS = Global assessment Scale (Endicott *et al.*, 1976).

2. Endogenous depression appears to respond to cognitive therapy as well as to medication.

3. Cognitive therapy in combination with medication has sometimes been found to be superior to either treatment alone and sometimes of equivalent efficacy. It is noteworthy that the combination of the two treatments has never been found to be inferior to either treatment given in isolation (an inhibition effect), as was feared by earlier polemists (Ulenhuth et al., 1969).

The methodological problems which transpire are now discussed.

DIAGNOSIS

Although most studies now make use of structured interviews (e.g. the Schedule for Affective Disorders and Schizophrenia, SADS, Endicott & Spitzer, 1978, or the Present State Examination, PSE, Wing et al., 1974,) and of Research Diagnostic Criteria (RDC) (e.g. Feighner et al., 1972; Spitzer et al., 1978), a diagnostic group such as major depression is known to be very overinclusive (Nelson et al., 1978; Feinberg et al., 1979). Notwithstanding the advantage of recognized and explicit diagnostic criteria in the selection of patients for studies, we have to guard against the current tendency of reifying these criteria.

Patients satisfying criteria for major depression vary in severity of illness which, as Elkin et al.(1989) demonstrated in their large NIMH study, can discriminate between drug and cognitive therapy response. These authors found no difference in response between active medication plus clinical management, placebo plus clinical management and two types of psychotherapy (interpersonal and cognitive) in the less severely depressed (scores <20 on the Hamilton Rating Scale for Depression, Hamilton, 1960). However, in the more severely depressed (scores ⩾20), medication was clearly superior to placebo plus clinical management, while cognitive therapy was not differentiated from placebo plus clinical management (see Table 10.1). It must, however, be noted that other researchers (McLean & Taylor, 1992; Hollon et al., 1992; Thase et al., 1991) have contested these findings and challenged the comments following the NIMH study in two editorials (Freedman, 1989; Gelder, 1990) that drug treatment is the first line treatment for the most severe depressive disorders. McLean and Taylor re-analysed the data from the McLean and Hakstian (1979) study and found no severity by treatment interactions. Similarly, Hollon et al. (1992) and Thase et al. (1991) found no effect for severity of depression on response to cognitive therapy.

In addition to severity of illness, patients may vary in symptoms profile. As mentioned earlier, the criteria for endogenous depression from the RDC (Spitzer et al., 1978) are not very exclusive (Klein, 1974). They may or may not include a subtype of depression labelled 'melancholia', in which biological dysfunction has been shown to be particularly implicated and which may be resistant to psychotherapy (DSM-III-R) (American Psychiatric Association, 1987). Different criteria for endogenous depression are not highly concordant, indicating the confusion attached to the term. In a current study in Edinburgh we found a Kappa coefficient of 0.19 between the RDC for endogenous depression and the Newcastle Depression Index (Carney *et al.*, 1965) for endogenous and neurotic depression. Thus, although several studies from Table 10.1 (Rush *et al.*, 1977; Blackurn *et al.*, 1981; Teasdale *et al.*, 1984) have found that an endogenous pattern of symptoms does not predict differential response to medication and to psychotherapy, these results are ambiguous and not very informative.

To conclude this section on the problems of diagnosis in relation to drug and psychotherapy effects, Costello's (1992) arguments in favour of allocating more research time to the study of symptoms rather than, or as well as, of syndromes are to be supported. I am not arguing against the use of diagnostic systems with explicit criteria, instead I am making a plea for researchers to go one step further. No progress can be made in the elucidation of response predictors to drugs and psychotherapy unless we study the effect of these two modes of treatment in detail on different classes of symptoms, perhaps using statistical methods such as time series analysis.

PLACEBO CONTROL

Placebo is a *sine qua non* in the methodology of controlled drug trials to distinguish between the specific and non-specific effects of treatment. Does psychotherapy, in particular cognitive therapy, need a placebo control to add credence to the evidence shown in Table 10.1? Parloff (1986) makes an elegant plea against the inclusion of placebo-control psychotherapy in outcome research. The main problems regarding placebo psychotherapy are as follows.

1. Good psychotherapy—of whatever theoretical orientation—must contain the so-called non-specific elements of warmth, genuineness and understanding. It is considered essential that the psychotherapist encourages hope, confidence and expectation of change in his/her patient. These characteristics are not considered inert in the same way that a placebo is inert.
2. Placebo psychotherapy cannot be standardized, as what may be considered as non-specific by one tye of psychotherapy may be seen quite differently by other therapies.
3. Psychological treatments, by definition, aim at psychological changes unlike pharmacotherapy which aims at physiological changes, so that a

placebo pill is expected to produce psychological changes (a subjective feeling of improvement), but not alter the physiological bases of the illness. The distinction between placebo psychotherapy and active psychotherapy can therefore be very blurred.
4. A placebo pill should be indistinguishable from an active pill to ensure that the patient and the therapist *believe* in the treatment. To devise a placebo psychotherapy which is entirely credible to both patient and therapist presents insurmountable problems.

The best procedure, therefore, in studying the effects of drug and psychotherapy is to compare a specified psychotherapy with an effective alternate treatment and/or a no-treatment condition (London & Klerman, 1982). In addition, studies such as that of the NIMH (Elkin *et al.*, 1989), which compared two psychotherapies with different theoretical assumptions and different procedures, are also extremely informative. The use of active therapies to serve as placebo for each other has been proposed before (Rosenthal & Frank, 1956).

SYSTEMATIZATION OF PSYCHOTHERAPY

Although there has been progress in the systematization of psychotherapy in the form of treatment manuals (Beck *et al.*, 1979; Blackburn & Davidson, 1990; Klerman *et al.*, 1984), the very nature of psychotherapy prevents complete systematization of its delivery to the same extent that it can be done in the administration of medication. The best attempts at monitoring the quality of psychotherapy in an outcome study to date has been the NIM study where therapists were continually monitored and given feedback. In spite of this, as would be expected, the quality of therapy varied and the results with cognitive therapy varied across study centres (Elkin *et al.*, 1989). De Rubeis and Feeley (1990) reported that cognitive therapists who use the focused, pragmatic aspects of cognitive therapy achieve a better and earlier outcome than therapists who use more abstract and less focused methods.

A moot point, therefore, in psychotherapy research is how much training should therapists have and who should be trained? A problem which is also generally ignored is the dyadic nature of therapy. Some therapists are better with some types of patient than others and vice versa. Usually, when patients do not respond to treatment, we examine the characteristics of the non-responders in terms of illness variables and demographic and psychosocial factors. However, at least one reviewer (Beckham, 1990) has emphasized the role of the therapist in non-response, not only in terms of expertise, but also in terms of personal characteristics, gender and attitudes.

COMPLIANCE

Compliance is another problem area which bedevils outcome studies, in particular the interpretation of psychotherapy effects. Studies involving the measurement of drug effects have now developed various means of monitoring compliance to medication. These involve pill counts, specially constructed pill distributors, measurement of plasma or urinary level of the active ingredients of the medication. Compliance with psychotherapy cannot be measured so directly or so accurately. However, studies should include some minimal attempt at monitoring compliance with psycho-therapy, for example, number of therapy sessions missed, turning up late for appointments, non-compliance with homework assignments and simple questionnaires for feedback about the client's view of therapy. As far as I am aware, none of the studies listed in Table 10.1 has followed such procedures.

In summary, when looking at drug and psychotherapy effects, although studies may show an equivalence effect, there remains the possibility of a Type II error (accepting the null hypothesis when it is in fact false) because of the weaknesses of outcome research, as discussed above. Some of these weaknesses are inherent in this type of research, but some can be corrected.

WHEN TO PRESCRIBE MEDICATION AND WHEN TO REFER TO PSYCHOTHERAPY?

The key question for the general practitioner or the psychiatrist when faced by an individual depressed outpatient in an outpatient clinic is, What would best benefit this patient? As discussed above, research to date can offer only limited answers. There is a plethora of antidepressant medi-cation of known efficacy, varying in their modes of action and in cost (Wright, 1993). The choice of drug will depend on the patient's profile of symptoms, his/her physical condition, previous drug profile and, in these cost-conscious times, on the actual cost of a course of treatment. The choice of psychotherapy is more ambiguous (Robinson et al., 1990) and the choice between drug and psychotherapy will often depend on patient's pre-ference, non-drug response in the past, physical health contra-indicating the use of antidepressants, availability of psychotherapists and, again, cost. There is no doubt that medication is much cheaper than psychological treatment, even when it is short-term and of proven efficacy (Scott & Freeman, 1992).

However, since the highly recurring nature of affective illness is now widely recognized (Kupfer, 1992), the question of relative efficacy of different treatments and of costs involved relates not only to the treatment of acute episodes of illness, but primarily to the long-term effects of treatment. The effects of maintenance medication in depression are well known (Glen *et al.*, 1984; Prien *et al.*, 1984). Nonetheless, these two studies showed that even patients on active medication had less than 50% probability of remaining well for 2–3 years. More recent studies of the prophylactic effect of maintenance medication, using full doses instead of reduced doses of antidepressants, have shown a far superior effect (Frank *et al.*, 1990—3-year controlled study; Kupfer *et al.*, 1992—5-year controlled study).

The evident problem of compliance with high doses of medication over long periods of time, if not for a lifetime, makes the long-term or prophylactic effect of psychotherapy a very important research question both in terms of clinical outcome and of economics.

Unfortunately, the prophylactic effect of psychotherapy has not been studied in controlled trials. Table 10.2 lists six studies of the long-term effect of cognitive therapy which hold a promise that recurrence of illness may be reduced with this type of psychotherapy.

There are several methodological problems with these studies.

1. They are naturalistic (except for Evans *et al.*, 1992) rather than controlled and, therefore, it is not possible to say what further treatment the patients in the different groups may have sought during the follow-up period.
2. The results are based on relatively small numbers, although the fact that all the data indicate the same outcome reduces the likelihood of false positives (Type I error).
3. Most studies report follow-up data only on patients who have recovered at the end of acute treatment. The long-term results are, therefore, selective as no information is available on the total original groups which had been randomly allocated to different treatments.
4. No studies have looked at the maintenance effect of cognitive therapy post-treatment. On the other hand, another type of psychotherapy, interpersonal psychotherapy (IPT) was compared, as a maintenance treatment, with imipramine and found to be less effective (Frank *et al.*, 1990). In a further study, the same group of researchers (Frank *et al.*, 1991) found that the quality of IPT sessions was highly related to outcome. Patients with high quality psychotherapy had a median survival time before recurrence of symptoms of 2 years and patients with lower quality psychotherapy had a survival time of less than 5 months.

In summary, research into the effects of drug and psychotherapy does not help the physician and the therapist in deciding which treatment mode would best benefit an individual patient. In deciding how to reduce the likelihood of recurrence of illness, recent research indicates that high doses of maintenance medication over long periods of time reduce recurrences

Table 10.2. Long-term effects of psychological treatment (% recurrence)

Cognitive Therapy (CT)	12 months	18 months	24 months
Kovacs *et al.* (1985)			
(N = 35)			
CT	33	–	–
TCA	59	–	–
Beck *et al.* (1985)			
(N = 33)			
CT	42	–	–
CT + TCA	9	–	–
Simons *et al.* (1986)			
(N = 79)			
CT	20	–	–
TCA	66	–	–
CT + TCA	43	–	–
CT + P	18	–	–
Blackburn *et al.* (1986)			
(N = 36)			
CT			23
TCA			78
CT + TCA			21
Shea *et al.* (1992)			
(N = 61)			
CT	–	41	–
TCA + CM	–	61	–
IPT	–	57	–
Evans *et al.* (1992)			
(N = 44)			
TCA (no continuation)	–	–	50
TCA (no continuation)	–	–	32
CT	–	–	21
CT + TCA	–	–	15

CT = cognitive therapy.
IPT = interpersonal psychotherapy.
TCA = tricyclic antidepressant.
CM = clinical management.
P = placebo.

markedly. The problems of compliance and the desirability of long-term medication increase the necessity of improved methodology in research in the long-term effect of psychotherapy. In particular, randomized post-treatment maintenance may be indicated and larger groups of patients have to be followed-up. The quality of therapy may be as important as the dosage of medication in the prevention of relapse and recurrence of illness.

THE PROCESS OF ACTION OF DRUGS AND PSYCHOTHERAPY

Very few studies have addressed this aspect of treatment. The NIMH study (Elkin *et al.* 1989) found a more rapid effect for antidepressants than for cognitive therapy or interpersonal psychotherapy (Watkins *et al.*, 1986). Rush *et al.*, (1981) found that with cognitive therapy, improvements in hopelessnes, view of self and mood generally preceded changes in vegetative and motivational symptoms, while no consistent pattern of change over time was found with pharmacotherapy. Rush *et al.* (1982) found greater improvement in hopelessness and self-esteem with cognitive therapy than with medication. Other researchers (Simons *et al.*, 1984; Blackburn & Bishop, 1983) did not come to the same conclusion, finding, instead, that changes in different classes of symptoms were commensurate with the general efficacy of the treatment, rather than more pronounced in one treatment relative to the other. More recently, DeRubeis *et al.* (1990) reported that changes in cognitive variables (attributional style, dysfunctional attitudes and hopelessness) at mid-treatment predicted overall improvement at the end of treatment with cognitive therapy, but not with medication. If this effect is replicated in future studies, these author's conclusion, that cognitive constructs play a mediational role in cognitive therapy but that this role is not sufficient, will be an important one in the understanding of the effects of drug and psychotherapy. A recent theoretical paper (Persons, 1993) considered whether cognitive therapy changes basic schemata, which are presumed to be vulnerability factors, or teaches compensatory skills. The two models generate different predictions regarding the *timing* of change (which would occurr earlier in the schema change model) and the *generalizability* of what is learned in therapy (the compensatory skills model providing more general skills which should provide more protection against future episodes of depression). The two models are empirically testable and it is hoped that some answers will be provided in the near future.

The understanding of the process of change in psychotherapy is of central importance to improve our therapeutic methods and, possibly, explain the equivalence effect of different therapies. However, as mentioned in a previous section of this paper, only detailed studies using sophisticated statistical techniques will be able to shed more light on the action of these two different treatment modes which, nonetheless, appear to have equivalent effects. It is also likely that neuropsychological techniques employing brain imaging techniques may be able to clarify how a talking therapy and drugs act in brain systems to achieve the same effect.

FUTURE DIRECTIONS

This short review of the effect of drugs and psychotherapy has been very selective in terms of the therapy, the class of drugs and the disorder discussed. Even within these confines, the studies discussed and the methodological problems highlighted have not been exhaustive. However, the author believes that some of the main problems inherent in the comparison of drug and psychotherapy in general have been considered. The main issues raised have been the selection of patients, including both inclusion and exclusion criteria; a plea for the study of symptoms as well as of syndromes; the systematization of therapy as well as of drug treatment, as specified by the NIMH Collaborative Research Program; the training of therapists; better designed follow-up studies and more detailed process studies, using up-to-date biomedical technology.

In a recurrent and costly illness such as depression (West, 1992), the emphasis has rightly moved from short-term outcome to long-term outcome. The time may now have come to focus on the prevention of onset of the disorder. Many of the factors associated with the onset of depression are known, for example, the occurrence of major negative life events, having a depressed parent and the presence of dysthymic symptoms, being some of the factors which have been implicated. It has also been documented that there is a disturbing increase in the prevalence of depression in childhood and adolescence (Klerman & Weissman, 1989). We can keep patients on medication when they have recovered from a depression, but we cannot give antidepressant drugs to groups of people who may be vulnerable to depression, but who are not yet depressed. However, psychotherapy, in small doses, may be an acceptable and effective way of preventing the onset of depression.

REFERENCES

American Psychiatric Association (1987). *Diagnostic and Statistical Manual of Mental Disorders—DSM-III-R*, 3rd edn, revised. Washingtron DC.

Beck, A.T., Hollon, S.D., Young, J.E. Betrosian, R.C. & Budenz, D. (1985). Treatment of depression with cognitive therapy and amitriptyline. *Archives of General Psychiatry* **42**, 142–148.

Beck, A.T., Rush, A.J., Shaw, B.F. & Emery, G. (1979). *Cognitive Therapy of Depression*. Guildford: New York.

Beckham, E.E. (1990). Psychotherapy of depression. Research at a crossroads: directions for the 1990's. *Clinical Psychology Review* **10**, 207–228.

Beutler, L.E., Scogin, F., Kirkish, P., Schretlen, D., Corbishley, A., Hamblin, D., Meridith, K., Potter, R., Bamford, C.R. & Levenson, A.I. (1987). Group cognitive therapy and alprazolam in the treatment of depression in older adults. *Journal of Consulting Clinical Psychology* **55**, 550–556.

Blackburn, I.-M. (1988). An Appraisal of Comparative Trials of Cognitive Therapy for Depression. In *Cognitive Psychotherapy. Theory and Practice* (C. Perris, I.-M. Blackburn & H. Perris, eds). Heidelberg: Springer-Verlag, pp 160–178.

Blackburn, I.-M. & Bishop, S. (1983). Changes in cognition with pharmacotherapy and cognitive therapy. *British Journal of Psychiatry* **143**, 609–617.

Blackburn, I.-M., Bishop, S., Glen, A.I.M., Whalley, L.J. & Christie, J.E. (1981). The efficacy of cognitive therapy in depression: a treatment trial using cognitive therapy and pharmacotherapy, each alone and in combination. *British Journal of Psychiatry* **139**, 181–189.

Blackburn, I.-M. & Davidson, K.M. (1990). *Cognitive Therapy for Depression and Anxiety. A Practitioner's Guide.* Oxford: Blackwell Scientific Publications.

Blackburn, I.-M., Eunson, K.M. & Bishop, S. (1986). A two-year naturalistic follow-up of depressed patients treated with cognitive therapy, pharmacotherapy and a combination of both. *Journal of Affective Disorders* **10**, 67–75.

Carney, M.W.P., Roth, M. & Garside, R.F. (1965). The diagnosis of depressive syndromes and the prediction of E.C.T. response. *British Journal of Psychiatry* **111**, 659–674.

Costello, C.G. (1992). Research on symptoms versus research on syndromes: arguments in favour of allocating more research time to the study of symptoms. *British Journal of Psychiatry* **160**, 304–308.

Covi, L. & Lipman, R.S. (1987). Cognitive behavioural group psychotherapy combined with imipramine in major depression. *Psychopharmacol Bulletin* **23**, 173–176.

DeRubeis, R.J., Evans, M.D., Hollon, S.D., Garvey, M.J., Grove, W.M. & Tuason, V.B. (1990). How does cognitive therapy work? Cognitive change and symptom change in cognitive therapy and pharmacotherapy for depression. *Journal of Consulting and Clinical Psychology* **58**, 862–869.

DeRubeis, R.J. & Feeley, M. (1990). Determinants of change in cognitive therapy for depression. *Cognitive Therapy and Research* **14**, 469–482.

Dunn, R.J. (1979). Cognitive modification with depression-prone psychiatric patients. *Cognitive Therpy and Research* **3**, 307–317.

Elkin, I., Shea, M.T., Watkins, J.T., Imber, S.D., Sotsky, S.M., Collins, J.F., Glass, D.R., Pilkoniz, P.A., Leber, W.R., Docherty, J.P., Fiester, S.J. & Parloff, M.B. (1989). NIMH Treatment of depression collaborative research program: general effectiveness of treatments. *Archives of General Psychiatry* **46**, 971–982.

Endicott, J. & Spitzer, R.L. (1978). A diagnostic interview: the schedule for affective disorders and schizophrenia. *Archives of General Psychiatry* **35**, 837–844.

Endicott, J., Spitzer, R.L., Fleiss, J.Z. & Cohen, J. (1976). The global assessment scale: a procedure measuring overall severity of psychiatric disturbance. *Archives of General Psychiatry* **33**, 766–771.

Evans, M.D., Hollon, S.D., DeRubeis, R.J., Piasecki, J.M., Grove, W.M., Garvey, M.J. & Tuason, V.B. (1992). Differential relapse following cognitive therapy and pharmacotherapy for depression. *Archives of General Psychiatry* **49**, 802–808.

Feighner, J.P., Robins, E., Guze, S.B., Woodruff, R.A., Winokur, G. & Munoz, R. (1972). Diagnostic criteria for use in psychiatric research. *Archives of General Psychiatry* **26**, 57–663.

Feinberg, M., Carroll, B.J., Steiner, M. & Commorato, A.J. (1979). Misdiagnosis of endogenous depression with research diagnostic criteria. *Lancet* **I**, 267.

Frank, E., Kupfer, D.J., Perel, J.M., Cornes, C., Jarrett, D.B., Mallinger, A.G., Thase, M.E., McEachran, A.B. & Grochocinski, V.J. (1990). Three-year outcomes for maintenance therapies in recurrent depression. *Archives of General Psychiatry* **47**, 1093–1099.

Frank, E., Kupfer, D.J., Wagner, E.F., McEachran, A.B. & Cornes, C. (1991). Efficacy of interpersonal psychotherapy as a maintenance treatment of recurrent depression, contributing factors. *Archives of General Psychiatry* **48**, 1053–1059.

Freedman, D.X. (1989). Editorial note (especially for the media). *Archives of General Psychiatry* **46**, 983.

Gelder, M.G. (1990). Psychological treatment for depressive disorder. *British Medical Journal* **300**, 1087–1088.

Glen, A.I.M., Johnson, A.L. & Shepherd, M. (1984). Continuation therapy with lithium and amitriptyline in unipolar depressive illness: a randomized, double-blind, controlled trial. *Psychological Medicine* **14**, 37–50.

Hamilton, M.A. (1960). A rating scale for depression. *Journal of Neurological and Neurosurgical Psychiatry* **23**, 56–62.

Hollon, S.D., DeRubeis, R.J., Evans, M.D., Wiemer, M.J., Garvey, M.J., Grove, W.M. & Tuason, V.B. (1992). Cognitive therapy and pharmacotherapy for depression. Singly and in combination. *Archives of General Psychiatry* **49**, 774–781.

Klein, D.F. (1974). Endogenomorphic Depression. *Archives of General Psychiatry* **31**, 447–454.

Klerman, G.L. & Weissman, M.M. (1989). Increasing rates of depression. *Journal of American Medical Association* **261**, 2229–2235.

Klerman, G.F., Weissman, M.M., Rounsaville, B.J. & Chevron, E.S. (1984). *Interpersonal Psychotherapy of Depression*. New York: Basic Books.

Kovacs, M., Rush, A.J., Beck, A.T. & Hollon, S.D. (1981). Depressed out-patients treated with cognitive therapy and pharmacotherapy. A one-year follow-up. *Archives of General Psychiatry* **38**, 33–39.

Kupfer, D.J. (1992). Maintenance treatment in recurrent depression: current and future directions. *British Journal of Psychiatry* **161**, 309–316.

Kupfer, D.J., Frank, E., Perel, J.M., Cornes, C., Mallinger, A.G., Thase, M.E., McEachran, AS.B. & Grocholinski, V.J. (1992). Five-year outcome for maintenance therapies in recurrent depression. *Archives of General Psychiatry* **49**, 769–773.

London, P. & Klerman, G. (1982). Evaluating psychotherapy. *American Journal of Psychiatry* **139**, 709–717.

McLean, P.D. & Hakstian, A.R. (1979). Clinical depression: comparative efficacy of out-patient treatments. *Journal of Consulting Clinical Psychology* **47**, 818–836.

McLean, P. & Taylor, S. (1992). Severity of unipolar depression and choice of treatment. *Behaviour Research and Therapy* **30**, 443–451.

Miller, I.W., Bishop, S.B., Norman, W.H. & Keitner, G.I. (1985). Cognitive/behavioural therapy and pharmacotherapy with chronic, drug-refractory depressed in-patients: a note of optimism. *Behavioural Psychotherapy* **13**, 320–327.

Murphy, G.E., Simons, A.D., Wetzel, R.D. & Lustman, P.J. (1984). Cognitive therapy and pharamacotherapy, singly and together in the treatment of depression. *Archives of General Psychiatry* **41**, 33–41.

Nelson, J.C., Charney, D.S. & Vingiano, A.W. (1978). False positive diagnosis with primary affective disorder criteria. *Lancet* **ii**, 1252–1253.

Parloff, M.B. (1986). Placebo controls in psychotherapy research: a 'sine qua non' or a placebo for research problems? *Journal of Consulting Clinical Psychology* **54**, 79–87.

Persons, J.B. (1993). The process of change in cognitive therapy: schema change or acquisition of compensatory skills? *Cognitive Therapy and Research* **17**, 123–137.

244CURRENT ISSUES

Prien, R.F., Kupfer,D.J., Mansky, P.A., Small, J.G., Tuason, V.B., Voss, C.B. & Johnson, W.E. (1984). Drug therapy in the prevention of recurrences in unipolar and bipolar affective disorders. *Archives of General Psychiatry* **41**, 1096–1104.

Robinson, L.A., Berman, J.S. & Neimeyer, R.A. (1990). Psychotherapy for the treatment of depression: a comprehensive review of controlled outcome research. *Psychological Bulletin* **108**, 30–49.

Rosenthal, O. & Frank, J.D. (1956). Psychotherapy and the placebo effect. *Psychological Bulletin* **53**, 294–302.

Ross, M. & Scott, M. (1985). An evaluation of the effectiveness of individual and group cognitive therapy in the treatment of depressed patients in an inner city health centre. *Journal of the Royal College of General Practitioners* **35**, 239–242.

Rush, A.J., Beck, A.T., Kovacs, M. & Hollon, S. (1977). Comparative efficacy of cognitive therapy and imipramine in the treatment of depressed patients. *Cognitive Therapy and Reasearch* **1**, 17–37.

Rush, A.J., Beck, A.T., Kovacs, M., Weissenburger, J. & Hollon, S.D. (1982). Comparison of the effects of cognitive therapy and pharmacotherapy on hopelessness and self-concept. *American Journal of Psychiatry* **139**, 862–866.

Rush, A.J., Kovacs, M., Beck, A.T., Weissenburger, J. & Hollon, S.D. (1981). Differential effects of cognitive therapy and pharmacotherapy on depressive symptoms. *Journal of Affective Disorders* **3**, 221–229.

Rush, A.J. & Watkins, J.T. (1981). Group versus individual cognitive therapy: a pilot study. *Cognitive Therapy and Research* **5**, 95–103.

Scott, A.I.F. & Freeman, C.L. (1992). Edinburgh primary care depression study: treatment outcome, patient satisfaction and cost after 16 weeks. *British Medical Journal* **8304**, 883–887.

Shea, M.T., Elkin, I., Imber, S.D., Sotsky, S.M., Watkins, J.T., Collins, J.F., Pilkonis, P.A., Beckham, E., Glass, D.R., Dolan, R.T. & Parloff, M.B. (1992). Course of depressive symptoms over follow-up. *Archives of General Psychiatry* **49**, 782–787.

Simons, A.D., Garfield, S.L. & Murphy, G.E. (1984). The process of change in cognitive therapy and pharmacotherapy for depression. Changes in mood and cognition. *Archives of General Psychiatry* **41**, 45–51.

Simons, A.D., Murphy, G.E., Levine, J.E. & Wetzel, R.D. (1986). Cognitive therapy and pharmacotherapy for depression. Sustained improvement over one year. *Archives of General Psychiatry* **43**, 43–48.

Spitzer, R., Endicott, J. & Robins, E. (1978). Research diagnostic criteria: rationale and reliability. *Archives of General Psychiatry* **35**, 773–782.

Stravynski, R. & Greenberg, D. (1992). The psychological management of depression. *Acta Psychiatrica Scandinavia* **85**, 407–414.

Teasdale, J.D., Fennell, M.J.V., Hibbert, G.A. & Amies, P.L. (1984). Cognitive therapy for major depressive disorders in primary care. *British Journal of Psychiatry* **144**, 400–406.

Thase, M.E., Simons, A.D., Cahalance, J., McGreary, J. & Harden, T. (1991). Severity of depression and response to cognitive behavior therapy. *American Journal of Psychiatry* **148**, 784–789.

Ulenhuth, E.H., Lipman, R.S. & Covi, L. (1969). Combined pharmacotherapy and psychotherapy: controlled studies. *Journal of Nervous and Mental Diseases* **148**, 52–64.

Watkins, J.T., Leber, W.R., Imber, S.D. & Collins, J.F. (1986). *NIMH Treatment of Depression Collaborative Research Program: Temporary Course of Symptomatic Change.* Presented to the Annual Meeting of the American Psychiatric Association, Washington, DC.

West, R. (1992). *Depression*. London: Office of Health Economics.
Wing, J.K., Cooper, J.E. & Sartorius, N. (1974). *Measurement and Classification of Psychiatric Symptoms*. Cambridge: Cambridge University Press.
Wright, A. (1993). *Depression. Recognition and Management in General Practice.* London, UK: Royal College of General Practitioners.

11 Design and Methodological Issues in Setting Up a Psychotherapy Outcome Study with Girls Who Have Been Sexually Abused

JUDITH A. TROWELL*, M. BERELOWITZ[a] AND I. KOLVIN
The Tavistock Clinic, and [a]Royal Free Hospital, London, UK

ABSTRACT   Psychotherapy outcome studies for child sexual abuse are long overdue. A number of such studies are being undertaken in the adult mental health field. Some of the design and methodological issues in setting up an intervention study are discussed and the relevant practical problems are considered.

Sexually abused girls who were psychiatrically symptomatic were taken as the subject group because this constitutes a major mental health problem and the outcome of psychotherapy with this client group is of particular interest. The hypotheses are outlined and the rationale for the choice of a comparison design is provided; the important subject of the equivalence of psychotherapy is reviewed. The complexity of the assessment is described together with an account of the need for as complete a collection of data as possible; and this is paralleled by the need to keep the interview length within manageable proportions for both the child and her carer. The emergent practical problems are described and the operational solutions which were arrived at explained.

Child abuse (CA) and child sexual abuse (CSA) are among the major mental health problems of the last decade. Considerable resources are being put into identifying and investigating cases of abuse (Kolvin *et al.*, 1988a). However, there is concern about the lack of *treatment* resources, and also lack of knowledge as to treatment: which treatments are most effective; to whom should they be provided; which symptoms or problems are most likely to be affected?

---

* Correspondence address: The Tavistock Clinic, Child and Family Department, 120 Belsize Lane, London NW3 5BA, UK.

Research Foundations for Psychotherapy Practice. Edited by M. Aveline and D. A. Shapiro.
Copyright © 1995, Mental Health Foundation and Individual Contributors.
Published 1995 by John Wiley & Sons Ltd

WHY TREAT SHORT- AND LONG-TERM EFFECTS OF CSA?

The exact number of women sexually abused as children is not known. The rate was found to be as high as 19% in college students (Finkelhor *et al.*, 1990) but this may have been inflated by incidents of observation of exhibitionists. Lower rates of 12% using a MORI poll are given by Baker and Duncan (1985). There have recently been reviews of follow-up studies of children who have been sexually abused. Beitchman *et al.* (1991) looked at 42 studies for short-term effects. In brief, they concluded:

1. Victims of child sexual abuse are more likely to develop some form of inappropriate sexual or sexualized behaviour.
2. The frequency and duration of sexual abuse is associated with more severe outcome.
3. CSA involving force and/or penetration is associated with greater subsequent psychopathology.
4. Sexual abuse perpetrated by the child's biological or stepfather is associated with greater psychological problems.
5. Victims of CSA are more likely than non-victims to come from families with a higher incidence of marital separation/divorce, parental substance abuse and psychiatric disorder.

In an allied review, Beitchman *et al.* (1992), in relation to long-term effects, concluded that there are three major sets of sequelae.

1. In comparison with women not reporting CSA, women who do report a history of CSA more commonly show evidence of anxiety and fear, and depression and depressive symptoms, which may be related to force or threat of force during the CSA. They also show evidence of re-victimization experiences, suicidal ideas and behaviour.
2. There is insufficient evidence to show a relationship between CSA and a specific post-sexual abuse syndrome; however, there is evidence to show a link between multiple personality disorder and borderline personality in adults and a history of childhood physical and sexual abuse.
3. Looking at the relationship between facets of abuse and particular or specific outcomes: more evidence exists to support a traumatic impact of post-pubertal than pre-pubertal abuse; longer duration of abuse is associated with greater impact; the use of force or threat of force is associated with negative outcome; penetration (oral and vaginal) is associated with greater long-term harm; abuse involving father or stepfather is associated with greater long-term harm.

This review considers work up to 1987, more recent papers indicate that there may be more links emerging (Ogata *et al.*, 1990; Brown & Anderson,

1991; Zanarini *et al.*, 1989). This work has been summarized by Cotgrove and Kolvin (1993) who conclude that there are four main long-term associations with CSA

1. Psychological symptoms consisting of depression, anxiety, low self-esteem, guilt, sleep disturbance and dissociative phenomena.
2. Problem behaviours including self-harm, drug use, prostitution and running away.
3. Relationship and sexual problems—social withdrawal, sexual promiscuity and re-victimization.
4. Psychiatric disorders particularly eating disorders, sexualization, post-traumatic stress disorder and borderline personality disorder.

We recognize that the above problems are more likely to be reported by clients in long-term therapy or referrals to psychiatric or social services, and it remains unclear how widespread the associations are in non-selected community samples of sexual abuse victims.

VIGNETTES

This section describes three cases already assessed, which illustrate the complexity of the families and the co-morbidity.

A 7-YEAR-OLD GIRL ABUSED BY HER UNCLE

Mother was physically abused by her father and emotionally abused by her mother. She left home at 15 years. Later she married and has three children, the last of which is a daughter idolized by her father. However, both parents are preoccupied with their own stormy relationship, the little girl receives most affection from her uncle who masturbates her and then vaginally abuses her, silencing her with threats to kill her if she speaks about what is happening. The girl presents as a 'little princess' but is not learning, cannot sleep and will not speak. Her uncle has been totally extruded and eliminated from the family.

SISTERS AGE 6 YEARS AND 9 YEARS ABUSED BY STEPFATHER

A mother who was physically and sexually abused in her own childhood was preoccupied and depressed, frequently unavailable to her children, either physically or emotionally absent. Mother remarries and stepfather provides most of the nurturing; he sexually abuses the girls; neither girl tells what happens to them, but a cousin, one of many children abused by the stepfather, tells her parents. The younger girl talks to the Police and

Social Services Department, the cousin gives evidence in Court and step-father is sent to prison.

The older sister still has not spoken about her abuse, she is not learning, has suicidal thoughts, has made a suicidal attempt, she has problems with other children at school, and is depressed; she misses her stepfather. She has marked symptoms of post-traumatic stress disorder.

The younger sister weeps for her stepfather, is depressed and not sleeping, she has post-traumatic stress disorder, she has talked about the masturbation by her stepfather and later full vaginal penetration.

GIRL AGE 7 YEARS ABUSED BY FATHER AND FATHER'S FRIEND

The girl has a white mother and an Afro-Caribbean father. Mother was much younger than father; mother had isolated herself from her family of origin where she had been physically abused and neglected. The girl was masturbated and vaginally abused by father, and a number of father's friends also abused her. The girl is depressed, wishes she was dead, and has symptoms of post-traumatic stress disorder. However she is doing well at school and enjoys it.

EXPLANATORY THEORY

We would suggest that the origins of many of the adult women's symptoms result from the disruptive effect of the abuse on their capacity as girls to relate to their own mothers and to develop their adult femininity and mothering capacities. There is also the disruption of their capacity to learn, to concentrate, to think, to use their memory and this is possibly the basis of dissociation. Thus, it seemed justified to offer psychotherapy to a sample of girls. The aim will be to see if it is possible to ameliorate or prevent both the short- and long-term effects, and also to ascertain if it is possible to clarify which particular psychological symptoms respond to which form of intervention. We are also interested to try and monitor the therapeutic processes in the child and carer, to try and establish significant points, areas in the therapy where change occurs, from the records kept by the therapists and carers workers.

THE CURRENT PROJECT

AIMS

Given the body of literature on the effects of childhood sexual abuse, we wanted to offer treatment interventions that we hope will contribute to the amelioration of the short- and long-term effects. Individual, group and family approaches have been considered as relevant models for dealing with the trauma and subsequent disturbance or distress. Group therapy for mothers has been used frequently (Giaretto, 1981), but there is no substantial evidence of improvement.

The literature on treatment mostly consists of clinical anecdotes and single case studies (Brassard *et al.*, 1983), and little in the way of carefully designed and controlled treatment outcome studies using appropriate standardized measures (Sgroi, 1982). However, more recently attention has focused on cognitive behavioural treatment for sexually abused children, especially those suffering from post-traumatic stress disorder (PTSD) (Deblinger *et al.*, 1990). Although changes were demonstrated over time on the post-treatment measures, this was again an uncontrolled study.

It is highly relevant that recent research has demonstrated that the symptoms exhibited by children who have been sexually abused fall into certain categories: post-traumatic stress disorder; chronic stress and trauma; and in adults, severe personality difficulties (Deblinger *et al.*, 1989; Terr, 1991; Herman *et al.*, 1989). For example, in the sample of McLeer *et al.* (1988), about half the children who had been sexually abused met the full DSM-111-R criteria for PTSD, while many of the remaining children met at least some of the criteria.

On clinical and theoretical grounds it is therefore crucial that we should set up carefully designed trials of psychotherapy for children who have suffered *contact sexual abuse* and who are *symptomatic*. But it is equally important that there should be an evaluation of the efficacy of individual child psychotherapy, which is currently being offered to sexually abused children by a number of institutions with special expertise in this area. Clinical judgement suggests that this approach is of particular benefit, and is well suited to personal traumas which impinge on long-term character development. In addition many referrals of such patients specifically request this treatment. Furthermore it is thought to be capable of dealing with the widespread symptomatology that may occur, including anxiety, depression, phobias, conduct disorder, sexualizing behaviour and the presence or sequelae of post-traumatic stress disorder. However, the effectiveness has not yet been systematically ascertained. It is therefore timely that such an evaluation is undertaken.

DESIGN

Taking into consideration the design issues already discussed, it was decided to offer individual and group psychotherapy to about a 100 sexually abused girls between 6 and 14 years. The decision to focus on girls was because sexual abuse of girls is known to be more common than boys.

The ages of 6 to 14 years were selected because by 6 years most girls would be able to read or follow the required questionnaires. The oldest children in this subsample would still be available for follow-up 2 years after the initiation of therapy because of the school leaving age (16 years).

The treatments offered were brief focused individual psychoanalytic psychotherapy once weekly for up to 30 sessions, and group psychotherapy for up to 20 sessions. The group therapy had two components—psycho-educational and psychotherapeutic. The subjects were randomly allocated to individual or group. This was complemented by supportive work with the carers, which helped to ensure the child's attendance for therapy.

The children were assessed at the start, at the end of treatment and after 1 year. The carers were interviewed to gather as much information as possible about the child, but also to learn about the carers response to the abuse. Information about the significant carers social history and abuse background was simultaneously gathered.

During the therapy the therapists, both group and individual, maintained records of the predominant themes that emerged during all sessions, and the child or carers' affect and the difficulties for the therapist. During supervision, experienced supervisors collected similar data but also rated the therapist on skill and compatibility.

HYPOTHESIS

Our hypothesis is that individual psychotherapy as practised by a skilled child psychotherapist, under supervision and backed up by a specific manual, can make a major contribution to the psychological stability of those who have been sexually abused as children. So far this crucial hypothesis has not been addressed in Western European research and this deficiency needs to be redressed.

Hypothesis one

In comparison with psycho-educational group therapy, focused individual psychotherapy administered by trained psychotherapists, guided both by a manual consisting of agreed statements of the aims, objectives and techniques of the therapy, and by clinical supervision, will:

1. Reduce the frequency of sexualized and eroticized behaviour (Yates, 1982).
2. Improve social functioning.
3. Improve academic functioning.
4. Give rise to changes in the child's internal working models of themselves and their important attachment figures (care-givers).
5. Alleviate distress and promote normal emotional development.

Hypothesis two

Patients receiving individual psychotherapy will benefit most from the treatment in the long-term, while those receiving psycho-educational group therapy will, in some respects, benefit more in the short-term.

There is also an expectation that further understanding will be gained as to which intervention is most helpful for which associated disorder. There are certain factors which we hypothesize will be important, irrespective of the outcome of treatment. These include carer variables, and the case complexity and co-morbidity.

SOME DESIGN AND METHOD ISSUES

THE COMPARISON DESIGN

Previously when there was no sound evidence about efficacy of child psychotherapy there were good arguments in favour of the 'control' design (Kolvin et al., 1988b; Bell et al., 1989). However, this position is no longer tenable and there are powerful arguments in support of the 'comparison' design (Kazdin, 1986; Parloff, 1986; Knyschild, 1993).

1. On ethical grounds the comparison design is preferable as it avoids the dilemmas about randomization into control situations which have dubious clinical credibility.
2. Non-treatment controls are not welcomed by families (Parloff et al., 1986).
3. In the comparison design the alternate groups are equally credible.
4. Any non-specific placebo influences will be eliminated.

However, on the negative side, the effect size between alternate psycho-therapy groups will be reduced as compared to where the comparison is with an untreated control group.

RANDOM ALLOCATION

Hopefully pre-treatment differences will be minimized by random allocation. We have also ensured that we control for age. We have used a random numbers technique whereby there is a prior allocation to different cells so that the client can be informed at the end of the assessment which specific form of therapy will be available for her.

EQUIVALENCE OF PSYCHOTHERAPY

In adults, good evidence is available that therapy is more effective than no therapy but there is little evidence of specificity of effectiveness of different therapies (Stiles *et al.*, 1986). The same is true of group and behavioural therapies in children despite substantial differences in duration (Kolvin *et al.*, 1981/5). There are two main explanations for this.

1. Despite superficial technical diversity, all therapies share common core therapeutic processes (Stiles *et al.*, 1986).
2. The lack of differences are due to inadequately sensitive outcome procedures.

Some authorities suggest that underlying the above lack of specificity is an equivalence of content, processes and outcome. Yet others suggest that such an equivalence is a mirage and the solution is in identifying the differences. All suggest greater precision and specificity of theory and method. The above include specificity of diagnoses, detailed monitoring of therapy, and the utilization of diverse and sensitive outcome measures. This should be complemented by a detailing of the qualities of therapists from direct observations via videotaped material. But the latter is not achievable in this study, the best we can do is to obtain ratings of process and outcome by supervisors. In addition, we will obtain accounts from the therapists about processes in therapy. While bias cannot be excluded, it remains an important method of gathering information. We will also gather information about the nature of the therapeutic alliance (Kernberg *et al.*, 1991). All of these may provide evidence of process.

VARIATION IN DELIVERY OF TREATMENT

This could constitute a serious threat to validity. One solution for dealing with variation in delivery of treatment consists of the use of *manuals* defining the principles and detailing procedures of treatment. Manualization should be complemented by process analysis of recorded sessions (Hardy & Shapiro, 1985). However, such video-recording methods are not appropriate for sexually abused children who merit privacy and sensitivity in management and these qualities cannot be sacrificed for the sake of science. However, to ensure homogeneity of quality of treatment, committed therapists will be used who are skilled in their own therapeutic modality and very senior therapists will be providing supervision to all the therapists. The supervisors will be receiving consultation from the authors.

DURATION OF THERAPY

Improvement has been shown as being related to the duration of therapy; being proportionately greater in early sessions but increasing more slowly as the number of sessions grow (Howard *et al.*, 1986; Garfield & Bergin, 1986). This has led to the suggestion that when comparing two different forms of therapy which are intrinsically different, the duration of therapy should be controlled. However, such arguments are negated by the fact that different therapies are likely to have optimal effects at different points in time (Kolvin *et al.*, 1988b; Bell *et al.*, 1989) and the specification of numbers of sessions could be counterproductive for those therapies which appear to peak earlier (e.g. group therapy). Further it may well be that each form of therapy has its own momentum. We have consulted widely with very experienced therapists in finalizing the optimum therapy duration for each modality.

PRINCIPLES OF ASSESSMENT

The principles addressed in the choice of measures are as follows.

1. Sound psychometric qualities are essential.
2. Sensitivity to change. The measures need to be sensitive to change and treatment outcome. For instance, there is very little evidence that cognitive measures in childhood respond quickly (or even slowly) to psychotherapeutic interventions.
3. Specificity of measures to goals of treatment (Shapiro, 1989). While this is desirable, in child psychiatry the effects are often non-specific and close adherence to the above principle can lead one widely astray about outcome. Also, the outcome measures need to be multi-faceted, involving different aspects of the subject's functioning.
4. Common core measures. These are thought desirable when comparing alternative therapies. However, a total focus on these may be at the expense of measures relevant to each of the alternatives, and exclusive use of common core measures may mask outcome diversity (Stiles *et al.*, 1986). Our solution has been to choose a judicious combination of both, accepting that some measures will only provide evidence of within group change.
5. Intrapsychic versus symptomatic change. Finally, we needed to seek measures of outcome that go beyond symptoms. Some will be tapping intrapsychic functions; some will tap interpersonal functioning (Weiss *et al.*, 1985).

DELAYED EFFECTS OF THERAPY

Psychotherapy research over the last decade has suggested that therapy may set in motion processes that have effects long after termination, and that these may remain undetected if provision is not made for long-term follow-up. These constitute the so called 'sleeper effects' (Bell *et al.*, 1989; Kolvin *et al.*, (1988). Sleeper effects are found after therapy has ended one or more years later. On follow-up there may be considerable improvement in symptomatology and functioning despite no evidence of improvement at earlier points in time.

HOMOGENEITY OF PATIENT GROUP

This is desirable and we have attempted to ensure this by including only girls who are symptomatic and present with evidence of PTSD. But a confounding issue in child psychiatry is the extensive co-morbidity in psychiatric samples. Unfortunately there is little in the literature which suggests solutions to this problem.

SIZE OF SAMPLES (Power studies)

In a previous study of therapy with maladjusted children, using an appropriate criterion of outcome (Kolvin *et al.*, 1988b, 1981/5), good clinical outcome was found in about 75% of cases in the treated groups (group therapy and behaviour modification) and in 45 to 50% of cases in control groups. Based on the above, the differences in outcome for the current study is estimated to lie between 25% and 30%. Assuming a drop-out rate of 10–15%, then 45 patients per group are required to detect a 30% difference with 75% power using a 5% two-sided test of significance. However, practical factors have necessitated that each therapy group starts with six children and so the initial sample size has needed to be 48.

THE RESEARCH IN ACTION

SOME METHOD PROBLEMS

One of the major problems that confronted us was the potential vastness of the information that could be collected in order to cover the wide areas of dysfunction outlined in the literature; and at the same time we wanted to ensure that we were covering the core symptoms of CSA. Thus we had to focus on common core measures and those specific to each treatment regime. Furthermore, we went beyond behaviour and explored intra-psychic change.

The Final Assessment Schedule covers four broad areas.

1. Psychological symptoms, as reflected by behaviour and psychiatric symptoms (Kiddie-SADS, see Chambers *et al.*, 1985; post-traumatic stress disorder, see Orvaschel, 1989).
2. Intrapsychic phenomena as reflected by self-esteem (Harter, 1985) Internal World of Child (child version of the AAI) (Main, 1989).
3. Interpersonal as reflected by:
   (a) family systems—the family assessment device (FAD) (Epstein *et al.*, 1983).

    (b) index of family support
    (c) parental response to abuse disclosure (PRAD).
    (d) adult attachment interview (AAI) (Main, 1989).
4. Cognitive and academic functioning:
    (a) British picture vocabulary test (Dunn et al., 1989).
    (b) Teacher report (CBCL) (Achenbach and Edelbrock, 1983).

The adult attachment interview (AAI) is given to the main female carer. It is used to explore the emotional environment for the child and to try and assess the availability of the main female carer to listen to the child and spend time giving the child attention. In childhood sexual abuse there is very frequently a situation where the child is unable to speak to the mother/or female carer. Also, where there has been childhood sexual abuse of the mother (female carer) herself there seems to be as a consequence a higher chance of the child not being protected.

ASSESSMENT

The assessment process is working smoothly. We are quickly being educated about the psychopathology, for instance, family intergenerational abuse, violence and involvement with drugs is common. Many of the families have distressing and traumatic histories that mean the assessments themselves are emotionally difficult for the families and the assessment team. Other families are having to cope with distressing and difficult behaviour in their adopted and fostered children. We have found that there are more siblings than single girls, often with two sets of traumas in the same family. A parallel project at Great Ormond Street for adolescent boys (abuser and abused) is involved with some of the families. The above illustrates the extent and complexity of the psychopathology within the subjects and in their network. The parallel assessment of the carers— natural parents, adoptive parents/foster parents, residential care workers or social workers—often reveals very distressing material, and so the contact and support for the carers in order to enable them to sustain treatment and manage the child is being seen as increasingly significant.

ALLOCATION TO TREATMENT REGIMES

As is to be expected, assembling a group of girls to fit into an age group together is a slow process. We aim to have three stratified groups and an older-girls group has started. It is easy to allocate subjects to group therapy but it is less easy to ensure a sufficiency of cases per age stratum (6 to 8 years; 9 to 11 years; 12 to 14 years).

    There are fewer problems with individual therapy. However, given the number of siblings, and recognizing that carers will only be prepared to attend once a week, there are often logistic and practical problems which have to be solved on a case by case basis. Siblings are allocated randomly to group and individual therapy. However, the supervisors of group therapy decided that to have twin girls or sisters in a group would be unhelpful to the group (and we already have three sets of twins in the project).

Social workers see the carers but we still have to plan for the availability of simultaneous group and individual treatment for siblings.

SUPERVISION

This is one of the ways of ensuring consistency of therapy between therapists and across the centres. The therapists have a shared supervision. They receive emoluments for seeing patients but not for their supervision time. The supervisors meet for joint consultations. Similar arrangements have been organized for the group therapists but such arrangements have been modified to meet the different circumstances.

OUTCOME

The assessment at the end of treatment and the follow-up, 1 year later will only provide information about short-term effects. Further funding will be needed if the subjects are to be followed-up 5 years later or into young adulthood. Given the current interest in 'sleeper effects', i.e. delayed effects of treatment (Kolvin et al., 1988b), long-term follow up seems essential. We are also aware of the sleeper effects of the sexual abuse itself, i.e. effects that are not immediately obvious: there is evidence that some symptoms emerge in later years. Hence the case for long-term follow-up, despite the high costs, is strong.

CONSENT

The nature of the project is explained to each child and their carer, when possible the family will be presented with the particular treatment. The duration of the treatment offered and the nature of the assessment before and after treatment and at follow-up is explained. The child, if it is old enough, is asked to consent to the treatment package. If the legal carer, or child, do not wish to be part of the project they have access to the ordinary resources available at the clinics and hospitals.

EXCLUSIONS

These are few but merit description.

1. Those that are too disturbed to tolerate outpatient treatment; girls who show such disturbed and difficult behaviour that they are unable to remain in the treatment room, travel to the treatment or be managed where they now reside.
2. Where there is concern about the child's safety (is the child adequately protected?). The appropriate child protection measures need to be in place. We realize new abuse or recurrences may occur and we need to be constantly aware of this.

EMERGENT PRACTICAL PROBLEMS

REFERRALS

Three main problems have emerged.

1. Referrals tend to come in batches with peaks and troughs, and thus ensuring a steady stream is problematic. A part-solution has been to maintain close contact with our main referrers—Social Service Departments.
2. The wide geographical spread of the referrals with families needing to travel long distances, sustaining contact over a long period may be difficult.
3. Many cases come from vulnerable multi-problem families with considerable financial difficulties and problems in organizing themselves. They need ongoing support, advice and encouragement to sustain attendance, the support of the referrer, generally a social worker, seems to be crucial.

SITES

The project is located across several sites. In order to assemble a sample of sufficient size and have access to sufficient therapists. The main sites include The Tavistock Clinic and Royal Free Hospital in North London, the Camberwell Child Guidance Clinic and the Children's Department at the Maudsley Hospital and Guys Hospital in South London.

COMMUNICATION

The need to have regular meetings of research workers, therapists and supervisors to ensure consistency was anticipated, but the time needed makes very considerable demands and has required considerable organization. Additional demands continue, criminal injuries compensation claims, repeated explanations and liaison with the external network and internally with managers. Other unanticipated additional demands include requests for completion of criminal injuries compensation claims but the greatest amount of work has arisen from the re-organisation of the health services and social services. Thus constant changes in staff in these organisations have resulted in the need for repeated explanations and liaison.

ACKNOWLEDGEMENTS

This research is supported by grants from the Mental Health Foundation and the Department of Health.

REFERENCES

Achenbach, T.M. & Edelbrock, C. (1983). *Manual for the Child Behaviour Checklist and Revised Child Behaviour Profile.* Burlington, VT: University of Vermont Department of Psychiatry.

Baker, A.W. & Duncan, S.P. (1985). Child sexual abuse—a study of prevalence in Great Britain. *Child Abuse and Neglect* 9, 457–467.

Beitchman, J.H., Zucker, K.J., Hood, J.E., Da Costa, G.A. & Akman D. (1991). A review of the short-term effects of child sexual abuse. *Child Abuse and Neglect* 15, 537–556.

Beitchman, J.H., Zucker, K.J., Hood, J.E., Da Costa, G.A., Akman, D. & Cassavia, E. (1992). A review of the long-term effects of child sexual abuse. *Child Abuse and Neglect* 16, 101–118.

Bell, V., Lyne, S. & Kolvin, I. (1989). Play group therapy: processes, patterns and delayed effects. In *Needs and Prospects of Child and Adolescent Psychiatry* (M.H. Schmidt & H. Remschmidt, eds). Stuttgart: Hogrefe and Huber.

Brassard, M.R., Tyler, A. & Kehle, T.J. (1983). Sexually abused children: identification and suggestions for intervention. *School Psychology Review* 12, 93–97.

Brown, G.R. & Anderson, B. (1991). Psychiatric morbidity in adult inpatients with childhood histories of sexual and physical abuse. *American Journal of Psychiatry* 148, 55–61.

Chambers, W.J., Puig-Antich, J., Hirsch, M. , Paez, P., Ambroson, I.P.J., Tabrisi, M.A. & Davies, M. (1985). The assessment of affective disorders in children and adolescents by semi-structured interview. *Archives of General Psychiatry* 42, 696–702.

Cotgrove, A.J. & Kolvin, I. (in press). The long-term effects of child sexual abuse. *Hospital Update.*

Deblinger, E., McLeer, S.V., Atkins, M.S., Ralphe, D & Foa, E. (1989). Post-traumatic stress in sexually abused, phsyically abused and non-abused children. *Child Abuse and Neglect* 13, 403–408.

Deblinger, E., McLeer, S.V. & Henry, D. (1990). Cognitive behavioural treatment for sexually abused children suffering PTSD: preliminary findings. *Journal of the American Academy of Child and Adolescent Psychiatry* **29**, 747–752.

Dunn, L.M., Dunn, L.M. & Whetton, C. (1982). *British Picture Vocabulary Scale.* Windsor, UK: NFER-Nelson.

Epstein, N.B., Baldwin, L.M. & Bishop, D.S. (1983). The McMaster family assessment device (FAD). *Journal of Marital and Family Therapy* **9**(2), 171–180.

Finkelhor, D., Hotaling, G., Lewis, I.A. & Smith, C. (1990). Sexual abuse in a national survey of adult men and women: prevalence, characteristics and risk factors. *Child Abuse and Neglect* **14**, 19–28.

Garfield, S.L. & Bergin, A.E. (1986). *Handbook of Psychotherapy and Behaviour Change*, 3rd edn. Chichester: Wiley.

Giaretto, H. (1981). A comprehensive child abuse treatment program. In *Sexually Abused Children and their Families* (P.B. Mrazek & C.H. Kempe, eds). Oxford: Pergamon Press.

Hardy, G.E. & Shapiro, D.A. (1985). Therapist verbal response modes in prescriptive vs. exploratory psychotherapy. *British Journal of Clinical Psychology* **24**, 235–245.

Harter, S. (1985). *Manual for the Self-Perception Profile for Children.* Denver: University of Denver.

Herman, J.L., Perry, J.C. & Van der Kolk, B.A. (1989). Childhood trauma in borderline personality disorder. *American Journal of Psychiatry* **146**, 490–495.

Howard, K.I., Kopta, S.M., Krause, M.S. & Orlinsky, D.E. (1986). The dose–effect relationship in psychotherapy. *American Psychologist* **41**, 159–164.

Kazdin, A.E. (1986). Comparative outcome studies of psychotherapy: methodological issues and strategies. *Journal of Consulting Clinical Psychology* **54**, 95–105.

Kernberg, P.F. (1991). Therapist verbal interventions: a reliability study. *Proceedings in Research in Psychoanalysis, International Psychoanalytic Association's 1st Scientific Research Meeting (April)*, London.

Kingschild, P. (1993). Trials and errors. Alternative thoughts on the methodology of clinical trials. *British Medical Journal* **306**, 1706–1707.

Kolvin, I., Garside, R.F., Nicol, A.R., Macmillan, A., Wolsenholme, F. & Leitch, I.M. (1981/5). *Help Starts Here: the Maladjusted Child in the Ordinary School.* London and New York: Tavistock Publications.

Kolvin, I., Steiner, H., Bamford, F., Taylor, M., Wynne, J., Jones, D. & Zeitlin, H. (1988a). Child sexual abuse—some principles of good practice. *British Journal of Hospital Medicine* **39**, 54–62.

Kolvin, I., Macmillan, A., Nicol, A.R. & Wrate, R.M. (1988b). Psychotherapy is effective. *Journal of the Royal Society of Medicine* 261–266.

Main, M. (1989). *Adult Attachment Rating and Classification System.* Unpublished scoring manual. Berkley: Department of Psychology, University of California.

McLeer, S.V., Deblinger, E., Atkins, M.S., Foa, E.B. & Ralphe, D.L. (1988). Post-traumatic stress disorder in sexually abused children. *Journal of the American Academy of Child and Adolescent Psychiatry* **27**, 650–654.

Ogata, S.N., Silk, K.R. & Goodrich, S. (1990). The childhood experience of the borderline patient. In *Family Environment and Borderline Personality Disorder* (P. Links, ed.). Washington DC: American Psychiatric Press.

Orvaschel, H. (1989). *Kiddie SADS-E Section. Designed to assess PTSD.* Philadelphia, PA: Medical College of Pennsylvania.

Parloff, M.B., London, P. & Wolfe, B. (1986). Individual psychotherapy and behaviour change. *Annual Review of Psychology* (M.R. Rosenzweig & L.W. Porter, eds). Palo Alto, California.

Sgroi, M. (1982). *Handbook of Clinical Intervention in Child Sexual Abuse*. Lexington MA: DC Heath.

Shapiro, D.A. (1989). Outcome research. In *Behavioural and Mental Health Research: A Handbook of Skills and Methods* (G. Parry & F.N. Watts, eds). Erlbaum. pp 163–167.

Stiles, W.B., Shapiro, D.A. & Elliott, R.K. (1986). Are all psychotherapies equivalent? *American Psychologist* **41**, 165–180.

Terr, L.C. (1991). Child traumas: an outline and overview. *American Journal of Psychiatry* **148**, 10–20.

Weiss, D.S., DeWitt, K.N., & Kaltreider, N.B. (1985). A proposed method for measuring change beyond symptoms. *Archives of General Psychiatry* **42**, 703–708.

Yates, A. (1982). Children eroticized by incest. *American Journal of Psychiatry* **139**, 482–485.

Zanarini, M., Gunderson, J.G. & Marino, M.F. (1989). Childhood experiences of borderline patients. *Comprehensive Psychiatry* **30**, 18–25.

12 A Clinical and Service Evaluation of Group Therapy for Women Survivors of Childhood Sexual Abuse

ZAIDA HALL, MARK MULLEE[a] AND CHRIS THOMPSON[*]
University of Southampton Department of Psychiatry, Royal South Hants Hospital, and [a]University of Southampton Department of Medical Statistics and Computing, Southampton General Hospital, Southampton, UK

ABSTRACT Probably as many as 12% of women and 50% of psychiatric patients have been sexually abused in childhood. Effective and economic methods of treatment must be found for adult survivors that are informed by reliable research in clinically realistic settings.

Since most patients cannot properly be used as controls, in this study outcome is evaluated by two methods: by estimating a change in depressive ratings, and in the use of psychiatric and primary care services after analytic group therapy.

From 1986 to 1990, because of clinical pressure of referrals rather than as a research project, 94 women survivors of childhood sexual abuse were treated in a slow-open group for up to 6 months. Hamilton and Beck depressive inventories were scored for up to 7 years after the group. Their use of psychiatric hospital and general practitioner services before and after the group was recorded for up to 5 years. Six months after the group, their depressive ratings and their contacts with the psychiatric hospital and their general practitioners had decreased; this was maintained for between 2 and 7 years. The fall in depressive ratings was independent of antidepressant medication. Comments from the patients suggest that the improvement was not only in depressive symptoms but was global.

[*] Correspondence address: University of Southampton Department of Psychiatry, Royal South Hants Hospital, Graham Road, Southampton SO9 4PE, UK.

Research Foundations for Psychotherapy Practice. Edited by M. Aveline and D. A. Shapiro.
Copyright © 1995, Mental Health Foundation and Individual Contributors.
Published 1995 by John Wiley & Sons Ltd

The difficulties of collecting data on service utilization in a geographically unstable population are discussed.

INTRODUCTION

In the UK we first learned of the extent of child sexual abuse in the early '80s. Since then we have come to realize that while between 10 and 14% of women in the general population have a history of this (Baker & Duncan, 1985; Hooper, 1990), in the psychiatric population the figure may be as high as 50% (Palmer *et al.*, 1992). These figures are probably an under-estimate. With such an extent of morbidity it is imperative that effective and economical methods of treatment are found. In order that resources are not wasted, clinical practice must be informed by reliable research findings.

There are two main ways of demonstrating the effectiveness of psychotherapy.

1. By a decrease in symptoms following treatment.
2. By a decrease in the use by the patient of psychiatric services in hospital and general practice.

Despite the vast literature on childhood sexual abuse there has been little systematic research into the treatment of adult survivors. There are many papers, mainly from North America, describing group therapy for women survivors, but the groups have been of short duration, are few in number, or the total number of patients is small (Deighton & McPeek, 1985; Lubell & Soong, 1981, Gazarain & Buchele, 1988).

There are very few papers with any kind of quantitative analysis of outcome. Some authors have used questionnaires of various types to measure outcome (Tsai & Wagner, 1978; Kearney-Cooke, 1989; Carver *et al.*, 1989). The first paper evaluating group therapy for women survivors, which used the waiting list as controls, was by Alexander and her colleagues (1989). They assigned 65 women randomly to a 10-week interpersonal transaction (IT) group, a 10-week process group and a waiting-list condition, assessing them before and after and at 6-month follow-up on various questionnaires, including the Beck Depressive Inventory. They found that both the IT group and the process group were more effective than the waiting-list in reducing depression, and that this improvement was maintained at follow-up.

The second criterion of efficacy is a reduction in the use of hospital and primary care services. Some workers have demonstrated a fall in general hospital services following psychological interviews in medical patients (Follette & Cummings, 1967; Mumford *et al.*, 1984). In a pilot study of

10 patients, Knowles (1989), looking at psychiatric hospital, GP and Social Services, before and after brief individual therapy, reported a fall in utilization of services. McDonald (1992) undertook a more comprehensive investigation of hospital and GP service utilization in a region of Scotland with a geographically stable population. He obtained the general practice data by a questionnaire from the GPs. He compared the use of services for an equivalent length of time (1–55 months) before and after weekly individual therapy, and reported that the patients decreased their usage.

There seem to be no reports in the literature of a service evaluation with women survivors of child sexual abuse, though Dick (1975) and Dick and Wooff (1986) did use a service evaluation as one outcome measure following group therapy for neurotic patients. They recorded hospital attendances and, in a proportion of cases, GP attendances for 1 year before and 1 to 4 years after therapy, and reported that treated patients showed a marked decrease in use of services.

BACKGROUND TO THIS RESEARCH

In the autumn of 1985 there was an increasing number of women patients presenting with a history of child sexual abuse to the psychotherapy department of the Royal South Hants Hospital. One of the authors (ZH) conducted a slow-open group for women survivors, using the group-analytic model of Foulkes (1964), from January 1986 until August 1990 (4 years, 7 months) with a series of six male co-therapists. It met weekly for one-and a-half hours with a maximum membership of 12. Patients were expected to remain in the group for 6 months.

Although this was not intended as a research project, each group member was asked to complete Hamilton (1967) and Beck (Beck *et al.*, 1961), depressive rating scales before and after the group and at 6 months follow-up. There was thus a prospective assessment of outcome. The Hamilton, with the exclusion of the insight question, was used as a self-rating questionnaire for the first patients. Later the Beck was added.

These patients' common history provides a well-defined category of psychiatric morbidity. All but three were assessed, and all were treated in the group by the author (ZH), which made the treatment programme fairly consistent. An account of this has been published (Hall, 1992), where the importance of having a male co-therapist, and of the initial assessment interview, was emphasized.

During the assessment interview of 1.5–2 hours, details of the abuse were elicited as part of the therapeutic work. Dream rehearsal was taught if a patient had recurrent nightmares preventing her from sleeping (Marks, 1948). If specific sexual problems existed, a conjoint session with a patient's partner was given.

In order to be selected for the groups, a patient had to have a conscious memory of the abuse, and at least one of the following four symptoms:

1. obsessive ruminations of the abuse;
2. with or without flashbacks or nightmares;
3. bad self-image;
4. a disabling mistrust of men or serious sexual unresponsiveness.

A total of 94 patients joined the group. A 'comparison group' consisted of 34 patients considered suitable for the group but who refused or failed to attend and who did not receive alternative therapy. These were not randomly allocated control patients. There was no significant difference in demographic data between the two sets of patients.

The ages of the members ranged from 17 to 59 years. They had been mainly victims of male relatives or of mothers' or grandmothers' boyfriends. Some had been abused by friends of the family and some by strangers. One had also been abused by her mother.

Some had had recent inpatient or day-hospital treatment, two had had manic depressive illnesses, one had been psychotic. A few had had individual therapy. Many were severely disturbed and psychologically fragile, reacting to any implied criticism. A formal psychiatric diagnosis was not made, but several had characteristics of borderline personality disorder. Many had chaotic lives, with sick or disabled children, abusive partners, and repeated intercurrent life crises. Many had been physically as well as sexually abused, some to the extent of torture, and all had to some degree been the subject of psychological abuse—'Soul Murder' (Hall, 1987). Five were still being abused by their fathers and could not extricate themselves from the situation. Two, who had themselves been abused, had abused young children.

CLINICAL EVALUATION

The group members had been asked to fill in Hamilton and Beck ratings before and after the group and at a 6 month follow-up appointment. A final Hamilton and Beck questionnaire was sent by post to each member during the year 1992–93 giving final follow-up ratings for between 2 and 7 years. To these were added a questionnaire rating five elements of group satisfaction. There was also a space for comments.

SERVICE EVALUATION

The number of psychiatric hospital and GP contacts were recorded by searching through the psychiatric hospital notes, the psychotherapy notes and the GP practice notes for the year before the group started and up to 1, 2, 3, 4 or 5 years after (depending on length of follow-up) for the 94 group members, using the date of the last group as the start of the follow-up. For the 34 'comparison' patients, the same data were collected using the date of assessment as the cut-off point for the year before and the 1–5 years after.

Hospital contacts consisted of inpatient days, day-hospital days, outpatient attendance or domiciliary visits, and community psychiatric nurse contacts. The GP contacts consisted of visits to the GPs for both medical and psychological complaints (excluding antinatal visits or visits for courses of injections), and contacts with the health visitor or practice counsellor.

RESULTS

Because this was not planned as a research project, inevitably there are some data missing. The main loss is in GP notes. Southampton does not have a geographically stable population, so many patients had moved out of the area. Unless we could be sure the patient still lived in the area, mainly by checking with the GP, negative hospital contacts could not be validated.

CLINICAL EVALUATION

The total number of sessions for each member ranged from 1 to 45. Forty-five left prematurely, 17 left early after discussion with the group and 32 went on to a planned termination. Several members stayed for longer than 6 months because of illness, severity of symptoms, because they could not break away from their abuser, or because they had rejoined the group several times.

Figure 12.1 is a graph of the mean scores from pre-group to the 6-month follow-up in those patients who did complete the full four ratings (pre- and post-group 6-month and final follow-up). It shows a gradual decrease in depressive symptoms, greater during the course of the group, and more gradual in the 6-month follow-up. The reduction was maintained at the final follow-up. (The horizontal scale is not an accurate time-scale as there was wide variation in length of time in the group, some variation in the timing of the 6-month follow-up, and extremely wide variation in the time of the final follow-up of from 2 to 7 years.) The decrease in the Hamilton scores at each assessment was significant up to the point at which the graph flattened out after 6 months, when they ceased to be significant. The Beck scores followed a similar trend with the exception that the decrease

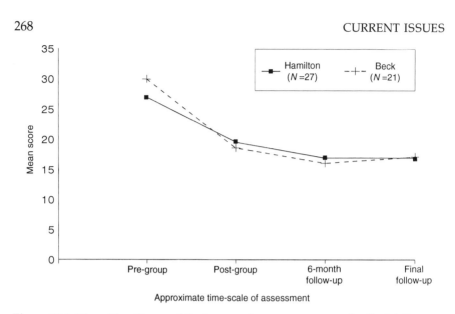

Figure 12.1 Mean Hamilton and Beck scores from pre-group to the final follow-up in patients completing the full four ratings

between post-group and 6-month follow-up failed to reach statistical significance.

It was possible that antidepressant medication from the GP might be responsible for any improvement in depressive symptoms. Inspection of the hospital and GP notes revealed that in only five of the patients with complete scores up to the final follow-up could medication, in a therapeutic dose, have been responsible for any improvement in depression. When the graph was repeated with those patients excluded, the curves were virtually identical. It can therefore be concluded that any improvement in depression following the group was independent of antidepressant medication.

In order to relate outcome to various factors we needed a clear-cut measure of outcome. From our experience of using scales we considered that a five-point change was likely to be significant and not due to random change. Taking, therefore, an alteration in Beck or Hamilton score of five or more as signifying change, Figure 12.2 shows the change in scores between the pre-group rating and the 6-month follow-up in those patients that completed both sets of ratings. It can be seen that both Hamilton and Beck scales showed an improvement of five or more points in 70% and 74% of patients, and a deterioration in 9% and 14%, with 20% and 12% showing no change, respectively.

In addition to the quantitative measures of outcome, there were subjective assessments of improvement by the therapists at the 6-month follow-up

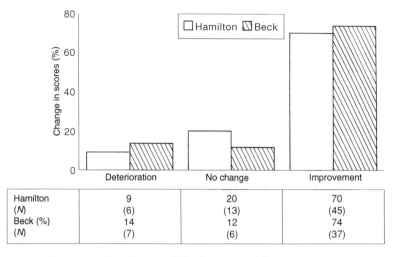

Figure 12.2 Change in Hamilton and Beck scores of five or more points between pre-group rating and 6-month follow-up

appointment (without knowledge of the depressive scores), and by the patients' views at the final follow-up: (1) of how helpful the group had been, and (2) their comments. The therapists considered that out of the 94 patients, 51 had improved, 41 had not changed and two had deteriorated.

Figure 12.3 (overleaf) compares what had actually happened to those patients in each category according to the Beck scores at 6 months in the therapists' opinion. The therapists' opinions were fairly congruent with the Beck changes, i.e. with the patients' assessments of themselves.

Outcome, surprisingly, did not appear to depend on the number of sessions attended, possibly because even one or two sessions can have the therapeutic effect of making the patient feel she is not alone in her experience of abuse.

SERVICE EVALUATION: HOSPITAL CONTACTS

Initial analysis of the data concentrated on the change of use of inpatient, outpatient and day-patient services. It was disappointing when analysis of 1 year before vs. 1 year after the group revealed no significant change in use of any of these services. This was in contrast to Dick (1975) and Dick and Wooff (1986) who detected a significant decrease for inpatient and outpatient usage for the year after therapy compared with the year before, but also found no significant decrease for day-hospital usage.

Evidence of decreases in usage observed by them and other researchers (MacDonald, 1992; Knowles, 1989) encouraged us to return to our data for further analysis. Our study of group patients revealed a fall in the usage

		Change in Beck score		
Therapist's opinion		Deteriorated	No change	Improved
of clinical	Deteriorated	1	0	0
outcome at 6-month	No change	5	5	6
follow-up	Improved	1	1	31

Figure 12.3 Comparison of therapists' opinion of clinical outcome with change in Beck scores of five or more points

of the combined inpatient, outpatient and day-hospital contacts from 744 (median = 1.0, range = 0–160), in the year before the group, to 525 (median = 0.0, range = 0–182), in the year following the group. This, however, narrowly failed to reach statistical significance ($n = 76$, $z = 1.84$, $p = 0.07$).

We had also collected data on the use of services 2, 3, 4 and 5 years after the group, and in the light of the findings of other workers we decided to study this data further. The percentage of patients using inpatient, outpatient and day-hospital services 1 year before the group was 15%, 56% and 8% respectively. The respective median (and inter-quartile range) for inpatient days was 0.0 (0.0–0.0); outpatient days was 1.0 (0.0–2.25); day-hospital visits was 0.0 (0.0–0.0). Therefore patients were classified into two categories, those who used the services and those who did not use the services, for each period of assessment (e.g. 1 year before group, 1 year after group, 2 years after group, etc.) and a comparison of the change in the proportion of patients using the services 1 year before the group with each year after was made.

Analysis of the combined services data revealed statistically significant decreases in the proportion of patients using the combined services 2, 3, 4 and 5 years after the group, but the decrease 1 year after failed to reach statistical significance. These results were mirrored by the analysis of the those patients who did or did not use the outpatient service (Figure 12.4). Identical analysis of the inpatient and day-hospital data revealed decreases in the proportion of patients using these services, but none were statistically significant (except the decrease of inpatient use, 3 years after, $n = 61$, McNemar's $\chi^2 = 5$, $df = 1$, $p = 0.03$; change in proportions $= -0.08$, 95% cl $= 0.15$–0.01).

We examined the data in an effort to discover why there appeared to be a lag of about 1 year before a significant decrease was detected in the proportion of patients using outpatient services. Figure 12.5 may help to explain this. The percentage of cases who recorded a decrease (i.e. reduction in the actual number of visits compared to 1 year before the group) remained relatively constant at about 43% at each of the five periods of assessment. The percentage of patients who recorded an increase in usage was also fairly constant (about 14%) for years 2–5 after assessment, but 24.4% of patients recorded an increase in usage 1 year after the group.

		Number of patients 1 year before		χ^2	df	p	Before – After	95% Confidence Interval
		Not used	Used					
1 year after	Not used	24	18	2.29	1	0.13	−0.10	−0.23 to 0.03
	Used	10	26					
2 years after	Not used	29	31	23.06	1	<0.001	−0.37	−0.49 to −0.24
	Used	3	13					
3 years after	Not used	22	21	15.70	1	<0.001	−0.32	−0.45 to −0.18
	Used	2	15					
4 years after	Not used	18	17	14.22*	1	<0.001	−0.38	−0.54 to −0.22
	Used	1	6					
5 years after	Not used	10	10	7.36*	1	0.01	−0.41	−0.65 to −0.17
	Used	1	3					

Figure 12.4 Group patients using outpatient visits. Patients were categorized into whether or not they used outpatients visits, at each stage of the study. McNemar's test was used to compare the proportions of patients making outpatient visits in the year before the group vs. the subsequent years following the group. (*Two out of four cells had expected values of less than 5)

GENERAL PRACTICE CONTACTS

Figure 12.6 shows a decrease over time in the median number of GP visits made for medical or psychological problems by the 53 group patients from 1 year before to 3 years after the group, of from nine to five, which was highly significant. The smaller number of 18 comparison patients whose notes were available showed no significant change. In those 19 group patients in whom there was a 5-year follow-up the decline at year 3 was maintained for years 4 and 5. (There were only 5 of the 34 comparison patients with a 5-year follow-up; the number was too small to be analysed.)

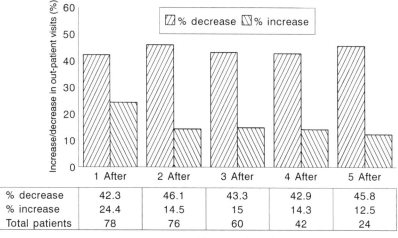

	1 After	2 After	3 After	4 After	5 After
% decrease	42.3	46.1	43.3	42.9	45.8
% increase	24.4	14.5	15	14.3	12.5
Total patients	78	76	60	42	24

After group (years)

Figure 12.5 Percentages of group patients whose outpatient visits increased or decreased during the 5 years following the group, compared with visits made during the year before the group. (The percentages of patients showing no change are not shown).

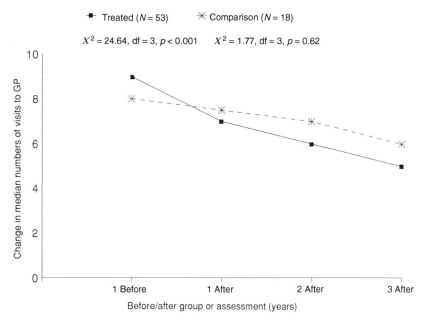

Figure 12.6 Change in the median number of visits to the GP by 53 group patients and 18 'comparison patients'. Medians were computed for 1 year before to 3 years after the group or after the assessment date. Friedman's test statistics are quoted for the two patient categories

Figure 12.7 shows the differing proportion of group patients prescribed any dose of antidepressants at each stage of the study. It can be seen that there is a decrease up to the 6-month follow-up, then a rise—though not up to the pre-group level—at the final follow-up. The decrease in the proportion of women prescribed antidepressants at the 6-month follow-up was significant. As many as 30% of patients consulted their GPs, again because of depression 2 to 7 years after the group, possibly because of intercurrent life events.

These findings do not give the flavour of the effect of therapy from the patient's point of view. Here are some extracts from the 12 adverse and 24 approving comments about the group:

> I would come in happy to the group then leave very unhappy. I got more depressed in the group than out of it, they were all doom, gloom and depression. (But her score fell dramatically.)

> I have met a wonderful man, and I think I may be—dare I say it?—in love!! And more significantly I have discovered that sex is great.

> I found the group very helpful, they helped me come to terms with what happened and I am no longer ashamed of it. I have now learned to drive, and I had the confidence to take lessons with a male instructor, which I had always thought would be impossible because of fears of men.

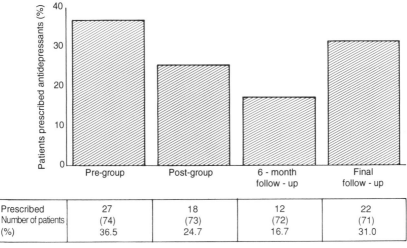

	Pre-group	Post-group	6 - month follow - up	Final follow - up
Prescribed Number of patients (%)	27 (74) 36.5	18 (73) 24.7	12 (72) 16.7	22 (71) 31.0

Approximate time-scale of assessment

Figure 12.7 Percentages of group patients prescribed antidepressants at each stage of the study.

DISCUSSION

Minz (1981) found that symptomatic improvement was as good an indi-
cator of early change as a more complex clinical judgement by an expert
clinician. However, decrease in symptoms does not necessarily occur as a
result of treatment. Patients might present with severe depression which
was going to remit anyway. (Equally, those who do not show any change
after treatment might have deteriorated without treatment.) It is possible
that this group of patients, with their particular type of post-traumatic
stress disorder, might have presented with a recent explosion of symptoms
shortly before assessment. Having often in adolescence suppressed their
memories of the abuse, these resurface in adulthood as a result of some
stimulus, and the resulting obsessive ruminations, flashbacks and nightmares produce a crisis in which the patient becomes agitated and
depressed. Such a crisis might possibly subside without treatment over the
next few months.

A way to establish causality would be to compare treated patients with
untreated controls but, as Candy (1972) reported, an adequately placebo-
controlled trial of psychotherapy is not really feasible. Luborsky (1981) has
pointed out that waiting-list patients cannot properly be considered as
controls as they are effectively minimally treated patients. The difficulty
here is that with severely damaged and distressed patients the possibility
that they might have to wait a considerable length of time for a treatment
already shown to be effective, or might not receive any treatment, would
be morally indefensible.

Shapiro and his colleagues (1991) state:

> There is a need for comparative outcome research that uses clinically realistic
> treatments with referred patients in clinically realistic settings. In addition,
> outcome research should pay attention to clinical as well as statistical
> significance in evaluating change and in presenting case outcomes.

Certainly in this study a standardized interview by the therapist with
multiple questionnaires could have interfered with therapist–patient
rapport and the therapeutic alliance. Trust might not have been built up,
the questions might have been felt to be intrusive, intimate details could
not so easily have been discussed and, in all probability, the therapy would
have been less effective.

The findings presented here form part of a larger study which includes
sociodemographic data on the whole population of 235 patients referred;
an attempt to predict outcome from factors such as age, initial depth of
depression and type of abuse; and an attempt to discover if factors, such
as age of onset of the abuse and its duration, influence the initial depth of
depression. It also includes an account of the fate of the 235 patients in

order to assess how economical it is in terms of time and finance to process those patients, assess 173, and treat 94 in order to benefit the 45 patients whose depressive ratings fell more than five points on the Hamilton Score.

THE CLINICAL EVALUATION

Depressive ratings alone are inadequate to reflect the global symptoms of anxiety and fear, lack of confidence and poor self-esteem that are found in these patients. Gorcey *et al.* (1986) found that patients with a history of child sexual abuse had much higher scores on the Beck, the State-Trait Anxiety Inventory and the Fear Survey Interview than had patients without such a history.

Considering that the Hamilton was originally intended to be an observer-completed instrument, it is interesting that in our study it gives similar results to the Beck, which was designed, and is valid, as a self-rating inventory. Whereas the Hamilton contains more items relating to vegetative symptoms, the Beck contains more items relating to self-esteem, particularly relevant to our patients (Thompson, 1989). The two inventories are therefore to some extent complementary. We suggest, however, that any further outcome studies should keep the Beck but omit the Hamilton. They should probably include an anxiety and a self-esteem questionnaire and the SCL (90R) (Derogatis *et al.*, 1974).

THE SERVICE EVALUATION

We found that during the second and succeeding years after therapy the proportion of patients reducing their use of (1) outpatient services and (2) combined hospital services progressively outweighed the percentage increasing their use. This supports Macdonald's (1992) results. In his service evaluation of psychiatric patients treated by individual psycho-therapy, he added together *all* types of hospital contacts—inpatient, day-patient and outpatient visits. He found that 35 patients made 744 hospital contacts in 1–55 months before therapy, and 173 contacts in the equivalent time-span after therapy. Knowles (1989) added the total hospital contacts and GP contacts together in her study of 10 patients.

Dick and Wooff (1975 and 1986), divided their patients into those who, in addition to the five assessment sessions, had received between 15 and 60 sessions ('treated cases') and those who had received less than 15 sessions whom they denoted as 'untreated'. They found that in the year following therapy 82% of their treated patients reduced their use of combined hospital services, while 17% increased their use, compared with the year before, and that the decrease in day- and outpatient usage was significant. In our study, however, we found no significant fall in use of hospital services in the first year, but only in the second and succeeding years. This may have been because Dick's therapy was more intensive, resulting in earlier improvement.

Statistical analysis of contacts with services is difficult when the proportion of patients using the service is small; use of the median is not appropriate. Moreover, in our particular study a comparison of the total contacts across the full 5 years would not have been valid because not all our patients completed a full 5-year follow-up. Comparison of contacts for each individual year after with the year before therapy seems an appropriate way to overcome the problem.

There are also considerable difficulties in attempting a service evaluation in regions such as Southampton, where there is much geographical movement of population. General practice records are essential in order to ensure that the patient still lives in the region and therefore that the recording of any negative hospital attendance is valid. Once the patients have moved out of the area their GP notes follow them, so neither the GP notes nor hospital data can be recorded.

For the service utilization data to have been collected more effectively it might have been wise, at the outset of the research, to ask patients to inform us of any change of address if they moved. We doubt, however, whether many of them would have responded when the time came. It might also have been helpful to have known the name of their new GPs.

General practitioners are at present overloaded with questionnaires of all kinds and are understandably reluctant to allow access to their notes without the patient's permission. The patients in this survey might well not have wished an outsider to read notes containing intimate personal histories. (Because of this, permission was requested from patients in only two instances, at the GP's insistence, one of whom acceded, and the other of whom did not reply.)

We doubt whether GPs from other areas would be prepared for a researcher to travel to look at the GP notes, (and the GPs we asked were not very ready to supply the information over the telephone or on a form we sent them). Any contacts with a psychiatric hospital would be recorded from the GP notes, so that with the relevant consultant's permission, the hospital notes could then be scanned for numbers of contacts. But patients sometimes consult hospitals without GPs being aware of it and often do not even register with their GP for some time after moving.

It seems more accurate therefore to exclude altogether those years following treatment in which a patient has moved out of the area. Ideally service evaluation research should, as we have said, take place in a geographically stable population.

It would be interesting to have accurate data on community psychiatric nurse, social worker and probation contacts. This would mean inquiring from the patient about these possible contacts regularly each year during the research period, as did Knowles (1989), and stressing to these workers the importance of keeping accurate records of contacts. Research of this sort is only as accurate as the records of the professional workers, and relies on the availability of all the relevant notes.

Since repressed memories may resurface shortly before referral and push up the number of contacts with services, it is possible that only 1 year of pre-group recordings may be too short a time to give a true base line. McDonald (1992) compared contacts in the same time period of up to 5 years before and after the group. Yet the percentage of group patients with a lifetime history of previous overdoses (40%), acts of deliberate self-harm (14%) and hospital admissions (27%) suggests that, even if memories were suppressed, there was still considerable morbidity.

It is possible that the recording of contacts with services is an under-estimate, as while some may be entered routinely by a computer, they may not always be entered in notes; but this discrepancy applied before as well as after therapy. Again, patients may have moved out of the area without the GP knowing, and be contacting services outside the area, thus making the figures appear lower than they should be.

These results are only preliminary and from a research point of view should ideally be confirmed by further studies using waiting-list controls or comparative group-treatment modalities. The use of reduction in symptoms as an outcome measure should include more global measures of social function.

CONCLUSIONS

In this study, analytic group therapy was accompanied by an alleviation of depression in women survivors. This confirms the findings of other group therapists treating women survivors (Alexander *et al.*, 1989; Carver *et al.*, 1989). We found a steady decline in depressive symptoms up to 6 months after the group which was then maintained for 2 to 7 years. The improve-ment was independent of antidepressive medication, and was confirmed by the subjective opinions of therapists and patients.

There was significant reduction in the proportion of patients using combined hospital and outpatient services in the second and succeeding years after the group.

Visits to the GP were significantly decreased from the year before the group to the third year after the group ended, and this decline was main-tained for up to 5 years. The comparison patients showed no significant change over the same time intervals.

The prcportion of patients prescribed antidepressants decreased signi-ficantly after the group up to 6 months.

It can therefore be concluded that analytic group therapy for up to 6 months for women survivors of child sexual abuse probably reduced depressive symptoms and the use of psychiatric hospital and primary care services.

Our finding that 6 months' weekly therapy for these women not only reduced their symptoms of depression but reduced both their attendance at hospital and at general practice and the prescription of antidepressants must reassure NHS managers of the value and cost-effectiveness of psychotherapy.

ACKNOWLEDGEMENTS

We should like to thank Jean Guion, Vivia Bashford, Demetra Onofriou and Anja Leydon-Keith for help in collecting data. We should also like to thank Glenys Parry for her initial encouragement in setting up the study and to the Wessex Regional Health Authority for supporting us with a grant.

REFERENCES

Alexander, P.C., Neimeyer, R.A., Follette, V.M., Moore, M.K. & Harter, S. (1989). A comparison of group treatments of women sexually abused as children. *Journal of Consulting and Clinical Psychology* **57**(4), 479–483.

Baker, A.W. & Duncan, S.P. (1985). Child sexual abuse: A study of prevalence in Great Britain. *Child Abuse and Neglect* **9**, 457–467.

Beck, A.T., Ward, C.H., Mendelson, M., Mock J. & Esbaugh, J. (1961). An inventory for measuring depression. *Archives of General Psychiatry* **4**, 53–63.

Candy, J., Balfour, F.H.G., Cawley, R.H., Hildebrand, H.P., Malan, D.H., Marks, I.M. & Wilson, J. (1972). A feasibility study for a controlled trial of formal psychotherapy. *Psychological Medicine* **9**, 345–362.

Carver, C.M., Stalker, C., Stewart, E. & Abraham B. (1989). The impact of group therapy for adult survivors of childhood sexual abuse. *Canadian Journal of Psychiatry* **34**, 735–757.

Deighton, J. & McPeek, P. (1985). Group treatment: adult victims of childhood sexual abuse. *Social Casebook* **66**(7), 403–410.

Derogatis, L.R., Lipman, R.S., Rickels, R., Uhlenhuthy, E.H. & Cori, L. (1974). The Hopkins symptom checklist (HSCL): self-report inventory. *Behavioural Science* **19**, 1–15.

Dick, B.M. (1975). A ten year study of out-patient analytic group therapy. *British Journal of Psychiatry* **127**, 365–375.

Dick, B.M. & Wooff, K. (1986). An evaluation of a time-limited programme of dynamic group psychotherapy. *British Journal of Psychiatry* **148**, 159–164.

Follette, W. & Cummings, N.A. (1967). Psychiatric services and medical utilisation in a pre-paid health plan setting. *Medical Care* **5**, 25–35.

Foulkes, S.H. (1964). *Therapeutical Group Analysis*. London: George Allen and Unwin. Reprinted (1984) London: Karnac.

Gazarain, R. & Buchele B. (1988). *Fugitives of Incest*. Conneticut: International Universities Press.

Gorcey, M., Santiago, J.M. & McCall-Perez (1986). Psychological consequences for women sexually abused in childhood. *Social Psychiatry* **21**, 129–133.

Hall, Z. (1992). Group therapy for women survivors of childhood sexual abuse. *Group Analysis* **25**, 463–475.

Hall, Z. (1987). Soul Murder. *Changes* **5**, 303–306.

Hamilton, M. (1967). Development of a rating scale for primary depressive illness. *British Journal of Psychiatry* **6**, 278–296.

Hooper, P.D. (1990). Psychological sequence of sexual abuse in childhood. *British Journal of General Practice* **40**, 29–31.

Kearney-Cooke, A. (1988). Group treatment of sexual abuse among women with eating disorders. *Women and Therapy* **7**(1), 5–21.

Knowles, J. (1989). Paper given to the Service Evaluation Group of the Society for Psychological Research. Uffcolme Clinic, Birmingham, November 1989 Conference Report.

Lubell, D. & Soong, W.T. (1981). Group therapy with sexually abused adolescents. *Canadian Journal of Psychiatry* **27**, 311–315.

Luborsky, L. (1981). Reply to letter from Loranger, A.W. on 'Outcome of Psychotherapy'. *Archives of General Psychiatry* **38**(3), 1070.

MacDonald, A.J. (1992). Training and outcome in supervised individual psychotherapy. *British Journal of Psychotherapy* **8**(3), 237–247.

Marks, I. (1948). Rehearsal relief of a nightmare. *British Journal of Psychiatry* **133**, 461–465.

Mintz, J. (1981). Measuring outcome in psychodynamic psychotherapy. *Archives of General Psychiatry* **38**, 503–506.

Mumford, E., Schlesinger, H.J., Glass, G.V., Patrick, C. & Cuerdon, T. (1984). A New look at evidence about reduced cost of medical utilization following mental health treatment. *American Journal of Psychiatry* **141**, 1145–1158.

Palmer, R.L., Chaloner, D.A. & Oppenheimer, R. (1992). Childhood sexual experiences with adults reported by female psychiatric patients. *British Journal of Psychiatry* **160**, 261–265.

Shapiro, D.A., Barkham, M., Hardy G.E., Morrison, L.A., Reynolds, S., Startup, M. and Harper, H. (1991). University of Sheffield Psychotherapy Research Programme: Medical Research Council/Economic and Social Research Council, Social and Applied Psychology Unit. In *Psychotherapy Research Programs* (L.E. Bentley & M. Crago, eds). Washington DC: American Psychological Association.

Thompson, C. (1989). Affective disorders. In *Instruments of Psychiatric Research* (C. Thompson, ed.). Chichester: Wiley pp. 5–6, 87–111.

Tsai, M. & Wagner N. (1978). Therapy for women sexually molested as children. *Archives of Sexual Behaviour* **7**(5), 417–427.

13 Evaluating the Benefit of General Practice-based Counselling Services

MICHAEL KING

Royal Free Hospital School of Medicine, London, UK

ABSTRACT   Over the past 20 years there has been a sharp increase in interest in the recognition and management of psychological disorder in primary medical care. This has been accompanied by an enormous expansion of counselling services in British general practice, with little attempt at evaluation. Counsellors have emerged as a profession with their own organizations, codes of ethics and recommended methods of working. Evaluation of counselling must address the questions: What type of intervention? What type of counsellor? What type of client? What type of emotional or social problem? What type of outcome? In this chapter these difficulties are discussed taking illustrations from the results of a pilot study. Our results indicate that patients referred by family doctors to counsellors are often seriously emotionally distressed and recovery is slow. Counsellors come from different backgrounds and use a variety of therapies. Although controlled research is feasible, in a definitive trial patients should be randomized in a stratified manner, according to severity, by the researcher after initial assessments have been made. With the widespread availability of counselling today, it is becoming less acceptable to clients or GPs that a randomization procedure bars people from the counselling intervention. This tension is increasing the difficulty of conducting classical controlled trials. Counsellors should have a recognized accreditation and preferably be employed for the trial to ensure uniformity of approach and avoid long waiting-lists. In practices where a counsellor is already in place, the arrival of a research counsellor may provide an opportunity to shorten the waiting list. There needs to be a consensus on outcome measures that will enable direct comparison between studies. Blind assessments of outcome are desirable but are not always feasible and reliance on patient self-report is important. Only controlled evaluations will provide us with a greater understanding of the efficacy of counselling in this setting.

Correspondence address: Department of Psychiatry, University of London, Royal Free Hospital School of Medicine, Pond Street, London NW3 2QG, UK.

INTRODUCTION

Interest in the recognition and management of psychological disorder in primary medical care has increased sharply over the past 20 years. The training of psychiatrists and general practitioners (GPs) has drawn closer together and to some extent begun to overlap (Burns *et al.*, 1991). Psychiatry and general practice have much in common; doctors in both specialties must deal with patients presenting vague, undifferentiated complaints, cope with considerable uncertainty and use psychotherapeutic skills. The proportion of work in general practice which is concerned with psychological or social difficulties is difficult to estimate largely because of problems of classification (Gray, 1988; Sharp & King, 1989). What one doctor may recognize as dyspepsia another diagnoses as anxiety. Nevertheless, it is estimated that about 14% of consultations in family practice are wholly or largely for psychological reasons. A further 7 to 10% are also for psychological reasons but are not readily recognized as such by the doctor (Shepherd *et al.*1966; Goldberg & Blackwell, 1970; Skuse & Williams, 1984). GPs show considerable variability in their rates of referrals to the psychiatric services (Horder, 1988) but on average only 1 in 20 patients, recognized as having psychological difficulties, is referred (Goldberg & Huxley, 1980). This cautious approach is probably justified as many of the so-called 'minor psychiatric' problems in general practice are self-limiting adjustment reactions, which have a relatively high rate of spontaneous resolution. Thus, the traditional rôle of the family doctor has been to manage most clinical and social problems, including such difficulties as marital strife, work problems, bereavement and behaviour problems in children.

Increased recognition of the prevalence of emotional problems has been accompanied by an enormous expansion of counselling services in British general practice. This did not occur on any planned basis or as a response to any particular government imperative. Rather, it has come about because of an increasing emphasis on counselling as a discrete skill. Counsellors have emerged as a profession with their own organizations, codes of ethics and recommended methods of working (Gray, 1988). This has led to the suggestion that GPs may need a specific training to counsel their patients (Rowland *et al.*, 1989), a proposal which has been regarded with scepticism by some doctors (Shepherd, 1989).

A government white paper, *Promoting Better Health* (1987) proposed increasing the opportunities for other professions to work in general practice, and funding became available for part reimbursement of salaries of attached staff such as counsellors. This provided funds for the payment of counsellors (Rowland & Irving, 1984; McLeod, 1988) but did little to clarify respective rôles, areas of expertise, or the ethical difficulties of access to medical notes by professionals other than the GP. More important still, remained the largely unanswered question of the efficacy of counselling. The caution to avoid 'promoting a large counselling service in general practice before establishing what benefit accrues from this service...' (Martin, 1988) went largely unheeded. In this chapter I will tackle the issue of assessing the efficacy of GP-based counselling services. Many of the more general issues that apply to psychotherapy research will be discussed only where they are of direct relevance.

DEFINING COUNSELLING

Before assessing the efficacy of counselling, the process must be defined. It is easier to describe what counsellors do rather than what counselling is; most writers make a sharp distinction between *counselling skills* and *counselling*. Counselling skills, which are regarded as the ability to listen, reflect and empathize, are not considered the exclusive domain of counsellors and are used by many professionals such as doctors, teachers and even sales people. Although incorporating ideas from several schools of psychotherapy, counsellors adopt a neutral stance of 'supportive listening'. They refrain from giving advice and instead enable clients to gain insight and understanding rather than use the directives of others (Rowland & Irving, 1984). Although counselling 'is more than the mere use of counselling skills' (Rowland, 1992) it is less easy to define. The British Association for Counselling (1979), states that '...the task of counselling is to give the client an opportunity to explore, discover and clarify ways of living more resourcefully and towards greater wellbeing', and makes no reference to suffering or illness in the client. Rowland has defined it as 'an ethical task in which the counsellor forms a therapeutic alliance with the client and uses a range of skills to facilitate the client's resolution of his or her problems' (Rowland, 1993). Such non-directive intervention may be rather different, however, from that applied in other medical settings (Miller & Bor, 1988) or by family doctors, many of whom regard counselling as including education and problem solving over the long term rather than in weeks or months (Shepherd, 1989).

There have been two main approaches to the evaluation of counselling. In much the more common one, the principles of audit are applied in examining the process of the work. A study is made of the numbers and backgrounds of the counsellors employed, types of clients seen, patterns of working with other members of the primary health care team and the management of issues of confidentiality and clinical responsibility. In the second, experimental conditions are applied to the evaluation of a stated hypothesis about counselling.

AUDIT

McLeod (1988) studied the work of counsellors in 14 general practices in England and highlighted the varied backgrounds of the professionals involved, including marriage guidance counsellors, general and psychiatric nurses, health visitors, social workers, psychologists, psychotherapists and 'general' counsellors. Sibbald *et al.* (1993) carried out a questionnaire survey of the work of counselling, provided as a distinct or separate service, within a large sample of general practices in England and Wales. The definition of counsellor used was very wide: 'someone who offers (formal) sessions to patients in which patients are helped to define their problems and enabled to reach their own solutions'. Thirty-one per cent of practices surveyed had a counsellor with no other task within the practice, but community psychiatric nurses and clinical psychologists constituted most of these, outnumbering practice counsellors two to one. The authors raised concerns about the varied qualifications of these professionals who were expected to cope with a wide range of often seriously affected patients. Possibly because of the broad definition of 'counsellor' used in the survey, it was found that just under half of the those identified had had no specific training in counselling. The authors called for a more rigorous accreditation of counsellors and for greater attention to the sorts of referrals which were appropriate. The Counselling in Primary Care Trust is a private charity set up in Britain with the aim of carrying out action research (another name for audit) into the rôle of counsellors in primary care. This organization has directed considerable funds into the study of counselling with the aim of establishing 'good practice' and assisting primary care teams to work with counsellors. However, they and many others have avoided any attempt to measure efficacy on the grounds that the field is not yet sufficiently developed for a controlled evaluation.

Uncontrolled studies of psychologists working in general practice have shown, if somewhat equivocally, that patients improve in their coping strategies or report reduced levels of distress (Freeman & Button, 1984; Milne & Souter, 1988) and that GPs appreciate these services (Jerrom *et al.*, 1983; Deys *et al.*, 1989). Some have reported positive changes in attendance rates and prescribing (Waydenfeld & Waydenfeld, 1980) while others have not (Martin & Martin, 1985).

CONTROLLED RESEARCH

In outcome research, experimental controls must be used to separate out treatment factors from other influences which may have brought about change. As Kendall and Norton-Ford (1982) have stated, 'Studies of intriguing therapeutic results that fail to pinpoint the effects that can be accurately attributed to the therapy provide at most speculative knowledge'. Controlled evaluations of counselling raise similar problems to those encountered in the evaluation of more established psycho-therapies (Wilkinson, 1984). The commonest opposition to quantitative, controlled trials of psychotherapy was the view that intrapsychic processes posed insurmountable problems of measurement and understanding. An early attempt to undertake a large, controlled trial of psychotherapy in Britain failed at the feasibility stage, largely because of difficulties in recruitment of patients considered suitable by the therapists (Candy et al., 1972). Nevertheless, there is now an extensive literature on controlled evaluations of psychotherapy, much of the work originating from the United States (Robinson et al., 1990). Arguably, evaluation of counselling may avoid some of the more complex issues in psychotherapy research. Supportive listening and non-directive reflection of clients' statements are simpler to define in terms extrinsic to the patient. Outcome may be more plainly defined in terms of problems solved, increased personal insight and patient satisfaction. Nevertheless, all counselling outcome research should address the questions: What type of intervention? What type of counsellor? What type of client? What type of emotional or social problem? What type of outcome? (Kendall & Norton-Ford, 1982).

Studies of psychologists working in primary care have most often taken a controlled approach (Earll & Kincy, 1982; Teasdale et al., 1984; Robson et al., 1984), but it is not always appropriate to regard their work as counselling. Many such studies have relied on attendance rates, levels of prescribing of psychotropic drugs or patient satisfaction, as measures of outcome. Although results have not indicated clear superiority of a psychologist over routine treatment from the GP, clinical improvement may be more rapid, or patient satisfaction greater, in those patients treated by psychologists. Robson et al. (1984) concluded that over a quarter of the salary of a senior psychologist working in primary care could be found from savings in the drug bill alone.

Increasing numbers of controlled studies of counselling have appeared over the past decade. In the 'The Leverhulme Project', conducted in a health centre in Southampton (Ashurst, 1981), patients with neurotic disorders were randomly assigned to a counsellor or to routine care by the GP. Counsellors employed a mixture of Rogerian, reflective techniques as well as assorted behavioural methods, including relaxation. Outcome was measured in terms of prescribing of psychotropic drugs, patient and GPs views of outcome and scores on the General Health Questionnaire (GHQ) (Goldberg, 1970) at 1 year. Although the evaluation was obscured by inclusion of patients who had declined the offer of counselling (42% of those offered it) in the control group, it appeared that counselling had a positive effect, at least on the level of prescribing of psychotropic drugs.

In an Australian study, brief psychotherapy was compared with care by the GP or a period of waiting in a control group (Brodaty & Andrews, 1983). Subjects were selected on the basis of persistently high scores on the GHQ. Although all three groups improved appreciably no matter what was done to them, large numbers of refusals and drop-outs during the selection and treatment phases, as well as a considerable amount of missing data, made interpretation of the results more difficult.

In a study of 80 women suffering from depression, Corney (1984) randomly allocated subjects to a social worker attached to the general practice or to routine care by the doctor. Although 60% improved with no significant differences between the two groups at 6 months, women suffering acute-on-chronic depression and who had major marital diffi-culties were more likely to benefit from the social work intervention.

Catalan et al. (1984) compared counselling by the GP to treatment with anxiolytic drugs in a study of 91 patients selected by the doctors with new episodes of minor affective disorder. Scores on questionnaire and struc-tured psychiatric interview at entry to the trial and at 1 and 7 months later showed no differences between the groups. Counselling, as perceived by both patient and doctor, comprised advice on coping, listening, explaining symptoms and reassurance. Only advice on coping, however, appeared to occur with greater frequency in the counselled group. Surprisingly, the doctors spent *less* time with the patients randomized to receive counselling only. Although the same doctors managing patients in both groups may have obscured differences between treatments, this study demonstrated that brief counselling by GPs was as at least as effective as psychotropic medication. Some would argue, however, that this study measured the application of counselling skills rather than counselling, which requires more specialized personal and professional training (Rowland et al., 1989).

Although primarily a study of the outcome of neurotic illness in general practice, Johnstone and Shepley (1986) followed up 84 patients detected initially by abnormal scores on the GHQ and randomly allocated to treat-ment by the GP or health visitor. Health visitors are trained nurses who work in the community and have special obligations to the care of mothers and young children and the elderly. Little difference in GHQ scores was found between the two groups of patients at 1 or 5 years after entry to the study. The authors concluded that even patients with quite severe levels of psychiatric morbidity could be treated satisfactorily by paramedical personnel in primary care. Health visitors were also utilized in a study of non-directive counselling of 60 women with post-partum affective disorder (Holden et al., 1989). Using structured psychiatric assessments by a blind psychiatric rater, the authors based clinical recovery on changes in psychiatric scores at an average of 13 weeks follow-up, a period perhaps too short to be meaningful. The health visitors taking part received very brief instruction which included a manual on post-natal depression and non-directive counselling and 6 hours of training in counselling skills. Counselling was reported to result in a better outcome than no intervention.

In a recent controlled evaluation of the work of community psychiatric nurses (CPNNs) based in general practice (Gournay & Brooking, 1992), patients who sought help from their doctor for a range of non-psychotic problems were randomized to remain with the GP or receive 'client-centred counselling' by a CPN. After the trial had begun a third group, a waiting-list condition, was added because of delays in patients seeing the CPNs. This considerably increased the complexity of the trial and high-lights the importance of piloting before beginning a definitive study. No differences between the groups in terms of symptomatology or use of services were found at the 3-month follow-up, although half the patients referred to the CPNs dropped out during the study. Patient satisfaction was mixed and bore little relationship to symptomatic or social outcome. The authors concluded that CPNs' work needs to be carefully targeted, although it was unclear what sort of client would benefit. Furthermore, they called for their work to be more problem orientated, implying that client-centred counselling was ineffective.

THE PROBLEMS OF CONTROLLED EVALUATIONS

The criteria for entry into studies of talking therapies vary enormously. Should all referrals be entered into the study or restricted to only those who might be most helped—and if so by what criteria? Although narrowly defined entry criteria may allow more exacting measurement of outcome, the results cannot be generalized to the range of psychological and social problems seen in general practice. Relaxing entry standards, however, imposes a need for greater numbers of subjects to allow later subcategorization into groups which are large enough for statistical comparison.

What the counsellor does is crucial. As already discussed, most counsellors use a non-directive approach which, for the purposes of research, can be refined and agreed upon to a reasonable degree of reliability. The training and experience of the counsellor, however, is less certain. Enormous variation in the length and content of training schemes for counsellors as well as differences in their personality, life experience and length of time in practice (McLeod, 1988), may all have differential effects on the outcome of therapy. Indirect evidence would suggest that ill-defined qualities in the counsellor, such as willingness to help, empathy and genuineness, may be as important in outcome as the theoretical assumptions he or she holds (Irving & Heath, 1989). Thus, a careful description is needed of the background of the therapists taking part.

Measurement of outcome of a heterogeneous group of conditions treated in a variety of ways is far from easy. How can behavioural change be balanced with increased insight and feelings of wellbeing? Although standardized evaluation of outcome in terms of patients' functioning can be used in a general practice setting, scales vary widely and simple improvement ratings may give a false impression when initial severity of the problem is not uniform. Effectiveness in terms of prescribing or surgery visits is not universally helpful as not all patients referred for therapy are high consumers of primary care services or prescribed drugs (Trepka & Griffiths, 1987) and in many instances such changes are not the central focus of counselling.

HOW TO DO IT

Despite these many obstacles we are moving ever closer to an under-standing of the function of counselling in primary medical care. Before further resources are committed, methodologically sound studies must be undertaken to examine the efficacy of counselling (Corney, 1986; Fallowfield, 1993). It is not sufficient to conclude that such research is no longer required and that a review of all studies to date might provide an answer (Trepka & Griffiths, 1987).

I will now illustrate the difficulties involved in this work from experience of a pilot study in which we aimed to test the feasibility of a controlled comparison of counselling with standard treatment by the GP. In planning the study we made three methodological decisions:

1. All patients presenting with emotional difficulties would be entered into the trial.
2. Non-directive counselling as defined by the British Association of Counsellors (1979) would be used as it is the treatment most often provided in general practice.
3. Outcome measures would be based on scales with established validity and reliability.

Two group general practices took part; the counsellors involved were those already working in the practices. Patients experiencing acute or acute-on-chronic episodes of emotional disorder, for whom the GP con-sidered that counselling would be of benefit, were recruited. Patient pre-ference was taken into account in the randomization; it was first established whether the patient had a strong desire to see a counsellor or to remain with the GP. If strong preferences was expressed patients were given their choice; if not they were randomly allocated to either pro-fessional. This method (Gossop et al., 1986; Brewin & Bradley, 1989) was utilized to avoid the difficulties of strict randomization and to reduce patient drop-outs. Subjects randomized to the counsellor were taken on for 6–8 sessions over a 12-week period; the remainder received routine treatment by the doctor over the same 12 weeks.

With a week of recruitment, each subject was approached by a research doctor for an assessment using a standardized interview, the Clinical Interview Schedule (Goldberg et al., 1970), and four questionnaires: the 28-item General Health Questionnaire (Goldberg, 1970), the Beck Depression Inventory (Beck et al., 1961), the Social Problems Questionnaire (Corney, 1988) and a Self-Esteem Questionnaire (Robson, 1989). Subjects in both groups were assessed again after 12 weeks and 6 months. At 12 weeks, a random half of subjects were interviewed and the remainder contacted by post. At this point GPs and counsellors completed a standard form detailing the number of consultations or counselling sessions, the type of treatment or therapy given and their view of the patient's response. At 6 months all subjects were reinterviewed at which time they also completed a questionnaire on their perceptions of counselling and factors of greatest help to them during the study. Attendance at the practice and prescribing patterns from 12 months prior to entry to the trial were collected from the practice records.

RESULTS OF THE PILOT

Only a summary of the results will be given here; the general lessons from a pilot study are more important. Altogether, 24 patients were recruited, of whom 19 were referred to a counsellor and 5 remained with the GP. At 3 months 22 were successfully contacted and 19 at 6 months. The mean number of sessions for patients seeing a counsellor was 10 (range of 1 to 20) while the mean time spent with either the counsellor or GP was 6 hours (range 30 minutes to 20 hours).

Symptom ratings on the Clinical Interview Schedule were higher than expected and correlated most closely with levels reported for patients seen in psychiatric outpatient departments (Strathdee et al., 1990). The mean duration of psychiatric symptoms was 15 months (range 1 to 60 months). Duration of disorder, however, was not associated with interview or questionnaire scores and only one subject reported previous psychiatric treatment. Problems were varied but mainly involved mixed anxiety and depression over relationship or marital problems, psychosomatic symptoms or difficulties related to child birth or child care. Recovery was reasonable over the initial 12 weeks but slow thereafter. There was no association between time spent with the professional and change in psychiatric scores at 6 months. Duration of disorder at entry to the study was also not related to outcome. There was a trend for patients who remained with the GP to show greater improvements in their scores than those who were randomized to the counsellor, despite higher baseline scores in the latter group. There was also a trend for patients who scored below the threshold on the Clinical Interview Schedule at follow-up to have consulted the GP less often and to have received less psychotropic prescriptions over the 6 months.

PRESENTING PROBLEMS

When we discussed the severity of the presenting problems with the participating doctors, it appeared that limited access to local psychiatric services was a decisive factor in their referrals. The counsellor was frequently the only mental health professional available. As discussed earlier, however, the suitability of such referrals for counselling is questionable (Sibbald et al., 1993), and the slow recovery that occurred may have reflected the difficult therapeutic challenge afforded by these patients.

If subjects with a wide range of severity of emotional distress are to be entered into a substantive trial, a stratified randomization on the basis of severity would be necessary. Stratified randomization reduces the possibility of chance differences between groups and ensures that the interaction between severity and outcome can be examined. A ceiling on symptom scoring, above which potential subjects would be referred to the local mental health services, might also be appropriate.

THE DOCTORS

Although the doctors maintained their interest, they reported that they would have appreciated reminders of the study protocol as well as regular updates on the number of patients recruited. They sometimes misunderstood our intention that they provide routine care to patients randomized to them and attempted to counsel patients in a way that was too time-consuming. This is a difficult issue; in planning for a substantive study we have had discussions with a large number of GPs about what they should do with patients randomized to remain solely in their care. Counselling has become so much a part of the services offered in (or at least available from) primary medical care that doctors regard it as difficult or unethical to deny patients access to the service, even for short periods. There is a growing reluctance to regard counselling as an integral part of the GP's work; rather it is quickly becoming a specialist, essential service.

THE RANDOMIZATION

We hoped that GPs would feel more in charge if they randomized the patients but they felt uncomfortable in this rôle. Furthermore, the chosen method of randomization was not successful. Randomization is undertaken to reduce systematic error in group allocation but where patients are asked to undertake a time-consuming treatment or where they have strong preferences, it may be unacceptable to them. Counsellors are now more available than ever before, particularly in the private sector; thus it may be difficult to prevent subjects who are randomized to remain with the GP from seeking their own therapy elsewhere. One way to optimize motivation is to take account of patient preference in the randomization (Brewin & Bradley, 1989). The GPs found, however, that they could covertly suggest to patients that seeing a counsellor might be helpful and thus, patient 'choice' was not always objective. This resulted in 19 patients being referred to counsellors, while five remained with the GPs. Patients referred to the counsellors tended to have higher psychiatric scores than those who remained with the GPs, also indicating that the doctors might have been influenced by the severity of their patients' distress.

Randomized clinical trials in general practice challenge the identification of doctors as clinicians or researchers and may lead to difficulties in the doctor–patient relationship (Taylor, 1992). Although the doctors taking part in the pilot agreed that randomization was essential for a definitive study, they were concerned about having to carry it out. Ideally, someone other than the GP or researcher should randomize patients in order that the researcher who measures outcome remains unaware of the treatment received by the subject; this is particularly important where outcome is assessed by interview. Use of a third party, however, increases the cost and complexity of the study. Patients may also be more likely to remain in the study if, during the initial assessment, the researcher makes the randomized choice and advises them into which group they have been allocated. If this is done by a third person, the process is more complicated and subject to delay. In any case it is very difficult for the researcher to remain completely unaware of group allocation. Some reference to the treatment group may come out at interview, even when patients are asked to conceal which treatment they received. Assuming that a study is blind when it is only partially so, is arguably the greatest pitfall in controlled trials of this sort (Oxtoby *et al.*, 1989). Thus, reliance on *self-report* schedules is perhaps the safest option. There is also a growing body of opinion within psychology and psychiatry that self-assessments are more accurate than observer ratings, whether or not blindness is an issue (Lewis *et al.*, 1988).

THE COUNSELLORS

The counsellors attached to the practices came from diverse backgrounds. Two were generic counsellors, one a social worker and two community psychiatric nurses. Their training and experience of counselling made uniformity in treatment difficult to achieve. Study of their counselling records demonstrated that although non-directive, patient-centred counselling usually took place, cognitive-behavioural approaches were also employed. Not all used methods recommended by the BAC, nor were any accredited by the BAC. Long waiting-lists, which hindered prompt referral, were the norm.

There is concern in Britain about the variation in training and qualifications of counsellors employed not only in general practice (Sibbald *et al.*, 1993) but also in other medical services (Fallowfield & Roberts, 1992). The only way to ensure uniformity of approach in controlled evaluations is to employ counsellors with an accredited training speifically for the trial. Although increasing the costs of the research, the work of the counsellors would be standardized. Where a participating practice already has an attached counsellor, recruitment for the trial will help to reduce the waiting-list of the practice's counsellor and should not interfere with normal practice routine. Providing a research counsellor is also a strong incentive for those practices without counsellors to take part in the trial. In the contemporary world of general practice, with its emphasis on the rôle of market forces, it is becoming increasingly difficult to persuade practices to take part in research without some form of inducement. Unfortunately, there are also disadvantages to providing a research counsellor. They may appear as an unfamiliar professional or as simply a new person to staff in the participating practices. Unless handled sensitively, difficulties may arise in communication with the GPs, access to medical notes, and working relations with other counsellors, if present, in the practice.

THE CLIENTS

Although the majority of subjects found the intervention helpful, counselling was not without its side-effects. Two patients said that they wished they had never started. One woman who revealed to the counsellor that she was being physically assaulted by her husband and son, felt abandoned when only four therapy sessions were considered necessary by the counsellor. The views of patients are obviously of great importance and should never be ignored in trials of this nature. Even in the more familiar controlled trials of medical drugs the quality of life of subjects taking part, over and above the specific effects of the intervention, has generated increasing concern.

OUTCOME MEASURES

Outcome measures, which have been well standardized in the population studied, should be employed wherever possible. The difficulty lies in deciding on which outcome measures to choose. In this pilot we focused on emotional symptoms, social problems and self-esteem. Other outcomes, such as consulting behaviour and consumption of psychotropic drugs, are also clearly of importance. However, a more specific measure of *disability* is essential. In measuring outcome it is important not only to take account of change in symptomatology and a reduction in social problems but also to make some assessment of daily functioning. This should include work or household duties, ability to socialize with family and friends, and physical health status. In most trials of counselling in general practice little attention has been paid to physical health problems in patients taking part, despite the evidence that physical disorder is a crucial factor in psychological ill health (Goldberg & Huxley, 1992). Although the randomized nature of a clinical trial will help to reduce this confounding effect, within-group analysis to differentiate which patients do well under which conditions would be aided by such measurements. One example is the Short Form 36, a disability schedule which was developed in the United States and which has been recently subjected to psychometric assessment in the United Kingdom (Jenkinson *et al.*, 1993). This brief self-report schedule, which has been age-standardized, provides a measure of functioning and wellbeing from the patient's point of view.

THE WAY FORWARD

A common criticism of quantitative research in the field of psychotherapy is that individuals have unique ways of experiencing distress and restitution and that human detail is lost in controlled trials of this nature. One result has been the popularity of randomized trials in single patients, but these too are not without their disadvantages (Galassi & Gersh, 1991). Controlled trials remain the most robust way of establishing efficacy but they should be deployed with a greater emphasis on subjective symptoms, daily functioning and quality of life (Knipschild, 1993). The new marketplace in health care in Britain has important consequences for providers and purchasers of psychotherapy services. A call has recently been made to resist the demand for more counselling services until better evidence of efficacy and safety is available (Fahy & Wessely, 1993). Well-designed, controlled evaluations are urgently needed.

The following is a list of suggestions taken from our current assessment of counselling in general practice. It is not intended to be overly specific or exhaustive, but serve as a guide to the way forward.

1. The design of studies of counselling in general practice should mimic as closely as possible the natural setting in which counselling occurs.
2. The qualifications of counsellors should be clear and acceptable to current standards of accreditation as laid down by the British Association of Counselling.
3. Some measure of the form and content of the counselling intervention is essential if the results are to be generalizable.
4. Introducing counsellors for the purposes of research may be helpful, but requires careful introduction to the practices taking part.
5. Subjects should be randomized (stratified by severity of emotional symptoms) by a person who is not associated with the practices.
6. An economic assessment should be incorporated into the trial design, but should not be allowed to dominate the research.
7. Outcome measures should be brief and include an evaluation of psychological, and social function as well as a global measure of disability. Reliance on standardized, self-report measures is preferable to subjective assessment by observers. An assessment of attendance rates at the practice and consumption of psychotropic drugs may also be important.
8. The views of subjects, counsellors and general practitioners should be taken into account in the measurement of outcome.
9. Follow-up of subjects taking part should be of sufficient length to assess longer term change; 6 months is suggested as a minimum.

REFERENCES

Ashurst, P.M. (1981). *Counselling in General Practice*. Report to the Mental Health Foundation Conference, Oxford, October 1981.

Beck, A.T., Ward, C.H., Mendelson, M., Mock, J. & Erbaugh, J. (1961). An inventory for measuring depression. *Archives of General Psychiatry* 4, 561–571.

Brewin, C.R. & Bradley, C. (1989). Patient preferences and randomised controlled trials. *British Medical Journal* 299, 313–315.

British Association for Counselling (1979). *Counselling: Definition of Terms in use with Expansion and Rationale*. Rugby: British Association for Counselling.

Brodaty, H. & Andrews, G. (1983). Brief psychotherapy in family practice. A controlled prospective intervention trial. *British Journal of Psychiatry* 143, 11–19.

Burns, T., Silver, T., Crisp, A., Flute, P. & Freeling, P. (1991). *General Practice Experience for Psychiatric Trainees in SW Thames*. Report to the Department of Health, St George's Hospital, London.

Candy, I., Balfour, F.H.G., Cawley, R.H., Hildebrand, H.P., Malan, D.H., Marks, I.M. & Wilson, J. (1972). A feasibility study for a controlled trial of formal psychotherapy. *Psychological Medicine* 2, 345–362,

Catalan, J., Gath, D., Edmonds, G. & Ennis, J. (1984). The effects of non-prescribing of anxiolytics in general practice. *British Journal of Psychiatry* 144, 593–602.

Corney, R.H. (1984). The effectiveness of attached social workers in the management of depressed female patients in general practice. *Psychological Medicine* (Supple. 6).

Corney, R. (1986). Marriage guidance counselling in general practice. *Journal of the Royal College of General Practitioners* **36**, 424–426.

Corney, R.H. (1988). Development and use of a short self-rating instrument to screen for psychosocial disorder. *Journal of the Royal College of General Practitioners* **38**, 263–266.

Deys, C., Dowling, E. & Golding, V. (1989). Clinical psychology: a consultative approach in general practice. *Journal of the Royal College of General Practitioners* **39**, 342–344.

Earll, L. & Kincy, J. (1982). Clinical psychology in general practice: a controlled trial evaluation. *Journal of the Royal College of General Practitioners* **32**, 32–37.

Fahy, T. & Wessely, S. (1993). Should purchasers pay for psychotherapy? *British Medical Journal* **307**, 576–577.

Fallowfield, L.J. (1993). Evaluation of counselling in the National Health Service. *Journal of the Royal Society of Medicine* **86**, 429–430.

Fallowfield, L.J. & Roberts, R. (1992). Cancer counselling in the United Kingdom. *Psychology and Health* **6**, 107–117.

Freeman, G.K. & Button, E.J. (1984). The clinical psychologist in general practice: a six year study of consulting patterns for psychosocial problems. *Journal of the Royal College of General Practitioners* **34**, 377–380.

Galassi, J.P. & Gersh, T.L. (1991). Single case research in counselling. In Watkins, C.E. & Schneider, L.J. (eds), *Research in Counselling*. Lawrence Erlbaum, New Jersey, pp. 119–161.

Goldberg, D.P. (1970). *Detection of Psychiatric Illness by Questionnaire*. Oxford: Oxford University Press.

Goldberg, D. & Blackwell, B. (1970). Psychiatric illness in general practice. A detailed study using a new method of case identification. *British Medical Journal* **2**, 439–443.

Goldberg, D.P., Cooper, B., Eastwood, M.R., Kedward, H.B. & Shepherd, M. (1970). A standardised psychiatric interview for use in community surveys. *British Journal of Preventive and Social Medicine* **24**, 18–23.

Goldberg, D. & Huxley, P. (1980). *Mental Illness in the Community: The Pathway to Psychiatric Care*. London: Tavistock.

Goldberg, D. & Huxley, P. (1992). *Common Mental Disorders*. London: Routledge.

Gossop, M., Johns, A.R. & Green, L. (1986). Opiate withdrawal: Inpatients versus outpatient programmes and preferred versus random assignation to treatment. *British Medical Journal* **293**, 103–104.

Gournay, K. & Brooking, J. (1992). *An Evaluation of the Effectiveness of Community Psychiatric Nurses in Treating Patients with Minor Mental Disorders in Primary Care*. Report to the Department of Health, London.

Gray, D. (1988). Counsellors in general practice. *Journal of the Royal College of General Practitioners* **38**, 50–51.

Her Majesty's Government (1987). *Promoting Better Health*. London: HMSO.

Holden, J.M., Sagovsky, R. & Cox, J.L. (1989). Counselling in a general practice setting: controlled study of health visitor intervention in treatment of postnatal depression. *British Medical Journal* **298**, 223–226.

Horder, J. (1988). Working with general practitioners. *British Journal of Psychiatry* **153**, 513–520.

Irving, J. & Heath, V. (1989). *Counselling in General Practice: a Guide for General Practitioners*, revised. Rugby: British Association for Counselling.

Jenkinson, C., Coulter, A. & Wright, L. (1993). Short form 36 (SF36) health survey questionnaire: normative data for adults of working age. *British Medical Journal* **306**, 1437–1440.

Jerrom, D.W.A., Simpson, R.J., Barber, J.H. & Pemberton, D.A. (1983). General practitioners' satisfaction with a primary care clinical psychology service. *Journal of the Royal College of General Practitioners* **33**, 29–31.

Johnstone, A. & Shepley, M. (1986). The outcome of hidden neurotic illness treated in general practice. *Journal of the Royal College of General Practitioners* **36**, 413–415.

Kendall, P.C. & Norton-Ford, J.D. (1982). Therapy outcome research methods. In *Handbook of Research Methods in Clinical Psychology* (P.C. Kendall & J.N. Butcher, eds). New York: Wiley, pp. 429–460.

Knipschild, P. (1993). Trials and errors. *British Medical Journal* **306**, 1706–1707.

Lewis, G., Pelosi, A.J., Glover, E., Wilkinson, G., Stansfield, S.A., Williams, P. & Shepherd, M. (1988). The development of a computerized assessment for minor psychiatric disorder. *Psychological Medicine* **18**, 737–745.

Martin, E. (1988). Counsellors in general practice. *British Medical Journal* **297**, 637.

Martin, E. & Martin, P.M.L. (1985). Changes in psychological diagnosis and prescription in a practice employing a counsellor. *Family Practice* **2**, 241–243.

McLeod, J. (1988). *The Work of Counsellors in General Practice*. Occasional paper 37. London: Royal College of General Practitioners.

Miller, R. & Bor, R. (1988). *AIDS. A Guide to Clinical Counselling*. London: Science Press.

Milne, D. & Souter, K. (1988). A re-evaluation of the clinical psychologist in general practice. *Journal of the Royal College of General Practitioners* **38**, 457–460.

Oxtoby, A., Jones, A. & Robinson, M. (1989). Is your 'double-blind' design truly double-blind? *British Journal of Psychiatry* **155**, 700–701.

Robinson, L.A., Berman, J.S. & Neimeyer, R.A. (1990). Psychotherapy for the treatment of depression: a comprehensive review of controlled outcome research. *Psychological Bulletin* **108**, 30–49.

Robson, P. (1989). Development of a new self-report questionnaire to measure self esteem. *Psychological Medicine* **19**, 513–518.

Robson, M.H., France, R. & Bland, M. (1984). Clinical psychologist in primary care: controlled clinical and economic evaluation. *British Medical Journal* **288**, 1805–1807.

Rowland, N. (1992). Counselling and counselling skills. In *Counselling in General Practice* (M. Sheldon, ed.). Exeter: Royal College of General Practitioners, pp. 1–70.

Rowland, N. (1993). What is counselling? In *Counselling in General Practice* (R. Corney & R. Jenkins, eds). London: Routledge, pp 17–30.

Rowland, N. & Irving, J. (1984). Towards a rationalisation of counselling in general practice. *Journal of the Royal College of General Practitioners* **34**, 685–687.

Rowland, N., Irving, J. & Maynard, A. (1989). Can general practitioners counsel? *Journal of the Royal College of General Practitioners* **39**, 118–120.

Sharp, D, & King, M.B. (1989). Classification of psychosocial disturbance in general practice. *Journal of the Royal College of General Practitioners* **39**, 356–358.

Shepherd, M., Cooper, B., Brown, A.C. & Kalton, G.W. (1966). *Psychiatric Illness in General Practice*. London; Oxford University Press.

Shepherd, S. (1989). Can general practitioners counsel? *Journal of the Royal College of General Practitioners* **39**, 304.

Sibbald, B., Addington-Hall, J., Brenneman, D. & Freeling, P. (1993). Counsellors in English and Welsh general practices: their nature and distribution. *British Medical Journal* **19**, 513–518.

Skuse, D. & .Williams, P. (1984). Screening for psychiatric disorder in general practice. *Psychological Medicine* **14**, 365–377.

Strathdee, G., King, M.B., Araya, R. & Lewis, S. (1990). A standardised assessment of patients referred to primary care and hospital psychiatric clinics. *Psychological Medicine* **20**, 219–224.

Taylor, K.M. (1992). Integrating conflicting professional roles: physician participation in randomised clinical trials. *Social Science and Medicine* **35**, 217–224.

Teasdale, J.D., Fennell, M.J.V., Hibbert, G.A. & Amies, P.L. (1984). Cognitive therapy for major depressive disorder in primary care. *British Journal of Psychiatry* **144**, 400–406.

Trepka, C. & Griffiths, T. (1987). Evaluation of psychological treatment in primary care. *Journal of the Royal College of General Practitioners* **37**, 215–217.

Waydenfeld, D. & Waydenfeld, S.W. (1980). Counselling in general practice. *Journal of the Royal College of General Practitioners* **30**, 671–677.

Wilkinson, G. (1984). Psychotherapy in the market place. *Psychological Medicine* **14**, 23–26.

Part IV

CONCLUSION

14 Building Research Foundations for Psychotherapy Practice

MARK AVELINE*, DAVID A. SHAPIRO[a], GLENYS PARRY[b]
AND CHRIS FREEMAN[c]

*Nottingham Psychotherapy Unit, [a]University of Sheffield, [b]Department of
Health, London and [c]Royal Edinburgh Hospital, UK*

INTRODUCTION

This chapter is based on the closing session of the conference that gave birth to this book. We aim to facilitate the development of psychotherapy research by setting research priorities and giving methodological guidance on attaining them. We hope that readers of this book will gain, both from the preceding chapters and this more condensed treatment of key issues, a clearer and more thorough appreciation of the opportunities and challenges facing psychotherapy research.

For research to inform practice in a productive manner, it is vital to ensure that it addresses questions of pressing concern to providers, purchasers and policy-makers. We begin by identifying questions important to these perspectives. Then, we consider issues of research strategy, design and method. These latter issues were the focus of lively debate at the conference. We have attempted to incorporate this discussion into our review of these issues. Accordingly, we acknowledge the contributions of the following colleagues: Anthony Mann (Chair), Chris Brewin, Sidney Crown, Richard Evans, Kenneth Howard, Joan Kirk, Frank Margison, Eugene Paykel, Paul Salkovskis, Varda Shoham and Peter Tyrer.

In an important paper, Horowitz (1982) recommended that investigators should 'game out' in advance the various possible patterns of data that a study might obtain. The purpose of this is to maximize the yield of the study by ensuring that no possible outcome is uninformative. The contributions to theory, practice and further research of each possible pattern of data should be carefully considered. 'Gaming out' offers scope for the

*Correspondence address: Nottingham Psychotherapy Unit, 114 Thorneywood Mount, Nottingham NG3 2PZ, UK.

Research Foundations for Psychotherapy Practice. Edited by M. Aveline and D. A. Shapiro.
Copyright © 1995, Mental Health Foundation and Individual Contributors.
Published 1995 by John Wiley & Sons Ltd

research questions to be refined in light of the likely yield of the available methods, as well as for the methods to be modified to help better answer the questions.

PROVIDER PERSPECTIVE

From the provider's perspective, there are broadly two kinds of objective that can be met by research. The first is concerned with advancing the provider's own understanding of treatment and its effects. The second is concerned with establishing and enhancing the effectiveness of treatment in order to promote the development of services.

Psychotherapy is provided in the NHS in specialist departments and as part of the range of interventions offered by generalist staff in mental health. Formerly these services were provided mainly from secondary and tertiary care but increasingly counselling and psychotherapy are being delivered at primary care level. At all levels, providers need to be able to match therapy to individual patient need and ensure that the most cost-effective use is made of limited, skilled therapeutic resource. The best way to achieve these objectives is through robust relevant research. From the perspective of a clinician providing a broad range of specialist NHS psychotherapy service, what are the *desiderata for psychotherapy research*?

DESIDERATA: THE SEVEN Rs OF PSYCHOTHERAPY RESEARCH

From the provider's perspective, psychotherapy research should be:

1. Representative

Research should be representative of real practice, i.e. patients, practitioners and therapy. Too much research is conducted on student or minimally disturbed participants, so that the results are limited in their generalizibility. While the fashionable manualized therapies control for therapy variability and facilitate between-therapy comparisons, they are unlikely to represent everyday practice. Indeed, they may minimize essential creative elements in psychotherapy.

2. Relevant

Research questions need to reflect the concerns of the interested parties in the provision of service. In today's NHS, outcome, efficacy and cost-effectiveness are the key areas.

3. Rigorous

In order to ensure that results are robust, the scientific methods used have to be of the highest possible quality. Pilot studies should precede the main study. Sample size must be based on power calculations that ensure the detection of clinically significant effects. The randomized controlled trial (RCT) remains the strongest method of demonstrating relative effects. One caveat has to be entered. The rigour of the RCT should not be allowed to so distort the psychotherapy under study that it is no longer representative of real practice.

4. Refined

Psychotherapists set very high standards of proof that relevant change has occurred. Symptom change is not enough. Alteration in self-concept and quality of personal relationships are often more important; these changes are more difficult to achieve and demonstrate. We should use measures which cover the three areas of symptoms, self-esteem and quality of relationships, and which are sensitive to significant change as determined by the patient, those important in his or her social world and the clinicians involved in his or her care. Measures need to take account of the fact that deterioration during psychotherapy may be a positive index of change as may the ending of a problematic relationship betoken a step forward in personal development. Sometimes leaving well-alone will be the optimum outcome when intervention might provoke decompensation: the medical principle of *non tocere* still holds. Outcome should be measured in the long-term as well as pre-, during and immediately post-therapy. Measures should take account of the broad health costs incurred and saved through successful psychotherapy, not just the immediate costs to a mental health service but to primary care, education, social services and welfare. Finally, we need to relate the severity of the problems of patients and their health gain through psychotherapy, delivered at any of the three levels of health care, to the problems of patients in medicine in general and the health gains that their therapies deliver.

5. Realizable

Given a particular design, a service conducting research must have sufficient patients or clients of the right type(s), therapists with the requisite skills and the time and motivation to complete the study within a reasonable time frame.

6. Resourced

The successful conduct of research crucially depends on the sustained commitment of the researchers and the researched. Sufficient staff and financial resource needs to be allocated to the project to ensure proper preparation, secretarial support, data entry and analysis. One should guard against underestimating the duration and likely complications of the study.

7. Revelatory

Ideally results need to be unambiguously worthwhile to all involved in obtaining them. This is the counsel of perfection but the purpose of research is to lead to improvements in clinical practice.

While these *desiderata* may vary in their relevance for particular projects, several priority steps stem from them. Providers urgently require tools to measure key parameters in the description of therapy and its effects. Consensus is needed on common measures to enable information to be pooled across centres, therapies and treatment populations; the measures need to be sensitive, relevant and not too laborious to complete. Three priority areas can be identified.

The first priority is the development and field-testing of measures of the effects of treatment. These include a 'core battery' of key outcome variables, including measures of general health status, quality of life, role functioning, self-esteem, interpersonal relationships and quality of life alongside more traditional measures of psychiatric symptoms. For each outcome measure, objective criteria of clinically significant change should be developed. It is especially important to place psychotherapy patients on one of these measures on the same spectrum as general medical disorders in terms of health disability. The now well-standardized self-report American measure, the SF-36, is promising; it provides scores on eight health constructs including limitations in social activities because of physical or emotional problems and general mental health (psychological distress and wellbeing) (McHorney *et al.*, 1993; Ware & Sherbourne, 1992).

A second priority concerns the description of therapy. Here, the most urgent need is for methods for assessing the competencies of therapists, both those that are general across treatments and those that are more specific to a given intervention. Outcome of therapy is the result of a complex interaction between the effects of the therapy, therapist, patient and social context. Patient and context factors include motivation, psychological-mindedness, age, complexity of disorder, fragility of self-structure, the extent to which the patient is gripped in restrictive social and family systems, general psychosocial issues such as unemployment and poverty, and the strength of the support network. These contextual factors, alongside therapist competencies, are among the most sorely neglected aspects of psychotherapy. Combining patient and therapist factors, one would expect highly skilled therapists to achieve better results with difficult patients and disorders more often than less skilled therapists. Before this proposition can be tested, measures of competence need to be agreed. As noted by Waltz et al. (1993), it is much more costly to assess competence than mere compliance with a treatment protocol.

A third priority relates to both of the above: consensus is required on methods of obtaining economic data on the costs and benefits of psychotherapy, in order to guide the development of services and to demonstrate their value to purchasers and policy-makers in terms of reduced health, social and economic costs.

Turning to more specific research questions to which answers are required by providers, there are several well-established domains of practice that lack research foundations. Accordingly, it is difficult to justify these to commissioners of services (Denman, Chapter 8). These practice domains depart from the single, brief treatment episodes of relatively mild disorder that have predominated in research. For example, much clinical practice is devoted to the treatment of disorders, such as borderline personality disorder, that carry high social and economic costs, concerning which the predominant professional opinion is that these patients require highly skilled treatment extending over several months or even years. It would be worthwhile to develop protocols for such extended, developmental therapy, and evaluate its effectiveness in comparison with 'maintenance' therapies involving less frequent contact, but nevertheless believed to protect patients from costly breakdown through long-term support and containment. In this context, it would be important to evaluate clinicians' beliefs concerning the ability of individuals to benefit from developmental therapy with a view to empirically-grounded assignment to developmental vs. maintenance treatment.

Thus, a controlled trial could be conducted of developmental therapy (weekly for 2 years) versus maintenance with planned episodes of developmental therapy (2-monthly contact plus, in each year, 3 months of weekly therapy). Both treatments would be administered by skilled dynamic therapists. All therapists would administer all three treatments. The patient population would be defined by a cut-off on the Inventory of Interpersonal Problems (IIP)(Horowitz et al., 1988) identifying substantial interpersonal problems, and/or DSM-III-R diagnostic criteria for Borderline Personality Disorder. Data would be collected on health care utilization as well as change in interpersonal problems. Outcome would be measured at 6 month intervals, with at least 1 year of follow-up, plus a monthly one-page questionnaire of reduced scales.

Such a study would not seek to evaluate 'long-term psychotherapy' as such, but rather to establish which of two treatment models is more cost-effective in bringing specific gains to specific groups of patients. Following Howard (Chapter 1), we would expect different goals to take different lengths of time to achieve: one or two sessions might suffice to increase subjective wellbeing, whereas real-life job and interpersonal functioning would take a long time to show change. This type of research would help to establish the place of psychotherapy within realistic models of long-term care. Just as with renal disease or diabetes, individuals with long-standing personality problems may be expected to require help of different kinds over many years.

Another interest of providers is in the development of new treatment approaches. They would like to see small-scale studies of speculative clinical innovations in which large effects may be identified in a pre-liminary way for subsequent testing in refined, controlled trials. These could include multiple simple, 'lo-tech' observations of process within each case (e.g. session impact ratings, patient accounts of sessions). Such observations would help identify probable effective components or features of the intervention.

Recognizing that many mental disorders are episodic and/or persistent, providers have an interest in tests of both immediate and longer-term effects of specific, active intervention or maintenance work undertaken at primary, secondary and tertiary levels of care.

Studies of treatment processes (Elliott, Chapter 2) promise substantial gains in understanding by developing and evaluating practitioners' theories concerning the factors contributing to successful outcomes.

PURCHASER AND POLICYMAKER PERSPECTIVES

When identifying issues for purchasers and policy-makers, it is worth noting that they are not solely preoccupied with containing costs, as is sometimes thought. Although purchasers have the difficult task of investing public money in the best way to improve the health of their population, the most cost-effective treatment is not the same as the cheapest. The striking feature of the recent NHS Research and Development strategy described by Professor Peckham (Chapter 5) is the commitment to commission services on the basis of evidence of cost-effectiveness. Cost-effectiveness research in psychotherapy is in its infancy, as is the economic evaluation of psychotherapy services, but both need to grow up quickly.

One of the main problems for purchasers wishing to design service specifications on the basis of commissioning therapies of proven effectiveness for different patient groups is the gap between a treatment's demonstrated efficacy in a controlled clinical trial and its clinical effectiveness as delivered. Purchasers have to be interested in both, and in maximizing the local effectiveness of a therapy which has only been demonstrated to work under controlled conditions. This efficacy/effectiveness gap is not specific to psychotherapy—it is also a problem for pharmacotherapy—but in commissioning psychotherapy services, it is vital to be aware of the issue, since one of the reasons for the gap is failure to ensure integrity and quality of the delivered treatment.

Peckham also notes a general tendency for health research effort to be focussed at the less distressing or disabling end of any continuum. Whether or not this is true for psychiatry, psychological and social sciences, there is a need to research psychotherapists' contribution to the care of more disabled populations. This concern converges with those of the chapters by Aveline and by Hall: many of those referred to psychotherapy services are severely disabled or burdened, exhibiting co-morbidity of borderline personality or other Axis II problems with Axis I disorders, often with histories of traumatic abuse. Given the Health of the Nation targets on suicide, and the problems of parasuicidal patients (Salkovskis, Chapter 9), any psychotherapeutic efforts to reduce or prevent suicide are a high priority to the policy-maker.

Nearly all research evaluation has been of one-shot therapeutic encounters, whether focal, time-limited or longer, open-ended therapy. That serves as a poor guide to the effective delivery of a care programme approach over time which integrates psychotherapy into the efforts of other mental health providers in a coordinated way. This is a more realistic and appropriate model of long-term care for many NHS patients who have severe and enduring problems. Given the continuing aspiration to managed care that characterizes health care policy in the UK, such realistic models are essential if research is to inform the development and implementation of UK policy.

One-off interventions may be appropriate for less severely disordered individuals, but to use Howard's terminology of remoralization, remediation and rehabilitation (Chapter 1), all three stages are necessary for adequate treatment of many NHS patients. Psychotherapy interventions in this case form part of a trajectory of planned care over time. For example, a developmental model of care offers intermittent brief therapeutic contact tailored to the individual's needs after case review over a number of years. The containing and continuation function is taken by the psychotherapy department and brings close liaison with other agencies, families and carers. Well-functioning psychology and psychotherapy departments can provide this continuity and co-ordination in cases where the treatment options are primarily psychosocial. A clinical example illustrates this process:

> A 19-year old man was referred to a multi-disciplinary NHS psychotherapy
> department by his consultant psychiatrist with a history of social withdrawal,
> depression and obsessional compulsive disorder. He also met DSM-IIIR
> criteria for borderline and narcissistic personality disorders. Many agencies
> had been involved (GP, GP counsellor, psychiatrist, social worker) and the
> prognosis seemed poor, with no response to antidepressant medication. He
> had no friends or social network, but a hostile and stormy relationship with
> his parents, on whom he depended for his care. The referral followed
> incidents of attacking his father then himself with a knife. After a 25-session
> cognitive analytic therapy (Ryle, 1990), self-injurious behaviour and violence
> had disappeared and the patient had greater insight and more mood stability.
> At review in the Department, 6 months after therapy ended, he was offered
> behavioural treatment for his obsessive compulsive disorder. One year later,
> following re-assessment, he joined an 18-month psychodynamic group. Over
> this whole period, the Department liaised with other health and social agen-
> cies and enabled a coherent and co-ordinated approach to care. At 3-year
> follow-up, he was much improved, less socially isolated, having a girlfriend
> and a place at University.

Research on patterns of service delivery for such patients would complement the example of a controlled trial of developmental therapy vs. brief
intermittent intervention described above; providing evidence of where
cost-effective patterns of delivery can be demonstrated in practice.

There are also important questions about the value of psychotherapy in
work with older adults, people with physical and learning disabilities, and
those with chronic health problems. Such research could examine cost-
effectiveness and clinical significance in terms of quality of life and
reducing the burden of ill-health to the individual, their family and the
health care system.

Another important question for purchasers is the service user's perspective. Most research does not really address this wholeheartedly. For
example, how many of us consult with a panel of users or ex-users about
our research designs at the point of designing them (cf. Hardy, Chapter 4)?
Users are an important stakeholder group whose aims and goals should be
aligned with those of the research. To take an example, if we design
studies to look at waiting list management, do we really check out with
groups of users which people prefer to receive some or any attention
quickly, and who would prefer to wait for the best available treatment?
Individual users assuredly do hold views and preferences of this kind.

Of course, those commissioning services must be concerned with 'value
for money': does the benefit outweigh the cost, including the opportunity
cost, of providing a given service? Can research address efficiency issues
whilst retaining clinical validity? To take an example, one could ask whether
it is efficient for the most skilled therapists to be offering supportive work

over many years. If what is required is 'safe containment', to use Peter Fonagy's analogy, the psychological equivalent of insulin to maintain the health of a diabetic, to sustain an individual's psychological functioning over many years, does this fit better with the social support paradigm rather than requiring the approach and skills of the experienced psychotherapist? That is a researchable question with important policy implications.

Balancing costs with quality and the competence of psychotherapists is a legitimate concern of policy makers. It is vital that the professionals, the psychotherapists, have ownership of such work, that it should be seen as part of good professional practice as reflective practitioners rather than simply imposed by purchasers (Parry, 1992). It is essential that psychotherapists be seen to take the lead in efforts to determine and improve the cost-effectiveness of their services. It should not be only the purchaser's concern to make the best use of public funds and to get the most benefit for patients from a given resource. Were clinicians to be uninterested in costs, it would set up a defensive, hostile or paranoid process between purchasers and providers that would be psychologically unhealthy and to be avoided.

Finally, purchasers are starting to be very interested in seeing evidence of effective *protocols* of care using the best evidence from controlled trials and service evaluations. Can one design optimum protocols of care, particularly across the primary/secondary care interface, for specific groups, e.g. people with bulimia, panic disorders or treatment-resistant depression? What can (and equally important, what cannot) be achieved by primary care workers being well supported by specialist staff ? How would a protocol giving guidance for intervention across the primary/secondary and secondary/tertiary service boundaries be developed?

DESIGN GUIDELINES

GENERAL RECOMMENDATIONS

In this section we offer general guidance to those contemplating a psychotherapy research study. The following features, in our judgement, will increase the probability that an application for research funds will succeed, by persuading those reviewing it that the research has a high probability of being successfully completed and of making a contribution to the field.

The proposal should show systematic attention to organizational factors to be considered in ensuring successful completion of the research (Hardy, Chapter 4).

The proposal should show active consideration of all the design features discussed in this chapter. Where these are not present, the omission should be explicitly justified. Justification may take the form of showing that desiderata are incompatible for the particular study, and justifying the choice made. The recommended design features will change as the field advances (e.g. the availability of growth curves and of multi-level modelling). In low-budget research, it may be possible to scale down the demands of the desiderata, whilst retaining some of their benefits (e.g. measuring outcomes three times is better than once or twice, despite falling short of what would be desirable for growth curve analysis).

A notorious pitfall in clinical research is the surprising shortage of patients meeting study requirements during the project period. It is often the behaviour of colleagues that determines the availability of patients, and this may both inform and reflect the researcher's current interests. A thorough and prolonged audit of the practice settings involved should be done whilst the study is being designed, to ensure continuing availability of patients.

The research proposal should be open and realistic about weaknesses and difficulties anticipated, rather than seeking to gloss over these. Reviewers of the proposal will have extensive practical experience of doing this kind of research. Thus, a proposal that is honest about 'dirt' rather than pretending to be 'squeaky clean' will be more favourably reviewed. For example, exclusion criteria should be exhaustively listed, numbers of drop-outs should be estimated, and steps to evaluate the effects of attrition on data interpretation proposed.

Excessive complexity should be avoided. A common fault is the excessive number of dependent variables in psychotherapy studies. Statisticians differ in the maximum ratio of variables to participants that they recommend. A conservative ratio would be one outcome variable for every 10 participants; a more realistic limit might be one variable to five. In many studies, the design includes more outcome variables than there are participants (Freeman & Tyrer, 1992).

The research proposal should specify at the outset the populations and settings to which its results are to be generalizible. This has major implications for the choice of methodology. For example, patients seen in general clinical practice may differ from groups assembled for studies of specific disorders. In general clinical practice, diagnosis may be less important than other patient characteristics.

Outcome measurement should encompass both the relevant specific problem domains and general parameters such as general wellbeing, quality of life, and role competencies. Whilst patient and therapist perspectives inevitably predominate, these are vulnerable to the charge of over-estimating treatment gains. Measures tapping more therapy-neutral perspectives (such as intimates and work mates of the patient) and objective data (such as performance tasks) are a priority for development.

Nearly all the problems treated by psychotherapy show variation over time, whether or not treatment is undertaken. In addition, evidence is growing that the immediate effects of treatment can be very different from its longer-term effects. From a practical point of view, it is vital to measure the time-scale of any benefits. For example, Freeman and Scott (1993) found that the greater effectiveness of prompt specialist treatment over GP care in reducing health care costs only emerged over a 2-year period. Accordingly, long-term follow-up is necessary. In the case of episodic disorders such as depression, this needs to monitor the patient's state or health care costs continuously, as 'snapshots' such as a 6-monthly current status report may miss inter-current episodes.

The proposal must include a convincing demonstration of adequate statistical power (Shapiro *et al.*, Chapter 7). This must include a defensible estimate of the size of effect that it is appropriate to seek to detect, based on practical, theoretical and empirical considerations. This applies equally to correlational as to experimental designs. Studies looking for different effects under different conditions (typically requiring tests of interactions or moderator effects) should provide power analyses for the more complex statistical tests these require. Cohen (1977) provides the necessary guidance for all types of design.

As discussed by Shapiro (Chapter 7), recommended sample sizes to compare treatments are usually much larger, around 64 per group, than those required to determine whether treatment is better than nothing, around 20. To address more than a single question, the 2^N factorial design, of which the 2×2 is the simplest case, gives most power for main effects (answering N separate questions at the same time) although the power of the interaction tests (seeing how the effects of the N factors are related one to another) is lower.

The choice of control or comparison group(s) depends on the specific question(s) to be addressed in the study. It is also constrained by what is ethical and feasible. Cost-effectiveness evaluation of a specialist service or specific intervention may require comparison with the more routine 'treatment as usual' that patients would otherwise receive. If so, the procedures followed with the comparison group may need to be standardized or at least measured more fully than would be the case in the absence of a research study. The costs of both specialist and standard treatments should be assessed fully; the latter may be surprisingly expensive for some groups of patients (e.g. borderline personality disorder).

RCT OR NOT RCT? THAT IS THE QUESTION

The randomized, controlled trial (RCT) enjoys unique standing in clinical research. Scientifically, it claims unrivalled power to sustain causal inference. It was developed for physical treatments across medicine, and was first used in psychiatry for the evaluation of psychotropic medication. It involves assigning individual patients at random to one of two or more standardized treatment conditions. This means that each individual is equally likely to be assigned to any one of the groups, and that the treatment received can be replicated.

If the groups of patients differ only in terms of the nature of the treatment to which they were assigned by the investigator, then it appears reasonable to attribute any difference in the groups' clinical outcomes to the treatments. Because of the strong causal inference it permits, the RCT is generally considered the most persuasive form of evidence concerning the effectiveness of a clinical intervention. All regulatory bodies will only permit new drugs to be used that have been shown safe and effective via RCTs. Ethical dilemmas (such as those relating to informed consent, and the withholding from individuals of treatments that are probably but not yet definitely the most effective) have been thoroughly addressed, especially in such literally life-and-death arenas as cancer treatment. Purchasers of mental health services, strongly influenced by public health physicians' endorsement of RCTs as the 'gold standard', may be reluctant to buy any service whose treatments are not so supported. There is therefore a strong argument for viewing RCTs as a *sine qua non* of the scientific foundation for psychotherapy services.

However, countervailing arguments have been put, to suggest that the RCT may be an insufficient or misleading basis for evaluating psychotherapy in practice:

The *prematurity* argument states that we are not yet ready to carry out RCTs in psychotherapy. It claims that we have not yet achieved sufficient consensus concerning operational characteristics and theoretical rationales of the treatments to be evaluated for meaningful RCTs to be carried out. In the domain of physical medicine, RCT testing of new drugs follows extensive development work in the laboratory and clinic, which builds empirical data and theoretical understanding sufficient to suggest that specific substances are likely to prove effective in a given disorder. The development of a new drug progresses through clearly defined phases. It is only in Phase III that drugs are tested in RCTs. Phase I and Phase II studies often involve open uncontrolled trials. There are many developments in psychotherapy that have not progressed beyond these early stages.

In response to the prematurity argument, it may be claimed that psychotherapists have already had 100 years to establish a theoretical basis sufficient for development of RCTs. Declining to do so will keep psychotherapy at the margins of psychiatric research and practice.

The *feasibility* argument states that RCTs are so costly and technically difficult that an adequate scientific basis for psychotherapy practice could never be built upon them. The National Institute of Mental Health (NIMH) established its Treatment of Depression Collaborative Research Program (TDCRP) in the 1970s, in a conscious effort to test the feasibility and yield of state-of-the-art RCT methodology in this field. At least 10 years and ten million dollars were spent generating somewhat equivocal findings concerning the comparative effectiveness of just two psychological treatments and one drug treatment in the treatment of a single disorder (Elkin *et al.*, 1989). This raises the thorny question of whether a similar level of research resources distributed over a variety of smaller projects would have contributed more to scientific understanding of psychotherapy. It also suggests that the cost of RCTs as a basis for psychotherapy practice will be prohibitive.

Consideration of RCTs' cost invites consideration of the relative resources available for research on drugs and psychotherapies. The scientific base of drugs is largely built at the expense of the pharmaceutical industry, for which research and development are appropriate investments, to be rewarded by the marketing of effective products. No comparable commercial incentives apply to psychotherapy research. The purchaser or policy-maker looking to the scientific literature for evidence of effective treatments will find more support for drugs than for psychotherapies, if only because more drug research is done since more money is available to support it. If psychotherapy research as a whole were as well supported financially as is drug research, the costs of RCTs would be less prohibitive, and less threatening to the resource base of other research methods. It falls to governmental and charitable agencies to level the playing field.

The technical difficulties of RCTs are legion. As discussed by Shapiro *et al.*, (Chapter 7), the 'perfect' study cannot be achieved, and necessary design decisions often reflect compromise between conflicting requirements. Although this is probably equally true of drug RCTs, there is a suspicion that publication standards may in practice be tougher for psychotherapy studies. For example, psychological issues (e.g. patients' capacity to differentiate active from placebo treatments) may be less searchingly examined by those reviewing drug studies than by reviewers of psychotherapy trials.

Some technical difficulties bring fundamental, conceptual challenges to the viability of the RCT. For example, the attrition problem discussed by Howard *et al.* (Chapter 1) violates the assumptions, and hence threatens the validity, of the RCT because individuals are lost to different groups within the trial for systematic, non-random reasons.

The *representativeness* argument contends that RCTs are, by their nature, unrepresentative of everyday clinical practice, so that their results cannot be assumed to hold for such practice. For example, the standardization and uniformity imposed within the RCT, in order to ensure that its results be replicable, produce a set of 'purified' treatment operations and client characteristics never seen in routine practice. Again, the knowledge of all parties that research procedures are in operation introduces psychological influences upon every step of the process. These difficulties, whilst present even for physical treatments, are particularly troublesome when the interventions under study are psychological in nature. The best answer to the representativeness argument would appear to be to call for empirical tests of the generalizability of findings established in RCTs to everyday practice.

The *informativeness* argument suggests that RCTs can only confirm or disconfirm the effectiveness or comparative effectiveness of specific interventions for specific disorders. Their theoretical importance is confined to the confirmation or disconfirmation of quite globally-framed hypotheses about effective interventions. Although incorporation of process studies (Elliott, Chapter 2) within RCTs can enable development and testing of theoretical propositions about treatment mechanisms, the scope for this may be limited by the constraints of RCT design. For example, the contribution to treatment effects of the characteristics and behaviours of therapists may be obscured by restricted numbers and types of therapists and by the uniformity of treatment delivery imposed by standardization. Comparison with drug research confirms that it is unreasonable to expect RCTs alone to advance our knowledge of treatment effects and mechanisms.

How might the conflicting positions, for and against RCTs, be resolved? What ways forward can be recommended?

It seems clear that RCTs alone are an insufficient basis for building the science of psychotherapy. Throughout medicine, they form only a part—although in some eyes the culmination—of the scientific endeavour. At the very least, the above arguments against reliance upon RCTs, taken together, make a strong case that other research methods are also required. RCTs are very costly, and are not the best way to build theory, devise new forms of treatment or describe clinical effects in everyday practice.

On the other hand, despite the difficulties of RCTs, it seems unlikely that psychotherapy researchers will succeed in weaning purchasers and policy makers from their attachment to this form of scientific evidence. The success of the Cochrane Database in Obstetrics and Gynaecology in discovering, by systematic reviews of RCTs, what was already known but hidden in the literature—what was known but we thought we did not know—has further strengthened the grip of the RCT on medical research. The establishment of Cochrane Centres throughout the world and the extension of the Cochrane collaboration into many areas of psychiatry will ensure its robust longevity and influence in psychiatric journals.

The most promising resolution of these issues appears to be a research strategy combining RCTs with other modes of investigation. There is merit in differentiating the task of establishing scientific foundations for psychotherapy practice into distinct but interlinked components. Furthermore, within a given domain, the different components may be best deployed in sequential order.

Salkovskis (Chapter 9) advances an 'hourglass' metaphor to describe one such sequence. According to this model, RCTs belong at an intermediate stage of development represented by the narrow stem of the hourglass, subsequent to broader theoretical developments and closely observed case series, but prior to broadening out again to more naturalistic studies in everyday service settings.

The first part of this sequence comprises basic theoretical work on mechanisms of disorder and models of treatment, together with small-scale clinical studies in which it would be hoped to observe large effects whose detection would not require large samples or sensitive method-ology. The basic theoretical work could be viewed as contributing more broadly to behavioural science as a discipline as much as to psychotherapy *per se*. This corresponds to the neurochemistry that precedes development of drug treatments. The initial clinical studies would test the theory by seeking large effects in a relatively informal way.

The intermediate, narrow stem of the hourglass involves RCTs, whose objective is to confirm (or disconfirm) the promise of the clinical studies in a more rigorous, definitive and convincing fashion. An RCT might estab-lish, for example, the extent to which a new treatment is an advance on a currently standard treatment of the disorder in question. In the absence of an existing standard treatment, it might seek to demonstrate the extent to which the effects of the new treatment are attributable to its theoretically derived key features by comparing it with similar methods lacking these.

The final, broader portion of the hourglass tests the generalizability of RCT findings to everyday clinical practice. The rigorous control and strong causal inference of the RCT are relaxed, in favour of representativeness of psychotherapy services as delivered. Methods range from experimental designs with weaker controls than required for rigorous RCTs to naturalistic, descriptive studies (cf. Howard *et al.*, Chapter 1). In present circumstances, naturalistic studies may usefully 'piggyback' on clinical audit: the addition of repeated administration of standard outcome measures to an audit package is a highly cost-efficient means of collecting naturalistic change data.

The hourglass metaphor requires strategic, policy-informed thinking to establish what stage a given field of investigation is at, and hence what kind of research is to be prioritized. Within large or interlinked research programmes, work of different kinds can be articulated together. This might take the form of separate studies organized programmatically. Alter-natively, a single study can incorporate work at different levels, to maximize the synergy between them. For example, the Second Sheffield Psychotherapy Project (Shapiro *et al.*, Chapter 7) is designed as an RCT, but also incorporates theoretical work analysing patient verbal material to test the Assimilation Model as an account of cognitive and emotional processing during psychotherapy (Field *et al.*, 1994).

The hourglass metaphor also implies different evaluative criteria for research at different stages. For example, small, relatively inexpensive clinical studies of hopefully large effects should not be judged by the standards of larger, more costly RCTs required to demonstrate effects more definitively. These different criteria should be explicitly shared among all parties to the grant award and research publication peer review processes.

This approach is appealing. However, it may not prove a panacea. There is already scepticism concerning generalization to practice of RCT findings from the middle of the hourglass. This problem is probably best addressed by those with primary commitments to service delivery undertaking audit-based, naturalistic research.

ENHANCING THE INFORMATIVENESS OF OUTCOME RESEARCH

In this section, we recommend features that are likely to enhance the informativeness of an outcome study, at relatively little additional cost. Their feasibility in a given study will depend upon circumstances, but each should be considered.

Stratification

The inferential statistics used in clinical research were originally designed for larger samples than are usually available to psychotherapy researchers. Accordingly, opportunities are rife for sampling error to compromise the results. So-called 'randomization failure' arises when individuals are assigned at random to one of two or more groups receiving different interventions, and the groups so constituted are found to be non-equivalent before treatment begins. This makes it very difficult to interpret post-treatment differences. For example, if a group that started higher ends up only slightly lower than another group, but the difference between groups is not statistically significant either before or after treatment, what can be inferred?

In addition to problems due to such random variation, systematic bias can occur (e.g. by differential attrition, whereby individuals may drop out of one treatment because their problems seem too severe for the treatment, and from another because their problems appear insufficiently severe to warrant the effort demanded by treatment).

These problems are best addressed by active steps to ensure pre-treatment equivalence of all groups formed for the study. Matching individuals on the basis of pre-treatment severity and other relevant parameters ensures greatest control, but is difficult to achieve in practice given the logistic constraints of recruitment to a treatment trial over a period of months or years. A more practical alternative is to stratify the sample, creating filrther design variables (e.g. severity of disorder, employment or relationship status), prior to randomization. However, this can only be done for one, two or at most three variables given typical sample sizes. Much therefore hangs on the choice of stratification variables as ones likely to impact upon response to treatment (Aveline, Chapter 6).

Monitor patient reference and drop-out

Another problem of randomization designs occurs when patients have strong preference for a particular treatment, leading to refusal to enter the randomized trial or patients dropping out of an unfavoured treatment. One proposed compromise is to assign patients to their preferred treatment where they express a view and to randomly assign only those who have no strong preference (Bradley & Brewin, 1990). Alternatively, an important safeguard is to measure and report characteristics and outcomes for the whole sample it was intended to treat, including those who refuse consent for randomization and those who are dropped from the study for whatever reason. This does not prevent the problem but allows the impact of 'hidden failures' of randomization to be better understood.

Aptitude–treatment interactions

Formally, the above procedures are all about accounting for variation between individuals, or reducing the size of the 'error term' comprising unexplained variation. Some of this unexplained variation may reflect the fact that different kinds of people respond well to different kinds of treatment—'Different strokes for different folks' (Blatt & Felsen, 1993; Shoham & Rohrbaugh, Chapter 3). Accordingly, the choice of stratification variables should reflect the best available theoretical rationale for predicting such differential treatment response. For example, it was reasonable to expect greater advantage of 16 sessions over 8 sessions of treatment for severe as contrasted with mild or moderate depression (Shapiro *et al.*, Chapter 7).

Effects not effectiveness

It has long been argued, especially by behavioural and cognitive therapy researchers, that clinical outcome studies often rely upon overly general measures of outcome; 'outcome equivalence' (Stiles *et al.*, 1986) could be largely due to this lack of precision. On this argument, by the time a treatment effect has percolated through to a general measure of wellbeing, or even of depression, any specificity of the treatment's effects will have been lost. Accordingly, we recommend against reliance upon averaged indices of *'overall effectiveness'* from several measures, and in favour of careful selection of measures to include ones that have a good chance of tapping the *specific effects* of each intervention in the study (Stiles & Shapiro, 1989).

Clinical significance of change

To establish the practical importance of research findings requires demonstration that the changes and differences observed are sufficiently substantial to make a worthwhile difference to the clinical status, effective functioning, and health care resource consumption of individuals. Jacobson and Truax (1991) describe methods for identifying individuals showing change that is both statistically reliable (i.e. greater than could be accounted for by measurement error alone) and large enough to indicate movement from the population of those requiring treatment to a level of dysfunction within the normal limits of the general population or of those not requiring treatment. Differences between treatment conditions can be expressed in terms of differences in the percentages of patients showing such change (Aveline, Chapter 6).

This method makes research results more persuasive to clinicians, purchasers and policy-makers. It also enables one to bridge the gap between clinical anecdote and quantitative research, by identifying 'improving' and 'non-improving' individuals and using this information to describe factors associated with favourable response to treatment.

Multiple occasions of measurement

Most of the disorders treated by psychotherapy are unstable over time, showing variations in intensity of symptoms and disability related to a host of factors within and without the individual, of which treatment is just one. Sole reliance upon single occasions of measurement before and after treatment is likely to be misleading. Treatment response must be viewed in the context of variations over time. If limitations of time and resources entail a trade-off between multiple administrations of short instruments, and pre-post only administrations of more complex instruments, the former is likely to be more informative.

This strategy has other advantages. Multiple measurement occasions enable statistical treatment of changes shown by individuals, using growth curves (Rogosa & Willett, 1985); treatment response can be described in terms of trends over time such as linear improvement, and the negatively accelerated 'diminishing returns' of the quadratic trend (Barkham *et al.*, 1993). Statistical methods are under development that will enable us to build enhanced statistical power from multiple observations on a limited number of individuals. Most exciting, multiple measurement occasions can be used to develop and test theories concerning the change process (Shoham & Rohrbaugh, Chapter 3).

Cost-effectiveness

Decisions concerning allocation of limited resources are best informed by data on both costs and benefits of treatment. The investigator should consider how broadly it is appropriate to define each of these. On the cost side, if these are to be defined in terms of the NHS internal market, it may be relatively straightforward to derive costs per case or per treatment session from service agreements. However, other costs (e.g. time costs to the patient) would be excluded from this. Effects or benefits are generally more problematic, because each of a number of potential audiences of the research will bring different interests requiring different kinds of data and will define 'benefit' differently. From the perspective of the NHS purchaser, health care utilization data are crucial. From a broader social policy perspective, such factors as employment status and benefit claims are important.

A cost-utility analysis (Kamlet et al., 1992) evaluates an intervention such as psychotherapy for depression by comparing the incremental societal costs of a health intervention and the incremental health benefits that result from it. Typically, the outcome is expressed as a ratio, with the units being dollars or pounds per quality adjusted life year. Kamlet et al. (1992) describe the assumptions and decisions entailed by one such analysis, where the economic evaluation was built on to an existing study rather than designed as such from the outset.

Therapist competencies

Substantial evidence points to the impact of the therapist upon treatment outcome. That impact must reflect observable behaviours of the therapist. Accordingly, investigators should pay at least some attention to the therapist's contribution to treatment. Ideally, therapists' *adherence* to a defined treatment method should be demonstrated, together with their *skilfulness* in delivering the treatment, as well as more *general competencies* in their fulfilment of the role of therapist. Adherence can be assessed by non-therapists rating audiotapes using rating scales designed to tap the procedures specified in the treatment protocol (e.g. Startup & Shapiro, 1993). This is a costly procedure, but much less so than the assessment of skilfulness. Logically, it is hard to see how this can be assessed by judges who do not themselves possess the skill involved; accordingly, it can only be measured with the help of trained and experienced practitioners.

More work is urgently needed, involving both practitioners and researchers, to define criteria for both specific skills and general competencies. Meanwhile, outcome researchers should at the very least establish that the level and extent of therapists' professional training and experience are sufficient for their role in the study. For example, King (Chapter 13) uses counsellors accredited by the British Association for Counselling (BAC). Where resources do not permit formal adherence or skilfulness measures, the integrity of treatment can be assured to a useful extent via regular tape-based supervision.

DESCRIPTIVE NATURALISTIC STUDIES

There are alternatives to the randomized, controlled trial that may be equally informative, albeit in different ways, and considerably less difficult to execute. Descriptive studies of therapy as practised may enable the extent and nature of change, and its correlates, to be ascertained. For example, the contribution of therapists to outcome may be best studied in naturalistic research in which each of a substantial number of therapists (upwards of 20) treats a number of patients (upwards of 10 per therapist) drawn from a common pool of referrals. Such designs can be analysed using multi-level regression methods, developed for such applications as the study of the effects of individual schools or teachers within schools on pupil attainment. In general, careful conceptualization and appropriate measurement of change in one or more treatments can address important theoretical questions without the loss of external validity characteristic of tightly controlled designs (cf. Howard, Chapter 1).

Implementation of such research can be 'piggybacked' on audit. The audit data will help interpretation of the outcome data. If audit is required for other purposes, the researcher can add to it at relatively low incremental cost. The multi-site option is worth considering for naturalistic studies, provided there are explicit criteria for the selection of participating sites. We have to understand why the sites are in the study, and be confident that they will deliver representative and high-quality data. In the course of piloting, key organizational factors (Hardy, Chapter 4) must be assessed.

PROCESS STUDIES

The counterproductive divide between clinicians and researchers may be best bridged by a scientific approach that exploits the clinical situation as a major source of scientific theorizing and discovery, rather than as merely a practice domain to which scientific theories are applied. The ultimate, overarching goal of process research is to identify change mechanisms accounting for the effects of psychotherapy, with a view to enhancing these. Elliott (Chapter 2) highlights the interdependence of research and practice in psychotherapy, and the political necessity of collaboration between researchers and therapists. He accords priority to research questions over research methods; the method chosen must be appropriate to the question to be addressed.

In consequence, Elliott (Chapter 2) is able to recommend a wide range of methodologies from which to select, according to the research question and the level of available resources. Quantitative process studies can exploit the number of treatment sessions or events within them to increase the descriptive and statistical power of work with a given number of patients.

Contrary to stereotype, qualitative and case-study methods can serve critical purposes and are not intrinsically less rigorous than quantitative studies involving large groups of participants. This is borne out by the examples in Elliott's Tables 2.3 and 2.4.

Current trends towards quantification of complex process variables (e.g. Kerr *et al.*, 1992; Field *et al.*, 1994) and establishment of quality standards for qualitative research (e.g Stiles, 1993), should render clinically meaningful process studies increasingly fundable. However, it will remain vital not to get so 'buried in the process' that the central tasks and objectives of therapy are forgotten. Outcome-related constructs should always be included, in order to maintain clear relevance to clinical practice. Furthermore, process studies should focus on features of the intervention that are practically open to change in the light of the findings. For example, the task analysis by Agnew *et al.* (1994). of resolution of challenges to the therapeutic relationship has strong and practical implications for the training of therapists to deal effectively with such challenges. Despite its apparently microscopic focus and method, such research has immediate relevance to the task of improving the quality and effectiveness of psychotherapy services.

REFERENCES

Agnew, R.M., Harper, H., Shapiro, D.A. & Barkham, M. (1994). Resolving a challenge to the therapeutic relationship: a single case study. *British Journal of Medical Psychology* **67**, 155–170.

Barkham, M., Stiles, W.B. & Shapiro, D.A. (1993). The shape of change in psychotherapy: longitudinal assessment of personal problems. *Journal of Consulting and Clinical Psychology* **61**, 667–677.

Blatt, S.J. & Felsen, I. (1993). Different kinds of folks may need different kinds of strokes: the effect of patients' characteristics on therapeutic process and outcome. *Psychotherapy Research* **3**, 245–259.

Cohen, J. (1977). *Statistical Power Analysis for the Behavioral Sciences*, 2nd edn. New York: Academic Press.

Elkin, I., Shea, M.T., Watkins, J.T., Imber, S.D., Sotsky, S., Collins, J.F., Glass, D.R., Pilkonis, P.A., Leber, W.R., Docherty, J.P., Fiester, S.J. & Parloff, M.B. (1989). National Institute of Mental Health Treatment of Depression Collaborative Research Program: general effectiveness of treatments. *Archives of General Psychiatry* **46**, 971–982.

Field, S.D., Barkham, M., Shapiro, D.A. & Stiles, W.B. (1994). Assessment of assimilation in psychotherapy: a quantitative case study of problematic experiences with a significant other. *Journal of Counseling Psychology*, 41, 397–406.

Freeman, C. & Scott, A. (1993). *The Edinburgh Primary Care Depression Study: Long-term Follow-up*. Paper presented at the Annual Meeting of the Royal College of Psychiatrists, Scarborough, July 1993.

Freeman, C. & Tyrer, P. (1992). *Research Methods in Psychiatry*, 2nd edn. London: Gaskell.

Horowitz, L., Rosenberg, A.E., Baer, B.A., Ureno, G. & Villasenor, V.S. (1988). Inventory of interpersonal problems: psychometric properties and clinical applications. *Journal of Consulting and Clinical Psychology* 56, 885–892.

Horowitz, M.J. (1982). Strategic dilemmas and the socialization of psychotherapy researchers. *British Journal of Clinical Psychology* 21, 119–127.

Jacobson, N.S. & Traux, P . (1991). Clinical significance: a statistical approach to defining meaningful change in psychotherapy research. *Journal of Consulting and Clinical Psychology* 59, 12–19

Kamlet, M.S., Wade, M., Kupfer, D.J. & Frank, E. (1992). Cost-utility analysis of maintenance treatment for recurrent depression: a theoretical framework and numerical illustration. In *Economics and Mental Health* (R.G. Frank & W.G. Manning, eds). Baltimore: Johns Hopkins University Press, pp 267–291.

Kerr, S., Goldfried, M.R., Hayes, A.M., Castonguay, L.G. & Goldsamt, L.A. (1992). Interpersonal and intrapersonal focus in cognitive-behavioral and psychodynamic-interpersonal psychotherapies: a preliminary analysis of the Sheffield Project. *Psychotherapy Research* 2, 266–276.

McHorney, C.A., Ware, J.E., et al. (1993). The MOS 36-item Short Form Health Survey (SF-36): II. Psychometric and clinical test of validity in measuring physical and mental health constructs. *Medical Care* 31, 247–263.

Parry, G. (1992). Improving psychotherapy services: applications of research, audit and evaluation. *British Journal of Clinical Psychology*, 31, 3–19.

Rogosa, D.R. & Willett, J.B. (1985). Understanding correlates of change by modelling individual differences in growth. *Psychometrika* 50, 203–228.

Ryle, A. (1990). *Cognitive-analytic Therapy: Active Participation in Change*. Chichester: John Wiley.

Startup, M. & Shapiro, D.A. (1993). Therapist treatment fidelity in prescriptive vs. exploratory psychotherapy. *British Journal of Clinical Psychology* 32, 443–456.

Stiles, W.B. (1993). Quality control in qualitative research. *Clinical Psychology Review* 13, 593–618.

Stiles, W.B. & Shapiro, D.A. (1989). Abuse of the drug metaphor in psychotherapy process-outcome research. *Clinical Psychology Review* 9, 521–543.

Stiles, W.B., Shapiro, D.A. & Elliott, R. (1986). Are all psychotherapies equivalent? *American Psychologist* 41, 165–180.

Waltz, J., Addis, M.E., Koerner, K. & Jacobson, N.S. (1993). Testing the integrity of psychotherapy: assessment of adherence and competence. *Journal of Consulting and Clinical Psychology* 61, 620–630.

Ware, J. & Sherbourne, C.D. (1992). The MOS 36-item Short Form Health Survey (SF-36): I. Conceptual framework and item selection. *Medical Care* 30, 473–483.

Index

Index compiled by Campbell Purton